CONTEXTS OF LEARNING MATHEMATICS AND SCIENCE

Effective tuition of mathematics and science is crucial to schools everywhere. While the subjects are clearly central to a good all-round education, modern economies are also increasingly dependent on a good supply of well-educated mathematicians, scientists and engineers. International research has already identified many factors that can have a positive or negative impact on pupils' achievement. This comprehensive new text brings together the results of research on these factors from around the world, harvested from the Third International Mathematics and Science Studies (TIMSS), under the auspices of the International Association for the Evaluation of Educational Achievement (IEA).

Contexts of Learning Mathematics and Science makes a unique contribution to the field, by bringing together the work of researchers of 17 different countries from a variety of contexts across the globe. This provides a means of looking at country differences in sufficient detail to promote understanding and to inform policy-making in respect of curriculum, pedagogy and teacher supply. Many examples already exist where countries have used existing TIMSS data to focus on areas of particular concern and highlight necessary action – this collection will likewise prove of great value.

The book explores two different approaches to educational research. In the first, researchers in principle begin with an open mind, considering all possible context variables as possible explanatory variables. Applying multivariate and multilevel analysis techniques they try to identify which factors explain levels of mathematics and/or science achievement. In the second approach, researchers already have reasons to choose certain context variables and investigate their relationship with achievement. The emphasis here is on teaching aspects, classroom processes and/or school management variables and their effect upon achievement.

Educational and international comparative researchers, teacher educators, policy-makers, and anyone with a vested interest in promoting the improvement of maths and science teaching in today's schools and universities, will all find the richly detailed international comparisons contained in this book fascinating and valuable reading.

Sarah J. Howie is the Director of the Centre for Evaluation and Assessment and Associate Professor of Curriculum Studies at the University of Pretoria, South Africa.

Tjeerd Plomp is Emeritus Professor of Education at the University of Twente, The Netherlands.

CONTEXTS OF LEARNING
Classrooms, Schools and Society

Managing Editors:

Bert Creemers, *Faculty of Psychology, Education and Sociology, University of Groningen, The Netherlands*.
David Reynolds, *School of Education, University of Exeter, UK*.
Janet Chrispeels, *Graduate School of Education, University of California, Santa Barbara, USA*.

CONTEXTS OF LEARNING MATHEMATICS AND SCIENCE

LESSONS LEARNED FROM TIMSS

Edited by

Sarah J. Howie and Tjeerd Plomp

Routledge
Taylor & Francis Group

LONDON AND NEW YORK

MT

First published 2006 by Routledge
2 Park Square, Milton Park, Abingdon, Oxon OX14 4RN

Simultaneously published in the USA and Canada
by Routledge
270 Madison Ave, New York, NY 10016

*Routledge is an imprint of the Taylor & Francis Group, an informa
business*

© 2006 Sarah J. Howie and Tjeerd Plomp selection and editorial matter;
individual chapters, the contributors.

Typeset in Times by Keyword Group Ltd
Printed and bound in Great Britain by Biddles Ltd, Kings Lynn

British Library Cataloguing in Publication Data
A catalogue record for this book is available from the British Library

Library of Congress Cataloging in Publication Data
A catalog record for this book has been requested.

ISBN 10: 0-415-36225-3 (hbk)
ISBN 10: 0-203-01253-4 (ebk)
ISBN 13: 978-0-415-36225-2 (hbk)
ISBN 13: 978-0-203-01253-6 (ebk)

3/8/07

Dedicated to Albert Beaton and David Robitaille, who provided leadership in the early years of TIMSS.

Contents

List of Figures and Tables

Figures

Tables

Notes on Contributors

Editors

Sarah J. Howie is the Director of the Centre for Evaluation and Assessment and Associate Professor of Curriculum Studies at the University of Pretoria, South Africa. She is the co-National Research Coordinator for Progress in Reading Literacy Study (PIRLS 2006). She was the National Research Co-ordinator of TIMSS 1995 (from 1997) and TIMSS 1999 in South Africa when she worked for the Human Sciences Research Council in South Africa. Her major research interests are educational monitoring, evaluation and assessment of education at system level and pupils' performance in mathematics, science and language. In recent years she has been involved in various countries across the world in training related to national assessments.
Address: Centre for Evaluation and Assessment, Faculty of Education, University of Pretoria, Pretoria, 0002, South Africa. Tel: +27 12 420 4131/4175; email: sarah.howie@up.ac.za

Tjeerd Plomp is Emeritus Professor of Education at the University of Twente. He was the chair of the International Association for the Evaluation of Educational Achievement (IEA), the organization under whose auspices the TIMSS studies take place, from 1989–1999. He was also the chair of TIMSS 1995 and 1999, and is involved in IEA studies on ICT in education. Since 2004 he has been study director of the IEA SITES 2006, a survey of schools and teachers on pedagogical practices and the use of ICT.
Address: University of Twente, Faculty of Behavioral Sciences, Department of Curriculum, PO Box 217, 7500 AE Enschede, The Netherlands. Tel: +31 53 489 3595; email: t.plomp@gw.utwente.nl

Authors

Carl Angell has a background in physics. After many years of high school teaching of mathematics and physics, he came to the Institute of Physics, University of Oslo, in 1988, where he has worked since. He has played an important role in establishing contacts between the university and physics teachers in Norway. He was involved both at the national and international level in the TIMSS 1995 study, particularly in developing the test for physics specialists. He took his PhD in 1996 on students' conceptual understanding based on this test.
Address: University of Oslo, Department of Physics, PO BOX 1048 Blindern, 0316 Oslo, Norway. Email: carl.angell@fys.uio.no

Alka Arora is a research associate with TIMSS and PIRLS International Study Center at Boston College. She is also a fourth-year doctoral student in the Educational

Research, Measurement, and Evaluation program at Boston College. She has a master's degree in physics and education from Delhi University in India. She received the Delhi University Gold Medal for her master's work there. She had taught mathematics and science to secondary school students for several years in India (April 1992–August 2001). Her major interests include educational measurement, statistical and psychometric analysis.

Address: TIMSS and PIRLS International Study Center, Manresa House, Boston College, 188 Beacon Street, Chestnut Hill, MA 02467, USA. Tel: +1-617 552 6252; fax: + 1-617 552 1203; email: aroraal@bc.edu

Celia R. Balbin is a science education specialist and Chair of the Information Science Workgroup in the National Institute for Science and Mathematics Education Development, University of the Philippines. She served as Data Manager for the Philippines TIMSS 1999 and 2003.
Address: National Institute for Science and Mathematics Education Development, University of the Philippines, Diliman, Quezon City, Philippines. Tel: (632) 927-4276; email: celia.balbin@up.edu.ph

Kiril Bankov is a professor of mathematics and mathematics education at the University of Sofia, Bulgaria. He is the National Research Coordinator (NRC) of TIMSS 1995, TIMSS 1999 and TIMSS 2003 in Bulgaria. He is also a member of the international mathematics item review committee for TIMSS and the international expert panels for the IEA studies TIMSS and TEDS. His research interests include educational measurement and evaluation (focusing on students' achievement in mathe -matics), teacher education and development.
Address: Faculty of Mathematics and Informatics, University of Sofia, bul. James Boucher 5, 1164 Sofia, Bulgaria. Email: kbankov@fmi.uni-sofia.bg

Jürgen Baumert is Professor of Education and has been Director of the Center for Educational Research at the Berlin Max Planck Institute for Human Development since 1996. He is one of Germany's most influential educational researchers, having headed the German TIMSS study and the OECD PISA survey. Jürgen Baumert has authored numerous books, book chapters and journal articles. His major research interests include research on teaching and learning, cultural comparisons, large-scale assessment, cognitive and motivational development in adolescence, and educational institutions as developmental environments.
Address: Max Planck Institute for Human Development, Center for Educational Research, Lentzeallee 94, D – 14195 Berlin, Germany. Tel: +49 (0) 30 82406303/304; email: baumert@mpib-berlin.mpg.de

Klaas T. Bos was the National Research Co-ordinator of TIMSS 1999, the TIMSS 1999 Video Study and TIMSS 2003 for the Netherlands in the period between 1997 and early 2003. He also worked on TIMSS 1995. His major research interest is the design and analysis of international comparative studies in education. From February

2003 he has been working as inspector and senior researcher at the Netherlands Inspectorate of Education.
Address: Netherlands Inspectorate of Education, Hanzelaan 310, 8017 JK Zwolle, The Netherlands. Tel: +31 38 4695400; email: k.bos@owinsp.nl

Fe S. De Guzman is a science education specialist and Chair of the Research and Evaluation Workgroup in the National Institute for Science and Mathematics Education Development, University of the Philippines, Diliman, Quezon City, Philippines. She was in charge of the administration of instruments and data collection in Philippines TIMSS 1995, 1999 and 2003.
Address: National Institute for Science and Mathematics Education Development, University of the Philippines, Diliman, Quezon City, Philippines. Tel: (632) 927 4276; email: timss@upd.edu.ph

Kadriye Ercikan is an associate professor at the University of British Columbia specializing in measurement, evaluation and research methods. She has conducted studies and published widely on international assessment issues, specifically on comparability and translation issues.
Address: University of British Columbia, 2125 Main Mall, ECPS, Faculty of Education, University of British Columbia, Vancouver, BC, Canada V6T 1Z4. Tel: +1 (604) 822 8953; fax: +1 (604) 822 3302; email: Kadriye.ercikan@ubc.ca

Kelvin D. Gregory is a lecturer in research and statistical methods at the School of Education, Flinders University, Australia. Prior to joining Flinders University, he worked at the International Study Center on TIMSS 1995 and TIMSS 1999, and at the Educational Testing Service on the National Assessment of Educational Progress projects. His research interests include psychometrics, assessment of student learning, and educational improvement and accountability research with a special interest in mathematics and science teacher education.
Address: School of Education, Flinders University, Box 2100 GPO, Adelaide South Australia 5001, Australia. Tel: + 61 (0) 8 8201 5737; fax: +61 (0) 8 8201 5387; email: kelvin.gregory@flinders.edu.au

Dirk Hastedt is Co-director of the IEA Data Processing Centre in Hamburg, working in the field of data processing for the studies undertaken by the International Association for the Evaluation of Educational Achievement (IEA). He has worked on the data processing of the IEA Reading Literacy study, the TIMSS 1995, 1999 and 2003 cycles and the PIRLS 2001.
Address: IEA DPC, Mexikoring 37, 22297 Hamburg, Germany. Tel: +49 40 48 500 700; email: dirk.hastedt@iea-dpc.de

Ali Reza Kiamanesh is Professor of Research Methods and Evaluation at the Teacher Training University of Tehran. He was the National Research Coordinator (NRC) of TIMSS 1995, TIMSS 1999 and TIMSS 2003 (field test) in Iran from 1992 to 2002. His major research interests are 'educational evaluation' with a focus on curriculum

evaluation, evaluation of students performance in mathematics and science in elementary and secondary schools and implications of large-scale assessment for improving teaching and learning.
Address: Teacher Training University, School of Psychology and Education, 49, Mofatteh Ave., Tehran, 15614, Iran. Tel: + 98 21 883 02 28; email: kiamanesh@dpimail.net

Marit Kjærnsli has a master's degree in science education. Since 1991 she has been involved in international large-scale studies at the Institute for Teacher Education and School Development at the University of Oslo. She was involved in TIMSS 1995 and PISA 2000, and she is the National Project Manager for PISA 2003 and PISA 2006, and is also involved in TIMSS 2003. She has written many papers on students' understanding of and attitudes to science.
Address: University of Oslo, Department of Teacher Education and School Development, PO Box 1099, Blindern, 0317 Oslo, Norway. Email: marit.kjarnsli@ils.uio.no

Mareike Kunter is a research scientist at the Center for Educational Research at the Max Planck Institute for Human Development, Berlin, Germany. She has been involved in the OECD PISA study since 2000. In 2004, she finished her PhD dissertation on multiple objectives in mathematics classes which included a re-analysis of the German sample of the TIMSS 1999 Study and the TIMSS 1995 Videotape Classroom Study. Her research interests include research on instruction and learning, instructional quality and motivational development of students.
Address: Max Planck Institute for Human Development, Center for Educational Research, Lentzeallee 94, D - 14195 Berlin, Germany. Tel: +49 (0) 30 82406246; email: kunter@mpib-berlin.mpg.de

Pekka Kupari is Head of the research team on Assessing Learning Outcomes at the Institute for Educational Research, University of Jyväskylä. He was the National Research Co-ordinator of TIMSS 1999 in Finland. Since 1999 he has worked in the national group for OECD/PISA programme and is now the National Project Manager of PISA 2006 study. His major research areas are assessment of mathematics education and international comparison of learning outcomes.
Address: Institute for Educational Research, University of Jyväskylä, P.O. Box 35 (Keskussairaalantie 2), FIN-40014 University of Jyväskylä, Finland. Tel: +358 14 260 3278; email: pekka.kupari@ktl.jyu.fi

Vanessa Lapointe is a doctoral student in the School Psychology programme at the University of British Columbia. She has conducted research on the effects of student attitudes, and teacher and parental expectations on mathematics and science educational achievement.
Address: University of British Columbia, 2125 Main Mall, ECPS, Faculty of Education, University of British Columbia, Vancouver, BC, Canada V6T 1Z4; email: vanessar@interchange.ubc.ca

Min Li is Assistant Professor of Educational Psychology at the University of Washington, College of Education. Her research interests are educational measurement and classroom assessment.
Address: College of Education, Box 353600, University of Washington, Seattle, WA 98195, USA. Tel: +1 206 616 6305; email: minli@u.washington.edu

Svein Lie has a master's in physics and has a PhD in nuclear physics. After many years of high school teaching he came in 1991 to the Institute for Teacher Education and School Development at University of Oslo to act as national Project Manager for TIMSS 1995 and later for PISA 2000. He has been involved in international science expert committees in both projects. His main research interest has been students' understanding of science and how large-scale international studies can provide valuable insight in this area.
Address: University of Oslo, Department of Teacher Education and School Development, PO Box 1099, Blindern, 0317 Oslo, Norway. Email: svein.lie@ils.uio.no

Michael O. Martin is the Co-Director of the TIMSS and PIRLS International Study Center at Boston College. He has been with the TIMSS and PIRLS projects since their inception, co-directing all three cycles of TIMSS in 1995, 1999 and 2003 and both cycles of PIRLS in 2001 and 2006. A co-founder of PIRLS, a member of the IEA's Technical Advisory Committee since 1992 and a Research Professor at Boston College, he is an internationally recognized expert in designing assessment instruments, operations procedures, and quality assurance procedures. Before joining Boston College, he was a Research Fellow at the Educational Research Center at St Patrick's College, Dublin, where he directed Ireland's national surveys of student achievement and served as Ireland's national project representative for four major international student surveys. Martin received a MSc degree in computer applications from Trinity College, Dublin (1973) and a PhD in psychology from University College, Dublin (1980).
Address: TIMSS and PIRLS International Study Center, Boston College, 140 Commonwealth Avenue, Chestnut Hill, MA 02467, USA. Email: martinas@bc.edu

Tanya McCreith recently obtained her doctorate in the area of measurement, evaluation and research methods from the University of British Columbia. She has authored and co-authored publications on differential item functioning, intelligence testing in multiple languages, and comparability and translation issues in international assessments.
Address: 2254 Canterbury Lane, Campbell River, British Columbia, Canada V9W 7Y7. Tel: +1 (250) 923 0724; email: mccreith@telus.net

Martina Meelissen is a researcher at the Faculty of Behavioural Sciences of the University of Twente. Since February 2003, she has been the National Research Coordinator of TIMSS 2003. Before that she was involved in a national study on the

implementation of ICT in education. Her main interest is gender and education, especially regarding gender differences in achievement and attitudes in mathematics and ICT.
Address: University of Twente, Faculty of Behavioural Sciences, Universiteit Twente, PO Box 217, 7500 AE Enschede, The Netherlands. Email: meelissen@edte.utwente.nl

Ina V.S. Mullis is Co-Director of TIMSS. She has extensive management and technical experience in conducting large-scale international and national assessments. She co-directs the TIMSS and PIRLS International Study Center at Boston College, and has managed aspects of all three TIMSS cycles (1995, 1999 and 2003). She co-founded PIRLS and co-directs the 2001 and 2006 assessments. Prior to joining Boston College a decade ago, Mullis was Project Director of the National Assessment of Educational Progress (NAEP) at Educational Testing Service. She is a Professor in the Lynch School of Education in the Department of Educational Research, Measurement, and Evaluation. Mullis serves on several national advisory panels, including the NAEP Validation Studies Panel, and is the independent evaluation specialist for the Foundation for Teaching Economics. The contributing author of more than 60 reports and articles about assessment, she received a PhD in educational research from the University of Colorado (1978).
Address: TIMSS and PIRLS International Study Center, Boston College, 140 Commonwealth Avenue, Chestnut Hill, MA 02467, USA. Email: mullis@bc.edu

Yasushi Ogura is a senior researcher of the National Institute for Educational Policy Research where he has worked for the TIMSS main studies and video studies, OECD/PISA, and national assessment of educational attainment. His major research interests are science lesson study for professional development, students' attitudes toward science study, and science curriculum that emphasizes thinking skills.
Address: National Institute for Educational Policy Research (NIER), Curriculum Research Center, 6-5-22 Shimomeguro, Meguro-ku, 153-8681 Tokyo, Japan. Tel: +81 3 5721 5082; email: ogura@nier.go.jp

Marie-Christine Opdenakker is a member of the Centre for Educational Effectiveness and Evaluation at the Katholieke Universiteit Leuven. She is co-ordinator of a division of the 'Steunpunt LOA' (a policy-oriented research centre sponsored by the Flemish government) studying educational effectiveness and educational careers and has a research grant from the University of Leuven. She received her PhD for her research into the effect of student, class, teacher and school characteristics on mathematics achievement in secondary education. Her previous research and writing has been in the areas of educational career research, educational effectiveness, and multilevel and structural equation modelling.
Address: Katholieke Universiteit Leuven, Centre for Educational Effectiveness and Evaluation, Dekenstraat 2, 3000 Leuven, Belgium. Tel: + 32 16 32 62 60; email: Marie-Christine.Opdenakker@ped.kuleuven.ac.be

Constantinos Papanastasiou was the National Research Coordinator of Reading Literacy, TIMSS 1995, TIMSS 1999, CivEd, SITES (1st module), PIRLS and TIMSS 2003. He has been guest editor of the journals *Educational Research and Evaluation*, *Studies in Educational Evaluation* and *International Journal of Educational Research*. He has published the books *Statistics in Education*, *Methodology of Educational Research* and *Measurement and Evaluation of Education*. His research interests lie in Monte Carlo studies and international comparative research within the framework of IEA.
Address: University of Cyprus, Department of Education, PO Box 20537, 1678 Nicosia, Cyprus. Tel: +357 22 753 745; email: edpapan@ucy.ac.cy

Elena C. Papanastasiou is a visiting professor at the University of Cyprus. In the past, she has worked at the University of Kansas as an assistant professor. She earned her PhD in the field of measurement and quantitative methods from Michigan State University in 2001. Her research interests are in the area of measurement, adaptive testing, as well as in comparative international research in the subject areas of science and mathematics and civic education.
Address: University of Cyprus, Department of Education, PO Box 20537, 1678 Nicosia, Cyprus. Tel: +357 22 753 745; email: elenapap@ucy.ac.cy

Chung Park is a researcher at the Korea Institute of Curriculum and Evaluation (KICE) and has served as the Korean National Research Co-ordinator of TIMSS 1999 (from 2000) and TIMSS 2003. Her major research interests are Item Response Theory and a large-scale assessment.
Address: Korea Institute of Curriculum and Evaluation, Division of Educational Assessment, 25-1 Samchung-dong, Gongro-Gu, Seoul, Korea, 110-230. Tel: +82 2 3704 3586; Fax: +82 2 3704 3636; email: Chungpark@kice.re.kr

Doyoung Park is an English teacher of the Chungnam Technical High School in Korea. His major research interests are hierarchical linear modelling and a large-scale assessment.
Address: Chungnam Technical High School, Daejeon, Korea. Tel: +82 41 232 0417; email: pdywsk@netian.co.kr

Richard S. Prawat is a professor of educational psychology and teacher education and Chair of the Department of Counseling, Educational Psychology, and Special Education at Michigan State University. His area of specialization is teaching and learning for understanding; his current interest is in bringing Pierce and Dewey to bear on this issue.
Address: College of Education, 449 Erickson Hall, Michigan State University, East Lansing, MI 48824-1034, USA. Tel: +1 517 353 6417; Fax: +1 517 353 6393; email: rsprawat@msu.edu

María José Ramírez is the Head of the Data Analysis Unit of the National Assessment System (SIMCE) at the Ministry of Education, Chile. She was a research assistant in the TIMSS and PIRLS International Study Center (2000–2004). She received her PhD in educational research and measurement from Boston College (2004). She worked at the Chilean Ministry of Education on TIMSS 1999 (1997–2000). Her research interests include measurement, school quality and educational policies.

Address: Ministerio de Educación, Alameda 1146 B piso 7, Santiago, Chile. Tel: + 562 3904690; Fax: + 562 3904692; email: mariajose.ramirez@mineduc.cl

Maria Araceli Ruiz-Primo is a senior research scholar at the Stanford University, School of Education. Her research interests are educational assessment, alternative assessments in science, and educational evaluation.

Address: School of Education, 485 Lasuen Mall, Stanford University, Stanford, CA 94305-3096, USA. Tel: 650 725 1253; email: aruiz@stanford.ed

William H. Schmidt is a University Distinguished Professor and Co-Director of the Education Policy Center in the College of Education at Michigan State University. He specializes in international comparative research, educational policy, and statistics.

Address: 463 Erickson Hall, College of Education, Michigan State University, East Lansing, MI 48824-1034, USA. Tel: +1 517 353 7755; fax: +1 517 432 1727; email: bschmidt@pilot.msu.edu

Richard J. Shavelson is Professor of Education and Psychology at the Stanford University School of Education. His research interests include educational measurement and evaluation, policy and practice issues in assessment reform, psychometrics, statistics and accountability in higher education.

Address: School of Education, 485 Lasuen Mall, Stanford University, Stanford, CA 94305-3096, USA. Tel: 650 725 7443; email: richs@stanford.edu

Ce Shen was research associate and programmer for the International Study Center (TIMSS) at Boston College. His work was involved in analyzing the TIMSS 1995 and 1999 data. Currently he is an assistant professor at the Graduate School of Social Work at Boston College. His research areas include comparative education, comparative sociology, demography and social development.

Address: Graduate School of Social Work, Boston College, Chestnut Hill, MA 02467, USA. Tel: +1 617 552 6363; email: shenc@bc.edu

Vivien M. Talisayon is Dean (2004–2007) and Professor of Science Education of the College of Education, University of the Philippines in Diliman, Quezon City, Philippines. She served as the Director of the National Institute for Science and Mathematics Education Development in the same university in 1996–2002. She was the National Research Coordinator for TIMSS 1999 and 2003. She is an Associate Member and Newsletter Editor of the International Commission on Physics Education. Her research interests include educational measurement and evaluation, concept learning, and policy research.

Address: College of Education, University of the Philippines, Diliman, Quezon City, Philippines. Tel: (632) 929 9322; email: vivien.talisayon@up.edu.ph

Ann Van den Broeck is a member of the Centre for Educational Effectiveness and Evaluation at the Katholieke Universiteit Leuven. She is the National Research Coordinator of TIMSS 2003 (Grade 8). She was responsible for the national part of TIMSS 1999 and conducted (multilevel) analyses on this dataset.

Address: Katholieke Universiteit Leuven, Centre for Educational Effectiveness and Evaluation, Dekenstraat 2, 3000 Leuven, Belgium. Tel: + 32 16 32 57 39; email: ann.vandenbroeck@ped.kuleuven.ac.be

Jan Van Damme is Professor at the Department of Educational Sciences of the Katholieke Universiteit Leuven. He is Head of the Centre for Educational Effectiveness and Evaluation. His major research interests are educational effectiveness and the longitudinal study of the educational career of students.

Address: Katholieke Universiteit Leuven, Centre for Educational Effectiveness and Evaluation, Dekenstraat 2, 3000 Leuven, Belgium. Tel: + 32 16 32 62 45; email: jan.vandamme@ped.kuleuven.ac.be

Ludger Wößmann is Head of the Research Department on 'Human Capital and Structural Change' at the Ifo Institute for Economic Research at the University of Munich. He has also worked at the Kiel Institute for World Economics and spent research visits at Harvard University and the National Bureau of Economic Research in Cambridge, Massachusetts. He currently coordinates the European Expert Network on Economics of Education (EENEE), which is funded by the European Commission. As an economist, much of his research focuses on the microeconometric estimation of the effects of family background, schools' resource endowments and institutional features of the education systems on student performance on several international student achievement tests, including TIMSS 1995, TIMSS 1997, PISA and PIRLS.

Address: Ifo Institute for Economic Research at the University of Munich and CESifo, Poschingerstr. 5, 81679 Munich, Germany. Tel: (+49) 89 9224 1699; email: woessmann@ifo.de

Ruth Zuzovsky graduated in 1987 from the Hebrew University, Jerusalem. Since 1970 she has been working at the Science and Technology Education Center at Tel Aviv University. Her professional areas of interest are large-scale assessment of achievements, among them several international studies of the IEA, including the 2nd International Science Study (SISS) of the IEA, the TIMSS, the TIMSS Trend Studies in 1999 and 2003 and PIRLS 2001 and 2006. In the four last studies she acted as National Research Coordinator.

Address: Tel Aviv University, School of Education, Center for Science & Technology Education, Ramat Aviv, Tel Aviv 69978, Israel. Email: ruthz@post.tau.ac.il

Foreword

Seamus Hegarty – IEA Chair

Mathematics and science are central to education in schools everywhere. This is for both educational and vocational/economic reasons. It would be an odd conception of a well-educated person that did not include having a good grasp of mathematical and scientific principles and some facility in carrying out mathematical operations, while at the same time modern economies are increasingly dependent on a good supply of mathematicians, scientists and engineers.

However, educational systems and labour markets around the world vary considerably in the role attached to mathematics and science education and, partly as a consequence, in the supply of mathematics and science graduates. While some countries have a plentiful supply of mathematics graduates, for instance, others struggle to ensure that mathematics is taught by teachers with an adequate grounding in mathematics themselves.

A unique contribution of IEA (International Association for the Evaluation of Educational Achievement) studies, and TIMSS among them, is that it provides a means of looking at country differences in sufficient detail to promote understanding at system level and to inform policy-making in respect of curriculum, pedagogy and teacher supply. There are many examples where countries have used existing TIMSS data to focus on areas of particular concern and highlight necessary action. This has included examining the effect of different methods of teaching, monitoring differences between schools, understanding the interaction between home language and language of instruction, and identifying the errors that students typically make.

TIMSS is a major project of the IEA, comprising studies of cross-national achievement on a four-year cycle beginning in 1995. TIMSS 2003 encompassed approximately 400,000 students in nearly 50 countries and produced a plethora of reports at national and international levels. The previous TIMSS studies in 1995 and 1999 respectively have likewise resulted in numerous useful reports.

Despite all this publishing activity, there remains a significant opportunity. TIMSS constitutes a unique suite of datasets covering mathematics and science achievement over 10 years in school systems around the globe. There is much to be learned from secondary analyses of these datasets, as the volumes of proceedings from the IEA International Research Conference in Cyprus in 2004 richly demonstrate (Papanastasiou, 2004).

It is a great pleasure to be able to introduce a book that takes this great task further. This volume is an important addition to the literature. It explores the relationship between background variables and student achievement both within countries and across groups of countries. It contains detailed studies of the effect of variables relating to the curriculum, pedagogy and school management. And all of this is set within a context of capitalizing on existing data so as to improve understanding, inform policy and offer guidance to practice. In summary, the book provides a wealth of illuminating information that is valuable in its own right and in the stimulus it should give colleagues to capitalize further on the TIMSS datasets.

Reference

Papanastasiou, C. (2004) *Proceedings of the IRC-2004: TIMSS, Volumes 1 & 2.* Nicosia: Cyprus University Press.

Preface

Sarah J. Howie and Tjeerd Plomp – Editors

The International Association for the Evaluation of Educational Achievement (IEA) initiated a major programme of studies in the area of mathematics and science education in the early 1990s. Following up on earlier studies in either mathematics or science, the IEA decided in 1990 to conduct the Third International Mathematics and Science Study (TIMSS), a study in which both areas were combined. The study was designed to assess students' achievements in mathematics and science in the context of the national curricula, instructional practices and the social environment of students. One of the purposes of TIMSS was to allow researchers, practitioners and policy-makers to analyse and relate performances in both mathematics and science, and to study these performances in relationship to background and context variables. TIMSS was conducted in 41 countries across the world and in three populations: population 1 being the 3rd and 4th grades in most countries; population 2 being the 7th and 8th grades in most countries, and population 3 being the final year of secondary school. In most countries, data collection took place in 1995, the reason why this study is referred to as TIMSS 1995.

TIMSS was repeated for population 2 in 1998 and 1999 in 38 countries in Europe, Africa, North and South America, Asia and Australasia. The repeat study is called TIMSS-R or TIMSS 1999.

The IEA made the data collected in both TIMSS studies available not only for the participants in the study, but also for researchers from across the world, interested to study relationships between achievement and background and context variables. The fact that researchers of 17 countries from all continents have contributed to this book illustrates the wisdom of this decision.

The ideas behind this book originated a few years ago when we met with Marc Weide, at that time representing Swets & Zeitlinger International Publishers, and Bert Creemers, one of the editors of the series Contexts of Learning, published by Swets & Zeitlinger. During our conversation the idea for this book was born and we gladly accepted the challenge not realizing the magnitude of the project we were embarking on.

This book is the result of a collaborative effort to which many people have contributed. From the very beginning we wanted a book with contributions from around the world and we are very happy and grateful that so many researchers from all corners of the globe were motivated to write a chapter.

Almost all chapters in the book have been reviewed twice – with reviewers being colleague authors or other researchers who are in some way also involved in analysing the TIMSS data. Our special thanks go to these external reviewers: Peter Allerup, Albert Beaton, Barbara Brecko, Liv Sissel Grønmo, Ronald Heck, Dougal Hutchison,

Barbara Japelj, Leonidas Kyriakis, Xin Ma, George Marcoulides, Hans Pelgrum, David Robitaille, Heiko Sibbern, Mojca Straus and Zacharia Zacharias.

Editing and reviewing a book is one thing, but preparing the book for publication is another branch of professionalism. We are very grateful to Sandra Schele (University of Twente) for her contribution in preparing all the manuscripts according to the specifications of the publisher.

During this project, the wave of mergers among publishers also influenced our work. We heard in 2004 that Routledge (Taylor & Francis Group) would be the publisher of this book. We very much appreciate the flexible and collaborative way the staff of Routledge, especially Tom Young, guided and supported us in finalizing this book.

In retrospect, we are not only impressed by the efforts of the people mentioned above in making our project happen, but we realized that designing, developing, conducting and coordinating studies of the size of TIMSS and other IEA studies is indeed a gigantic operation. We therefore want to express our appreciation and admiration for those responsible for making TIMSS happen. It is obvious to express this to the International Study Center for TIMSS and the IEA Secretariat. However, we want to dedicate this book to two pioneers who provided leadership to TIMSS in the early years of its realization: Albert Beaton (Boston College) and David Robitaille (University of British Columbia).

We hope that you, the reader, will have as much satisfaction and pleasure in reading and using this book as we had in preparing it.

Part 1

Introduction to TIMSS and to the Book

1

Lessons from Cross-national Research on Context and Achievement: Hunting and Fishing in the TIMSS Landscape

Sarah J. Howie and Tjeerd Plomp

Introduction

This book brings together results of research from across the world that addresses questions on what background and contextual factors are related to students' achievement in mathematics and sciences. Researchers from 16 countries all utilized data from the Third International Mathematics and Science Studies (TIMSS), which was conducted in 1995 and 1999 under the auspices of the International Association for the Evaluation of Educational Achievement (IEA).

The TIMSS was conducted in 41 countries around the world between 1992 and 1999, with data collection taking place in 1995. The study was designed to assess students' achievements in mathematics and science in the context of the national curricula, instructional practices and the social environment of students. Testing was carried out in more than 40 countries and in three populations: population 1 being 3rd and 4th grade in most countries; population 2 being 7th and 8th grade in most countries; and population 3 being the final year of secondary school.

TIMSS was repeated for population 2 in 1998 and 1999 in 38 countries in Europe, Africa, North and South America, Asia and Australasia. The repeat study is called TIMSS 1999.

The purpose and the design of the TIMSS studies are presented in Chapter 2. This chapter provides the 'environment' of TIMSS as an international study of achievement in mathematics and sciences, as well as the scope and an overview of the book. First, a characterization of the IEA will be given, followed by a brief discussion of functions and goals of international comparative studies of educational achievement, of which TIMSS is an outstanding example. Then the organization of TIMSS will be described, which will be followed by the scope and structure of the book.

International Association for the Evaluation of Educational Achievement – IEA

The IEA is a non-governmental organization with more than 60 member countries that conducts international comparative studies in education. Founded in the early 1960s, the IEA's mission is to contribute to enhancing the quality of education through its studies. Its international comparative studies have two main purposes: to provide policy-makers and educational practitioners with information about the quality of their education in relation to relevant reference countries; and to assist participating countries in understanding the reasons for observed differences within and between educational systems (Plomp, 1998). To achieve these two purposes, the IEA strives in its studies to make two kinds of comparisons. The first consists of straight international comparisons of the effects of education in terms of scores (or sub-scores) on international tests. The second focuses on the extent to which a country's intended curriculum (what should be taught in a particular grade) is implemented in schools and is attained by students. The latter kind of comparison focuses mainly on national analyses of a country's results in an international comparative context. As a consequence, most IEA studies are curriculum-driven.

Over the years, the IEA has conducted studies in major school subjects, such as mathematics, science, written composition, reading comprehension, foreign languages and civics education. The IEA's most recent studies, in addition to TIMSS, are the Second Civics Education Study, the Progress in International Reading Literacy Study (PIRLS) and the Second International Technology in Education Study (SITES). In addition the Teacher Education and Development Study (TEDS) recently has commenced. With respect to TIMSS, the present IEA policy is to have a four-year cycle for TIMSS under the name 'Trends in Mathematics and Science Study' – the most recent TIMSS study was TIMSS 2003 (the results of which were released in December 2004), which will be followed by TIMSS 2007.

More information about the IEA, its studies and the international reports and databases can be found on the IEA web site (http://www.iea.nl). Chapter 2 provides the references of the TIMSS international reports.

International Comparative Achievement Studies

The development of large-scale international comparative studies of educational achievements dates back to the late 1950s and early 1960s when the IEA began its first international mathematics study. Large-scale international comparative studies were made possible by developments in sample survey methodology, group testing techniques, test development and data analysis (Husén and Tuijnman, 1994: 6). The studies involve extensive collaboration, funding and negotiation between partici-pants, organizers and funders, resulting in a long-term commitment of all those involved in a study. These types of studies usually have a variety of purposes, such as to compare levels of national achievement between countries; to identify the major determinants of national achievement, country by country, and to examine to what extent they are the same or different across countries; and to identify factors that affect differences between countries (Postlethwaite, 1999: 12). The functions of these studies have been analysed and described by Kellaghan (1996), Plomp (1998) and Postlethwaite (1999). Plomp et al. (2002) and Howie and Plomp (2005) list these functions as description (mirror) benchmarking, monitoring, enlightenment, under-standing and cross-national research.

The studies published in this book illustrate the potential and relevance of the last two functions. Various chapters contribute to the *understanding* of differences between or within educational systems, and provide information that is helpful in making decisions about the organization of schooling, the deployment of resources, and the practice of teaching (Kellaghan, 1996, in Howie and Plomp, 2005). A number of chapters can be characterized as *cross-national research* as they study variations between educational systems ('the world as a laboratory'), which may lead to a better understanding of the factors that contribute to the effectiveness of education.

The Organization of TIMSS

The IEA TIMSS studies, as well as the studies in reading (PIRLS), are coordinat-ed from the TIMSS and PIRLS International Study Center in the Lynch School of Education at Boston College (USA) that is responsible for managing the design and implementation of TIMSS and reporting the results for all countries interna-tionally.

TIMSS is a collaborative effort involving dedication, expertise, and participation from individuals and organizations worldwide. In particular, the project receives important contributions and advice from panels of specialists and experts from partic-ipating countries that work directly with the TIMSS and PIRLS International Study Center and IEA. The IEA Secretariat (located in Amsterdam, The Netherlands) provides overall support in coordinating TIMSS with IEA member countries and is directly responsible for fund raising, translation verification and organizing the international Quality Control Monitors. The IEA's Data Processing Center (in Hamburg, Germany) is responsible for the accuracy and comparability of the international data-files within and across countries. Statistics Canada (located in Ottawa, Canada) is responsible for designing and evaluating the samples in TIMSS, and helping participants to adapt the TIMSS sampling design to local conditions. Educational Testing Service (in Princeton, USA) provides support in scaling the TIMSS achievement data.

The TIMSS National Research Coordinators (NRCs) are responsible for imple-menting the study in their countries and work with international project staff to ensure that the study is responsive to national concerns, both policy-oriented and practical. NRCs review successive drafts of the frameworks, tests and questionnaires. They also provide considerable input about the feasibility of the design and procedures for survey operations. To ensure the relevance of the data, NRCs review the analyses and report at every step for accuracy and utility. Finally, the NRCs work nationally with the schools, teachers and students whose participation in each country is absolutely vital.

A number of organizations have provided support for this series of studies, including the World Bank, the US National Science Foundation, the US National Center for Education Statistics and the United Nations Development Programme (UNDP). This support made it possible for a number of countries to participate in the study that otherwise would not be able to do so. Examples of such countries represented in this book are South Africa, the Philippines and Bulgaria.

Scope and Structure of the Book

As stated, the research reported in this book addresses questions on what background and contextual factors are related to students' achievement in mathematics and science.

International research (with the size of the studies varying greatly from case studies to large-scale surveys) shows that there are many factors on school, class and student levels that have positive and negative effects on mathematics and science achievement (see Howie, 2002, for an extensive review of the literature). Most of these factors were included in secondary analyses, as they were included in the TIMSS instruments.

The factors on *student level* represented in the TIMSS database include: socioeconomic status; books in the home; parental education; parental press; pupils' attitudes to mathematics; family size; jobs in the home; pupils' aspirations; peer group attitudes; pupils' self-concept; self-expectations; enjoyment of mathematics; attitudes toward mathematics; gender; age; and time spent on homework. Some countries included a national option in their TIMSS instruments. For example, in South Africa, where a vast majority of the children are being taught in another language than the language of the home, a test of English language proficiency was included, and in Flanders, Belgium, a test of cognitive ability.

On *classroom level*, the following factors found in the literature were represented in the TIMSS databases: the learning environment; teachers' characteristics (including gender); streaming; computers; teachers' competence; teachers' confidence; education background; teachers' qualifications; teachers' methods; class size; time on task; disruptions in class; calculators; content coverage; and assessment.

Finally, on *school level* the following factors were included in the TIMSS databases: textbooks; time on task; leadership; decision-making; school size; and location.

This book is divided into five parts, of which Part 1, containing the background chapters, sets the context for TIMSS and for the book as a whole. In the analyses reported in Parts 2–5 two different approaches can be distinguished, which can be characterized by the metaphors of 'fishing' and 'hunting'.

In the 'fishing' approach, researchers in principle begin with an open mind, considering all available context variables as possible explanatory variables. Applying analysis techniques such as regression analysis, Lisrel, PLS (partial least square analysis), HLM (hierarchical linear modelling), and MLM (multi-level modelling), they try to identify which factors within their countries, or across a number of countries, explain (to a certain extent) mathematics and/or science achievement. These chapters form Parts 2 and 3, both labelled 'Background Variables and Achievement', with Part 2 (Chapters 6–12) focusing on within country analysis and Part 3 (Chapters 13–15) on across country analysis.

In the 'hunting' approach, researchers have reasons to choose certain context variables upfront and investigate their relationships with achievement. In Part 4 (Chapters 16–20), the authors discuss a number of curriculum related variables in relation to achievement, whereas in Part 5 (Chapters 21–24) the emphasis is on teaching aspects, classroom processes and/or school management variables and their effect upon achievement in several countries.

All the secondary analyses utilize TIMSS 1995 and/or 1999 data. In terms of the topics of the secondary analyses reflected in the analytical chapters (Chapters 6–24), seven studies involve data only from TIMSS 1995 and 11 studies utilize data only from TIMSS 1999 (see Table 1.1). Just one study analyses both 1995 and 1999 data. While most of the chapters (viz. 10) focus on mathematics achievement only, there are three studies on science achievement specifically. However, six of the chapters include both mathematics and science achievement.

In the following sections, a summary is given of the contents of the various parts of the book, including a brief abstract of each chapter.

Part 1: Introduction to TIMSS and to the Book

This part provides both the context to the TIMSS projects as well as setting the scene for the book. Apart from introducing the TIMSS study as an international comparative achievement study and the scope and structure of the book (Chapter 1 by the editors Sarah J. Howie and Tjeerd Plomp), this part provides the introduction to the TIMSS study (Chapter 2) and to the background and contextual variables included in the study (Chapter 3). To illustrate that investigating background and contextual variables in relation to achievement is not a trivial endeavour, two chapters addressing methodological and technical issues are included: Chapter 3, 2nd part, on how background data can be reported; and Chapter 4 about the reliability of data gathered by questionnaires. The final chapter of this part (Chapter 5) presents a 'cross chapter' analysis of the findings reported in this book.

Chapter 2, 'TIMSS: Purpose and Design', written by the Co-study Directors of the TIMSS project, Michael O. Martin and Ina V.S. Mullis (USA), provides an overview of the purpose and background of TIMSS, the types of information it provides, and how it is designed. In describing the various components of the study design, the focus is on TIMSS 1999 from the perspective of how the project is evolving over time. In particular, they discuss the frameworks and tests, sampling, booklet design, questionnaires, data collection, and analysis of the results in an attempt to provide readers with information that can serve as a reference for various chapters throughout the volume.

Table 1.1 Focus of the analyses reported in the book.

Chapter	Authors	Countries focused on in analysis	Mathematics (M) or science (S)	TIMSS 1995 or 1999 studies
6	Van den Broeck, Opdenakker and Van Damme (Belgium)	Belgium (Flanders)	M	1999
7	Ramírez (Chile)	Chile	M	1999
8	Papanastasiou and Papanastasiou (Cyprus)	Cyprus	M	1999
9	Kupari (Finland)	Finland	M	1999
10	Kiamanesh (Iran)	Iran	M	1999
11	Howie (South Africa)	South Africa	M	1999
12	Park and Park (Korea)	Korea	M and S	1999
13	Bos and Meelissen (the Netherlands)	Netherlands, Flanders, Germany	M	1995
14	Ercikan, McCreith and Lapointe (Canada)	Canada, USA, Norway	S	1995
15	Talisayon, Balbin and De Guzman (Philippines)	Philippines, 38 countries	M and S	1999
16	Gregory (USA) and Bankov (Bulgaria)	Bulgaria	M	1995 and 1999
17	Prawat and Schmidt (USA)	41 countries	M	1995
18	Angell, Kjærnsli and Lie (Norway)	41 countries	S	1995
19	Li, Ruiz Primo and Shavelson (USA)	USA	S	1999
20	Ogura (Japan)	Japan	M and S	1995
21	Kunter and Baumert (Germany)	Germany	M	1995
22	Wößmann (Germany)	18 countries	M and S	1995
23	Zuzovsky (Israel)	Israel	M and S	1999
24	Shen (USA)	38 countries	M and S	1999

Chapter 3, 'Using Indicators of Educational Contexts in TIMSS', by Alka Arora (USA), María José Ramírez (Chile) and Sarah J. Howie (South Africa), discusses how the vast amount of background information that aimed to provide a context to better understand education systems around the globe was designed and developed and a number of challenges that faced this process. It also discusses how these data can be reported using examples from the 1999 assessment. These examples illustrate the main concepts and steps involved in reporting background data, while presenting methods and statistical techniques that can be used for developing indices. Important criteria to evaluate the overall quality of the measures are introduced, paying special attention to their validity and reliability. The advantages of using indices are discussed both from a conceptual and a measurement perspective.

Chapter 4, 'Inconsistent Student Responses to Questions Related to Their Mathematics Lessons', by Dirk Hastedt (Germany), evaluates the problem of the reliability of lesson data gathered by student questionnaires. One expects that students within a class would report in a similar way about the frequency of, for example, how often computers were used in the mathematics lessons. Consequently, when analysing the variance for these variables it is expected that these variables show high between-class variance components and only a minor proportion of within-class variances. The research evaluates how the 20 different parts of question 26 of the student question-naire behave in terms of the variance components and possible reasons for high within-class variance components.

The final chapter in the first section, Chapter 5, 'Raising Achievement Levels in Mathematics and Science Internationally: What Really Matters', by Sarah J. Howie and Tjeerd Plomp, is a synthesis and reflection on the main findings of the research in this book.

Part 2: Background Variables and Achievement – Single Country Studies

Part 2 comprises seven chapters exploring and relating background variables from the various TIMSS questionnaires to mathematics and/or science achievement in specific countries ranging across Europe (Belgium Flanders, Cyprus, Finland), Africa (South Africa), Latin America (Chile) and Asia (Iran, Korea). Authors apply a variety of mul-tivariate and multi-level analysis techniques.

Chapter 6, 'The Effects of Student, Class, and School Characteristics on TIMSS 1999 Mathematics Achievement in Flanders', by Ann Van den Broeck, Marie-Christine Opdenakker and Jan Van Damme (Belgium), provides readers with an example of the richness of comparative international studies when enhanced by national additions to the international design. In Flanders (Flemish Belgium), the TIMSS 1999 sample design differed from the general design: instead of taking one class in each of the selected schools, two classes per school were selected. Other national options in Flanders were the extension of the student's, teacher's and princi-pal's questionnaires with additional questions, a parent's questionnaire, and a numer-ical and spatial intelligence test for the students. Because of this design, a multi-level model with an intermediate level could be implemented.

Chapter 7, 'Factors Related to Mathematics Achievement in Chile', by María José Ramírez (Chile), analyses the distribution of mathematics achievement among Chilean eighth graders and identifies variables that account for the achievement spread among the schools and among students in the same classes. Given the segre-gation in the Chilean school system it was considered important to analyse how social economic status would influence achievement. The author identifies a number of variables characteristic for the Chilean school system that contribute to the understanding of mathematics achievement. The study also identified a number of student variables as significant predictors of achievement.

In Chapter 8, 'Modeling Mathematics Achievement in Cyprus', Constantinos and Elena C. Papanastasio (Cyprus) investigated the achievement of mathematics of students in Cyprus. The reason for this study was that education is valued highly in Cyprus and the educational system was thought to work very effectively, but international comparative studies changed this belief drastically. Students involved in the study were eighth graders enrolled in the year 1998/9. Structural equation modelling was used, containing two exogenous constructs – the educational background and the school climate; and two endogenous constructs – student attitudes toward mathematics, and teaching. The results of this study contribute to the discussions about important issues and problems that the Cyprus educational system faces.

Chapter 9, 'Student and School Factors Affecting Finnish Mathematics Achievement: Results from TIMSS 1999 Data', by Pekka Kupari (Finland), starts from the observation that many assessments have shown that differences in mathematics performance between Finnish schools are quite small, that the variation of student performance is among the smallest internationally and that the availability and general conditions of mathematics teaching are fairly equal across the country. The key research question addressed is what factors most strongly explain achievement in this educational context. The conceptual framework in selecting the background variables is taken from the educational effectiveness research. In total, 24 background variables divided into five groups are explored in relation to mathematics achievement, by applying HLM techniques.

In Chapter 10, 'The Role of Students' Characteristics and Family Background in Iranian Students' Mathematics Achievement', Ali Reza Kiamanesh (Iran) seeks to identify the student background factors that contribute to mathematics achievement in Iran by utilizing TIMSS 1999 data. Where there are no reliable and valid data at the national or regional level, international studies such as TIMSS provide unique sources for evaluating Iran's educational system. Eight factors were identified utilizing principal component factor analysis and varimax rotation. The findings may serve as a guideline to train teachers to understand the importance of teaching and how to teach effectively; and for educational practitioners and curriculum developers to utilize educational methodologies that would help students improve their academic self-concept, beliefs and attitudes toward school subjects.

Chapter 11, 'Multi-level Factors Affecting the Performance of South African Pupils in Mathematics', by Sarah J. Howie (South Africa), seeks to identify various factors that could explain the performance of South African students in TIMSS 1999, which (as in 1995) was extremely low compared to the other countries in the studies. In both studies more than 70 per cent of the South African pupils wrote the achievement tests in their second or third language. A national option, an English test, was included in an attempt to study the effect of the pupils' language proficiency on their mathematics achievement. In this research PLS analysis was used to explore the relative contribution of various background variables. Multi-level analysis was also employed to investigate the main factors explaining mathematics achievement of South African pupils.

In Chapter 12, 'Factors Affecting Korean Students' Achievement in TIMSS 1999', Chung and Doyoung Park (Korea) aim to identify the main factors that affect Korean students' unusually high achievement in TIMSS 1999. Despite the exceptional performances in international assessments, recently there has been a growing concern within Korea about the lowering of academic achievement. Furthermore, Korea obtained very low scores on affective measures, and it is one of the countries that reveal a big difference in academic achievements between boys and girls. These phenomena seem to be inconsistent with Korean students' high academic achievement, which leads to the question of what the main factors are that affect Korean students' academic achievement. HLM was applied to address the research question.

Part 3: Background Variables and Achievement – Multiple Country Studies

Part 3 encompasses three chapters relating contextual data to achievement in mathematics and/or science in cross-national analyses including countries from Europe, America and Asia. The first two chapters each compare results across three countries, while the third chapter looks at the Philippine data against the context of a variety of high and low performing countries.

Chapter 13, 'Exploring the Factors that Influence Grade 8 Mathematics Achievement in the Netherlands, Belgium Flanders and Germany: Results of Secondary Analysis on TIMSS 1995 Data', by Klaas T. Bos and Martina Meelissen (the Netherlands), presents the results of explorative analyses on the TIMSS 1995 data of the Netherlands, Belgium (Flanders) and Germany. The TIMSS study showed that Belgium Flanders performed significantly better on the mathematics test than the Netherlands, while Germany performed significantly lower compared to the other two educational systems.

Applying HLM, the authors explore similarities and differences across the three systems regarding student and classroom/teacher variables, which could influence mathematics achievement of eighth grade students.

In Chapter 14, 'Participation and Achievement in Science: An Examination of Gender Differences in Canada, the USA and Norway', Kadriye Ercikan, Tanya McCreith and Vanessa Lapointe (Canada) examine, using the TIMSS 1995 data, the relationship between achievement in science and participation in advanced science courses and individual student attitudes, home related variables associated with socioeconomic status and expectations. These relationships are examined for males and females in Canada, the USA and Norway. While gender differences in achievement and participation have been documented in science as a unified subject area, an important aspect of this study is that it takes science as a discipline composed of many subcontent areas, such as biology, physics, chemistry and earth sciences. The findings indicate differences in levels of association between the factors and achievement and participation in science for gender groups and the three countries.

Chapter 15, 'Predictors of Student Achievement in TIMSS 1999 and Philippine Results', by Vivien M. Talisayon, Celia R. Balbin and Fe S. De Guzman (Philippines), aims to determine predictors of student achievement in science and mathematics in the Philippines in comparison to other countries. Indices of the student, teacher and school variables were calculated for the top five and the bottom five countries in student achievement in TIMSS 1999 and compared. Multiple regression analyses were conducted using these indices and class size as predictor variables for mathematics score for all participating countries, using country as unit of analysis. For the science score, the mathematics score was added as an independent variable. Similar multiple regression analyses were done on Philippine data with student as unit of analysis.

Part 4: Curriculum Related Variables and Achievement
Part 4 comprises five chapters relating curriculum variables to achievement in mathematics and science.

Chapter 16, 'Exploring the Change in Bulgarian Eighth Grade Mathematics Performance from TIMSS 1995 to TIMSS 1999', by Kelvin D. Gregory (Australia) and Kiril Bankov (Bulgaria), explores possible reasons for the decline in Bulgarian students' performance in TIMSS 1999 when compared to TIMSS 1995. While it was suspected that changes in sampling lay at the heart of the decrease, the existence of a substantial measurement error could not be ruled out. The authors investigate various possible sources of this measurement error, including the degree of alignment between the TIMSS assessments and Bulgarian curriculum goals, instructional and assessment practices. Furthermore, they research whether changes made in 1997 to the Bulgarian school system have contributed to the misalignment between TIMSS and Bulgarian mathematics education.

In Chapter 17, 'Curriculum Coherence: Does the Logic Underlying the Organization of Subject Matter Matter?', Richard S. Prawat and William H. Schmidt (USA) examine the issue of curriculum coherence and more specifically the logic behind the ordering of content. This emerged as a key variable in accounting for students' test performance in mathematics in international studies, most notably TIMSS. Specifically, they examine the possibility that more child-centred and thus more process-oriented countries, such as the USA (and Norway), may opt for a 'content in service of process' orientation, while high performing countries, such as Japan or the Czech Republic, employ a subtly different approach, best characterized as 'process in service of content'. These two differing logics, they hypothesize, account for the now well-known differences in curricular profiles between the USA and the top performing countries.

Chapter 18, 'Curricular Effects in Patterns of Student Responses to TIMSS Science Items', by Carl Angell, Marit Kjærnsli and Svein Lie (Norway), show how detailed item-by-item achievement results from TIMSS 1995 provide an opportunity for a closer investigation of the similarities and differences between groups of countries. Their analyses focus on two main issues: important similarities and differences between countries concerning student conceptions and to what extent these differences can be understood in the light of data on Opportunity to Learn (OTL) from teacher questionnaires. Two domains, 'Earth in the Universe' and 'Structure of Matter', were selected for investigating the latter. The chapter illustrates how item-by-item data from large-scale international assessment studies can contribute to our understanding of similarities and differences on these issues between countries.

Chapter 19, 'Towards a Science Achievement Framework: The Case of TIMSS 1999 Study', by Min Li, Maria Araceli Ruiz-Primo and Richard J. Shavelson (USA), implements a knowledge framework to illustrate the distinct aspects of science achievement measured by the TIMSS 1999 items and their different relations to students' instructional experiences. The framework conceptualizes science achievement as four types of knowledge: declarative, procedural, schematic, and strategic. The authors apply the framework to the TIMSS 1999 Booklet 8 science items and scores by logically analysing the test items and statistically modelling the underlying patterns of item scores. They conclude that their framework provides more meaningful interpretations of students' learning than other models.

In Chapter 20, 'Background to Japanese Students' Achievement in Science and Mathematic', Yasushi Ogura (Japan) explores the Japanese data from TIMSS 1995 in order to explain the exceptional achievement of their students. The system of entrance examinations for upper-level schools at the end of Grade 6, 9 or 12 pushes students to study hard and out-of-school education (*juku*) plays a significant role in improving students' level of achievement. However, because extrinsic motivation can undermine intrinsic motivation, the author was keen to investigate whether Japanese students continue to study autonomously after their examinations. The influence of this type of out-of-school education on student achievement in Japan is assumed to be significant, but it has never been investigated. This chapter studies the effect of attending *juku* on achievement in Japan using data from TIMSS 1995.

Part 5: Teaching Aspects/Classroom Processes, School Management Variables and Achievement

The last part of the book focuses on important issues such as teaching aspects, classroom processes and/or school management variables and their effect upon achievement in several countries. Four chapters are included here from Europe, the Middle East and the USA.

Chapter 21, 'Linking TIMSS to Research on Learning and Instruction: A Re-Analysis of the German TIMSS and TIMSS Video Data', by Mareike Kunter and Jürgen Baumert (Germany), examines the effects of classroom management strategies and a constructivist teaching approach on students' mathematical achievement and interest development within a multi-perspective framework, using data from the German sample of TIMSS 1995 and the TIMSS 1995 Videotape Classroom Study. In Germany, the international design was extended to a longitudinal design, with additional student measures being collected one year after the international study. The authors combined student questionnaire data with observational data from the TIMSS-Video study, and used multilevel regression analyses to investigate what variables were predicting math achievement and interest in Grade 8.

In Chapter 22, 'Where to Look for Student Sorting and Class-size Effects: Identification and Quest for Causes Based on International Evidence', Ludger Wößmann (Germany) addresses two interrelated research questions using the TIMSS

1995 data: the much-researched question of class-size effects and the much less researched question of the effects of student sorting. He developed a strategy of identification in order to disentangle the two questions which is necessary for answering the two questions. The author demonstrates that international evidence can inform the discussion of these questions substantially, although, due to the limited number of country observations and the multitude of potential underlying factors, such cross-country evidence can never be definitive. Still, the international evidence provides a helpful starting point when asking 'where to look' for the effects, thereby increasing our understanding of how school systems work.

In Chapter 23, 'Grouping in Science and Mathematics Classrooms in Israel', Ruth Zuzovsky (Israel) reveals from the TIMSS 1999 data that, in spite of a policy of social integration in schools, a significant extent of ability grouping exists within the lower secondary school in Israel. In this chapter she addresses a number of research questions such as to what extent different modes of learning organization (ability grouping) enhance the overall level of achievement in science and mathematics studies and act to widen or close achievement gaps associated with social stratification. She also looks at whether variables related to the organization of learning act to attenuate the effect of student background variables on students' learning outcomes, and whether there are any interaction effects between the organization of learning and other variables such as instruction that can explain the variability in outcome measures.

In Chapter 24, 'Factors Associated with Cross-national Variation in Mathematics and Science Achievement based on TIMSS 1999 Data', Ce Shen (USA) takes as a starting point that both TIMSS 1995 and TIMSS 1999 reveal substantial variations in student mathematics and science achievement cross-nationally, with a significant gap between the highest and lowest performing countries. Using multiple regression analysis for eighth grade students from 38 school systems, the author tests the effects of selected country-specific variables on achievement in mathematics and science such as the countries' economic development level, school and classroom environment, the literacy of students' families, and students' self-perceived easiness of the subjects of mathematics and science. The policy implications of these findings are discussed.

Conclusion

The purpose of this book is both for insight and reflection based upon empirical data gathered in one of the largest international comparative studies to date. We hope that the chapters stimulate debate and provide researchers in the field with inspiration for further research in school-based education. The great variation in methodology applied across the chapters may also serve as useful examples for new and up-and-coming researchers eager to apply similar research methods. Finally, given that the TIMSS 2003 data has recently been released in December 2004, we hope that the research contained within this book will encourage others to tackle the latest TIMSS data in a further effort to pursue and extend the frontiers of knowledge.

References

Howie, S.J. (2002) English language proficiency and contextual factors influencing mathematics achievement of secondary school pupils in South Africa. Doctoral dissertation, University of Twente.

Howie, S.J. and Plomp, Tj. (2005) International comparative studies of education and large-scale change. In N. Bascia, A. Cumming, A. Datnow, K. Leithwood and D. Livingstone (eds) *International Handbook of Educational Policy* (pp. 75–99). New York: Springer.

Husén, T. and Tuijnman, A.C. (1994) Monitoring standards in education: Why and how it came about. In A.C. Tuijnman and T.N. Postlethwaite (eds) *Monitoring the Standards of Education* (pp. 1–22). Oxford: Pergamon.

Kellaghan, T. (1996) IEA studies and educational policy. *Assessment in Education, Principles, Policy & Practice* 3(2): 143—60.

Plomp, Tj. (1998) The potential of international comparative studies to monitor the quality of education. *Prospects* XXVIII(1): 45–59.

Plomp, Tj. Howie, S.J. and McGaw, B. (2002) International studies of educational achievements. In T. Kellaghan and D.L. Stuffelbeam (eds) *The International Handbook on Evaluation* (pp. 951–78). Oxford: Pergamon Press.

Postlethwaite, T.N. (1999) *International Studies of Educational Achievement: methodological issues*. CERC Studies in Comparative Education. Hong Kong: Comparative Education Research Centre.

2

TIMSS: Purpose and Design

Michael O. Martin and Ina V.S. Mullis

Introduction

The purpose of the Trends in International Mathematics and Science Study (TIMSS)[1] is to inform educational policy in the participating countries around the world. TIMSS collects educational achievement data in mathematics and science to provide information about trends in performance over time. Regular monitoring of national performance in education is important to provide information for policy development, to foster public accountability and to identify areas of progress or decline in achievement. Most importantly, TIMSS is designed to collect information that can be used to improve the teaching and learning of mathematics and science for students around the world. TIMSS also collects extensive background information to address concerns about the quantity, quality, and content of instruction (see Chapter 3).

This chapter is intended to provide an overview of the background of TIMSS, the study design, the data provided and how the project is evolving across time. In an attempt to provide readers information that can serve as a reference for various chapters throughout the volume, we discuss the frameworks and tests, sampling, booklet design, questionnaires, data collection and analysis of the results.

Background

Planning for TIMSS, the most ambitious project ever undertaken under the auspices of IEA, began in 1990, with data collection taking place in most of the 41 countries in 1995. Subsequently, this first round of TIMSS has become known as TIMSS 1995. Mathematics and science achievement data were collected from more than 500,000 students in 40 countries around the world as well as from their teachers and principals. Five grade levels were covered, with the participating countries collecting data at the seventh and eighth grades (or the equivalents internationally) and some countries also collecting data at the third and fourth grades and/or the final year of secondary schooling. With such comprehensive testing, together with an ambitious curriculum analysis and a video study of mathematics teaching at the eighth grade in Japan, Germany and the United States, TIMSS 1995 collected an enormous amount of information of interest to educators, policy-makers and researchers (Robitaille *et al.*, 2000).

The second round of TIMSS data collection in 1999 involved 38 countries assessing students at the eighth grade. Of the countries participating in TIMSS 1999, 26 had participated in 1995 and were able to measure trends in mathematics and science achievement over time, while the remaining countries were participating for the first time in order to compare their students' performance internationally and establish baseline information for future studies. Since the 1999 eighth-grade cohort was the fourth-grade cohort in 1995, countries also could compare changes in the relative performance of students from fourth to eighth grade. TIMSS 1999 used the experience gained in the 1995 study to focus on those aspects of curriculum and classroom practice most closely related to achievement outcomes in mathematics and science. The 1999 study administered questionnaires to curriculum experts to collect information about the mathematics and science curriculum in each country, and questionnaires to students, mathematics and science teachers, and school principals to collect data on educational experiences, instructional practices and facilities and resources. In a related effort, the TIMSS 1999 Video Study studied instructional practices in mathematics and science using video recordings of mathematics and science lessons from seven countries (Hiebert *et al.*, 2003).

Continuing the regular cycle of studies at four-year intervals, TIMSS 2003 involved more than 50 countries and included the fourth and eighth grades so that participating countries could assess trends in relative progress across grades as well as across time. For TIMSS 2003, 35 countries had achievement results for their eighth-grade students from one or both the previous assessments conducted in 1995 and 1999, and 18 countries had results for three different points in time (1995, 1999 and 2003). Fifteen countries had results for their fourth-grade students at two different points in time (1995 and 2003). Based on discussions with the participating countries, TIMSS 2003 continued and improved the contextual data collected through the curriculum, student, teacher and school questionnaires.

Value of TIMSS Information

Educational policy is formulated and implemented by large numbers of decision-makers and practitioners at all levels of the educational system, from the highest governmental or ministerial level to the actions taken by teachers and students each minute in each classroom. Each participant in the vast and complex network of the educational process works extremely hard in delivering instruction to students, but to make teaching and learning as effective as possible, it is important to have solid factual information on which to base plans and actions.

First, TIMSS provides comprehensive information about what mathematics and science concepts, processes and attitudes students have learned. In an environment of global competitiveness and local accountability, it is important for countries to monitor the success of their mathematics and science instruction and TIMSS provides a vehicle for doing this with a degree of validity and reliability, while meeting the highest standards for technical quality. It provides achievement data in mathematics and science overall and in major content areas, permitting informed between-country comparisons of student achievement as well as changes over time. Countries can monitor strengths and weaknesses in student learning; both in the absolute sense of attaining standards and in relation to what is happening in other countries.

Understanding the contexts in which students learn enables a fuller appreciation of what the TIMSS achievement results mean and how they may be used to improve student learning in mathematics and science. In addition to testing students' achievement in mathematics and science, TIMSS enables international comparisons among key policy variables relating to curriculum and instruction. By administering a series of questionnaires, TIMSS collects a range of information about the contexts in which students study and learn mathematics and science and links achievement to a host of factors. The essential questions for TIMSS have to do with what it is about the curriculum, the instructional practices, the allocation of resources and the sociological environment in certain countries, that result in higher levels of student achievement.

TIMSS has the additional benefit of providing a vehicle for countries to address internal policy issues. TIMSS allows for comparisons among educational jurisdictions at the local, state, regional and national levels. Within countries, TIMSS provides an opportunity to examine the performance of population subgroups, for example, by race/ethnicity, gender or region, to ensure that concerns for equity are not neglected. It is efficient for countries to add questions of national importance (national options) as part of their data collection effort.

Educators from different national and cultural backgrounds can use the TIMSS results as a kind of mirror that reflects their education systems in an international perspective. While much of what TIMSS does is to describe 'what is' in mathematics and science instruction and learning, its power is most fully realized in considering the between-country perspective. Given the differences in the ways in which education is organized and practiced across cultures and societies, a comparative perspective such as that provided by TIMSS not only enables better understanding of the current situation but serves to expand horizons as to what is possible.

The TIMSS Curriculum Model

Building on earlier IEA studies of mathematics and science achievement, TIMSS uses the curriculum, broadly defined, as the major organizing concept in considering how educational opportunities are provided to students, and the factors that influence how students use these opportunities. The TIMSS curriculum model has three aspects: the intended curriculum, the implemented curriculum and the achieved curriculum (see Figure 2.1). These represent, respectively: (1) the mathematics and science that society intends for students to learn and how the education system is organized to facilitate this learning; (2) what is actually taught in classrooms, who teaches it, and how it is taught; and, finally, (3) what it is that students have learned, and what they think about these subjects.

Working from this model, TIMSS uses mathematics and science achievement tests to describe student learning in the participating countries, together with questionnaires to provide a wealth of information about the educational context. In particular, TIMSS asks countries about the curricular goals of the education system and how the system is organized to attain those goals. Teachers and school administrators are asked about the educational resources and facilities provided as well as about specific classroom activities and characteristics. Teachers are asked about their educational preparation, support in their teaching and opportunities for further development. Students are asked about home involvement and support for learning, experiences outside of school related to mathematics and science learning, and about classroom activities in mathematics and science. Finally, students and teachers are asked about the knowledge, attitudes and predisposition that they themselves bring to the educational enterprise. Chapter 3 provides more information about the analysis of the data from the questionnaires.

Measuring Trends across Time and Grades

To help countries monitor and evaluate the efficacy of their mathematics and science teaching, TIMSS is settling into a regular data collection pattern of assessments at the fourth and eighth grades every four years. This will permit countries to:

- Highlight aspects of growth in mathematical and scientific knowledge and skills from the fourth to eighth grades.

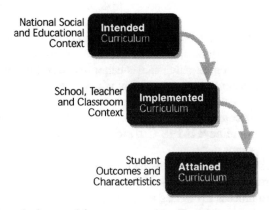

Figure 2.1 TIMSS curriculum model.

- Assess progress in mathematics and science learning across time for students at the fourth grade.
- Assess progress in mathematics and science learning across time for students at the eighth grade.
- Monitor the relative effectiveness of teaching and learning at the fourth as compared to the eighth grade, since the cohort of fourth-grade students will be assessed again as eighth graders.

With each cycle, TIMSS releases test questions to the public and then replaces these with newly developed questions. For example, a large portion of the items used in 1995 had not been released; they were kept confidential. The TIMSS 1999 test booklets consisted of the non-released items from the 1995 study supplemented by new items. Similarly, TIMSS 2003 included unreleased items from 1995 and 1999.

Frameworks and Tests for Assessing Mathematics and Science Achievement in an International Context

The TIMSS mathematics and science tests are based on comprehensive frameworks developed cooperatively by the participating countries. TIMSS framework development involves widespread participation and reviews by educators around the world. *The Curriculum Frameworks for Mathematics and Science* (Robitaille *et al.*, 1993) formed the basis for the 1995 and 1999 assessments. These frameworks encompassed a wide range of mathematics and science topics and several levels of cognitive functioning.

For TIMSS 2003, the frameworks were updated and focused on the fourth and eighth grades, although the content domains were basically the same as those of the previous TIMSS studies. Designed to permit the content assessed by TIMSS to evolve, the *TIMSS Assessment Frameworks and Specifications 2003* (Mullis *et al.*, 2003) reflect changes during the last decade in curricula and the way mathematics and science are taught. In particular, the 2003 frameworks were expanded to provide specific objectives for assessing students at the fourth and eighth grades. Figures 2.2 and 2.3 show, for mathematics and science, respectively, the content and cognitive domains assessed in TIMSS 2003, and the target percentages of the assessment devoted to each domain by grade level.

The frameworks do not consist solely of content and behaviors included in the curricula of all participating countries. The aim of extensive consultation and review is to ensure that goals of mathematics and science education regarded as important in a significant number of countries are included. The ability of policy-makers to make sound judgments about relative strengths and weaknesses of mathematics and science education in their systems depends on achievement measures being based, as closely as possible, on what students in their systems have actually been taught.

Based on the frameworks, the TIMSS tests are developed through an international consensus-building process involving input from experts in education, mathematics, science and measurement. After training in TIMSS item development procedures and

	Fourth Grade	Eighth Grade
Mathematics Content Domains		
Number	40%	30%
Algebra*	15%	25%
Measurement	20%	15%
Geometry	15%	15%
Data	10%	15%
Mathematics Cognitive Domains		
Knowing Facts and Procedures	20%	15%
Using Concepts	20%	20%
Solving Routine Problems	40%	40%
Reasoning	20%	25%

*At fourth grade, the Algebra content domain is called
Patterns, Equations, and Relationships.

Figure 2.2 Target percentages of TIMSS 2003 mathematics assessment devoted to content and cognitive domains by grade level.

	Fourth Grade	Eighth Grade
Science Content Domains		
Life Science	45%	30%
Physical Science	35%	*
Chemistry	*	15%
Physics	*	25%
Earth Science	20%	15%
Environmental Science	*	15%
Science Cognitive Domains		
Factual Knowledge	40%	30%
Conceptual Understanding	35%	35%
Reasoning and Analysis	25%	35%

*At fourth grade, Physical Science will be assessed as one
content area including both physics and chemistry topics.
Some understandings related to Environmental Science will
be assessed as part of the Life Science and Earth Science
content domains at fourth grade.

Figure 2.3 Target percentages of TIMSS 2003 science assessment devoted to content and cognitive domains by grade level.

provided with a manual detailing how to develop mathematics and science items for TIMSS, all participating countries that wish to do so submit items to the TIMSS International Study Center for review and possible inclusion in the tests. The items are reviewed extensively by panels of experts and by representatives from the participating countries. After field-testing with nationally representative samples, items are selected for the assessment. TIMSS 2003 included 696 items (194 mathematics and 189 science items at eighth grade as well as 161 mathematics and 152 science items at fourth grade).

The tests contain questions requiring students to select appropriate responses or to solve problems and answer questions in an open-ended format. From the early development of TIMSS, there has been widespread agreement that the item pool not be limited simply to multiple-choice items. In the beginning there was widespread support for authentic or performance testing, which has evolved into an emphasis on problem-solving and inquiry tasks. As part of the item development process, scoring guides (rubrics) are developed for each of the constructed-response items. Both for the field test and the actual assessment, considerable time and energy are devoted to collecting sample student responses, developing training materials and conducting training sessions for those responsible for coordinating the scoring of those items in their own country.

According to procedures specified by the TIMSS International Study Center, countries are responsible for collecting information about the reliability of the scoring within and across countries as well as across time. Most recently, the IEA Data Processing Center has been capitalizing on the capabilities of image processing to provide countries with files of students' responses to use in monitoring scoring reliability. The scoring reliability studies within and across countries to monitor the degree of agreement indicate high (typically 80 to 90 percent) agreement between independent ratings.

TIMSS development is conducted in English, necessitating translation of the items into the more than 30 different languages in use in the study. Country representatives are trained in translation procedures and provided with a manual detailing procedures designed to minimize the introduction of any systematic bias to the items in the translation process. National centers are encouraged to have at least two independent translations made for each instrument, and to put in place a process for reconciling differences between or among the translations. For both the field test and the actual assessment, countries must send their translated instruments to IEA for verification by a professional translation company. Implementation of suggested changes in translations is the responsibility of each country, but is monitored by IEA and the TIMSS International Study Center. Finally, as part of the data analysis process, a number of statistical checks are undertaken to identify problematic translations.

Target Populations

One of the challenges in making international comparisons of student achievement lies in deciding how to identify the population of students in each country to be compared. Essentially the choice is between a definition based on student age, for example, all students born in a particular year, or one based on grade level, for example, all students in the eighth grade of formal schooling. Age-level definitions lead to populations scattered across several grades, with resulting differences in exposure to the curriculum, as well as administrative difficulties in collecting data. Definitions based on grade level have the advantage that students all have had approximately the same amount of instruction, although the average age of students in a particular grade will not be the same in all countries because school starting age and admission practices vary across countries. Additionally, grade-level population definitions make it possible to sample intact classes, which are both more convenient administratively and facilitate linking teacher information to the students in the class.

Because TIMSS considers providing information about curriculum and instruction of central importance, and in almost all countries these are organized around grade levels, TIMSS chose grade as the basis of its population definitions. However, TIMSS also uses students' age to help countries identify the appropriate grades in primary and middle schools to be tested. For example, for the two younger populations assessed in TIMSS 1995, the formal definition of 'the two adjacent grades containing most 13-year-olds at the time of testing' led to the seventh and eighth grades in most countries and the definition of 'the two adjacent grades containing the most 9-year-olds' led to the third and fourth grades in most countries.

Keeping the target population the same for purposes of measuring trends, but focusing on the eighth grade only, TIMSS 1999 asked countries to choose the 'upper of the two adjacent grades with most 13-year-olds at the time of testing'. Where there was any doubt about the appropriate choice, countries were encouraged to choose the eighth grade. To measure trends, TIMSS 2003 continued with the focus on the grade representing eight years of formal schooling for students in middle or lower secondary school, and used a similar approach in determining the primary or elementary school population to be assessed (focusing on the grade representing four years of formal schooling).

Sampling Design

TIMSS is a cross-sectional survey of the mathematics and science achievement of students in participating countries. To be cost-effective, TIMSS uses sample survey methods to draw samples of schools and students representative of the populations in the target grade in each country. Conducting the assessment on samples rather than the whole student population greatly reduces the data collection burden, while the TIMSS sampling design ensures that the results are sufficiently precise to provide reliable measures of trends in student achievement. Statistics Canada develops the sampling design and procedures, and in consultation with members of the TIMSS collaborative, assists countries in adapting the general design to national circumstances and priorities.

The TIMSS quality standards require that the international population definition (the *international desired population* in IEA parlance) be implemented in each country in a way that includes all eligible students, and where this is not possible, that any omissions or exclusions be clearly documented. In TIMSS 1999, for example, this meant including all eighth-grade students and nearly all 38 participating countries defined their target populations accordingly (except Latvia and Lithuania restricting the study to Latvian-speaking and Lithuanian-speaking students, respectively).

While requiring that exclusions be kept to a minimum, TIMSS recognizes that a small percentage of otherwise eligible students in each country have to be excluded from sampling for a variety of reasons: because they attend a very small school or one in a remote area; because they attend a special school for mentally-handicapped students; because they do not speak the language of the test; or because they have a disability that prevents them from completing the assessment. The TIMSS 1999 countries all complied with the standards for inclusion, except for Israel, which excluded 16 percent of its eighth-grade population.

Recognizing that the participating countries vary considerably in size and in diversity and that the sampling strategy has to be tailored to some extent to each one, TIMSS uses a basic sampling design that is efficient yet flexible and straightforward to apply. Essentially, the design calls for sampling in two stages: first, a sample of schools containing students at the target grade, and then a sample of intact mathematics classes from the target grade in those schools. Countries adapt the basic design to local requirements, with support and supervision from Statistics Canada, the TIMSS sampling consultants.

Because sample survey methods inevitably introduce some unreliability in the results through variability of the sampling process, TIMSS determines the size of the sample in each country according to the precision requirements of the study. The TIMSS standard is that the margin of error should be no more than 5 percent for survey estimates of percentages.[2] To have samples large enough for analyses at school and classroom level, countries are requested to sample at least 150 schools from the target population. For example in TIMSS 1999, sampling 150 schools with one eighth-grade class per school (3,500 students, approximately) was sufficient in most countries to meet the TIMSS sampling precision guidelines. For TIMSS 2003, to broaden analysis possibilities, most countries had sample sizes of 4,000 to 5,000 students.

Because of the importance in the TIMSS design of presenting student achievement in the context of curricular goals and implementation and instructional practice, and since teachers are the primary source of much of this information, it is necessary to link the responses of students and teachers. In order to facilitate this student–teacher linkage, students are sampled in intact classes from participating schools rather than individually from across the grade level. Generally at the fourth grade, and sometimes at the eighth, students form the same class for both mathematics and science instruction, and so there is no difficulty in listing classes for sampling purposes. By the eighth grade, however, more often students are formed into different classes for mathematics and science. In this situation, TIMSS asks countries to sample mathematics classes, since these often reflect a more general classroom assignment in schools.

To assist countries in conducting sampling operations, Statistics Canada has developed detailed manuals showing step-by-step how to list the school population and draw the school sample, and actually selects the school sample for many of the participating countries. Since the TIMSS within-school sampling procedure can be an onerous task (for example at the eighth grade, sampling an intact eighth-grade mathematics class, listing all students in the class, and identifying their associated science teachers), the IEA Data Processing Center worked with Statistics Canada to provide software to reduce the burden on national centers and document the entire process.

TIMSS expects countries to make every effort to ensure that all sampled schools and all students in those schools participate in the study. For example, all countries in 1999 succeeded in meeting the minimum requirements of at least 85 percent of sampled schools and 85 percent of sampled students, although Belgium (Flemish-speaking part), England, Hong Kong SAR and the Netherlands did so only after including replacement schools in their samples.

Assessment Booklet Design

Ambitious goals for measuring student achievement in mathematics and science are an important characteristic of TIMSS. From the beginning in 1995, implementing the coverage goals as described in the *Curriculum Frameworks for Mathematics and Science* (Robitaille *et al.*, 1993) resulted in a large pool of mathematics and science items (more than 300 items) comprising many more hours of testing than was reasonable to administer to any one student. To overcome this problem, TIMSS arranges the items into student booklets according to a complex scheme, so that each booklet contains a subset of the item pool. The booklets are assigned to participating students according to a predetermined plan, so that each student receives just one booklet and the students responding to each booklet constitute equivalent random sub-samples of the entire test-taking student sample. Depending on the grade assessed, the booklets are to be completed in about 60 to 90 minutes, with a short break in the middle.

In 1995 and 1999 at the eighth grade, the items were arranged into eight separate 90-minute booklets with some items appearing in just one booklet and others in two, three or more booklets. Each booklet contained sufficient mathematics and science items to provide estimates of student proficiency in each subject. Given that each student booklet had to contain broad coverage of both mathematics and science items, include a mix of multiple-choice, short-answer and extended-response response formats, and be of approximately equivalent difficulty, a 'cluster-based design' was used (Adams and Gonzalez, 1996). Each item was assigned to one of 26 blocks or clusters, and the booklets then were assembled from the clusters in a way that maximized the linkages between items and students. Because TIMSS 2003 included even more items (more than 300 at fourth grade and almost 400 at eighth grade), the booklet design was modified so that each item is found in one of 14 mathematics or 14 science clusters. The 28 clusters are arranged into 12 booklets, with each cluster appearing in at least two booklets.

Because each student responds to just a portion of the mathematics and science items in the entire pool, TIMSS uses a form of psychometric scaling based on item response theory with multiple imputation (Yamamoto and Kulick, 2000) to estimate and report student performance. As well as reporting student proficiency in both mathematics and science overall, TIMSS reports achievement in mathematics and science content areas.

Context Questionnaire Design[3]

In all three data collections, TIMSS included a questionnaire for students. It also included questionnaires for their mathematics and science teachers as well as for the principals or heads of their schools. Furthermore, in 1999 and 2003 TIMSS began collecting curriculum information through questionnaires. Information about the intended curriculum in mathematics and science, that is how the curriculum is organized in each country and which topics were intended to be covered up to the eighth grade, was collected through mathematics and science curriculum questionnaires. The National Research Coordinator (NRC) in each country completed one questionnaire about the mathematics curriculum and one about science, drawing on the expertise of mathematics and science specialists in the country as necessary.

Information about how schools are organized and how well provided with resources to implement the intended curriculum was collected through the school questionnaire, which was completed by the school principal or administrator. In particular, principals were asked about staff levels and qualifications, instructional resources, including information technology and Internet access, instructional time for mathematics and science, student behavior, admission policies and parental involvement.

Ultimately, the curriculum is implemented by teachers working with their students, so TIMSS collected much of the information about the implemented curriculum from the teachers of the students in the assessment. There were separate questionnaires for each student's mathematics and science teachers, each in two parts. The first part dealt with teachers' background, experience, educational preparation, attitudes, and teaching and administrative duties. The second part focused on teaching mathematics or science to the class sampled for the TIMSS testing. It included questions on classroom instruction such as the areas of mathematics or science emphasized most; use of textbooks, computers and calculators; allocation of instructional time; resources for mathematics and science instruction; homework and assessment; and especially the extent of coverage of a range of mathematics and science topics included in the assessment.

Each student in the assessment was asked to complete a questionnaire about home background, attitudes and beliefs about mathematics and science, and experiences in mathematics and science class. To accommodate the diversity of the educational systems participating in TIMSS, there were two versions of the questionnaire: a general science version for use by countries where eighth-grade science is taught as a single subject, and a separate-science-subject version for countries teaching science subjects separately (biology, chemistry, physics, earth science). By asking students about what they do and what they study in mathematics and science class, the equipment they use, and the homework and tests they get, the student questionnaires supplemented the information provided by teachers on the implemented curriculum. Student questionnaires also complemented the information on the achieved curriculum from the TIMSS mathematics and science achievement tests by providing data on students' attitudes and self-concepts with regard to mathematics and science, and on their academic expectations.

Data Collection and Monitoring

Data collection for the fourth- and eighth-grade students occurs toward the end of the school year for each country. Thus, for the original TIMSS in 1995, data collection for these two grades in the northern hemisphere took place in the first half of 1995, and in the southern hemisphere was carried out in the latter part of 1994. Similarly, data collection for TIMSS 1999 and TIMSS 2003 was in the first half of those years, respectively, for the northern hemisphere and in the latter part of 1998 and 2002, respectively, for the southern hemisphere countries.

The TIMSS International Study Center provides detailed manuals and trains those responsible for overseeing data collection in their own countries in standardized test administration procedures. Each country is expected to monitor data collection procedures and document the results for review by the TIMSS International Study Center. Additionally, each country is asked to identify a person to serve as an IEA quality control monitor for their national study. The IEA Quality Control Monitors attend an international training session during which they are briefed on the kinds of activities they are expected to perform. These include visiting their national center and interviewing the NRCs, visiting a number of randomly selected schools participating in the study, and observing the data collection process in a number of classes. Each IEA Quality Control Monitor provides an independent report to the TIMSS International Study Center, and the reports of the quality-control monitors indicate the test administrators generally adhere to the procedures.

Finally, it should be noted that TIMSS conducts extensive training sessions in data entry methods. The IEA Data Processing Center in Hamburg, Germany works very closely with the data managers in the participating countries throughout the data entry process. Upon delivery of the data files to Hamburg, staff members painstakingly check the records from each country for within- and across-country accuracy and comparability.

Analysis and Reporting

In line with its mission of informing educational policy, the TIMSS reporting strategy is to provide clear, concise, and timely information about student achievement in mathematics and science in the participating countries. The 'International Reports' prepared for each subject and each grade are the major vehicle for releasing the results for all of the participating countries. Each country also works on a series of national reports.

The TIMSS international reports present student achievement in mathematics and science, respectively, from a variety of perspectives, including overall performance over time; relative performance in content areas such as algebra, geometry, life science, physics, and so forth; comparisons with international benchmarks of achievement; and results for individual items. The international reports also provide extensive information about the educational context in each country, including the organization and coverage of the mathematics and science curriculum, teacher preparation and classroom instruction, school organization, resources, and problems, and student home background and attitudes to mathematics and science. The results for TIMSS

1999 can be found in the *TIMSS 1999 International Mathematics Report* (Mullis, Martin, Gonzalez *et al.*, 2000) and the *TIMSS 1999 International Science Report* (Martin, Mullis *et al.*, 2000). The results for TIMSS 2003 can found in the *TIMSS 2003 International Mathematics Report* (Mullis *et al.*, 2004) and the *TIMSS 2003 International Science Report* (Martin, Mullis, Gonzalez *et al.*, 2004).

TIMSS also publishes technical information and the full database with each assessment cycle. All of the TIMSS publications, including the international databases, are available on the TIMSS web site (http://timss.bc.edu). For example, to foster understanding of the methods employed, the *TIMSS 1999 Technical Report* (Martin, Gregory and Stemler, 2000) and the *TIMSS 2003 Technical Report* (Martin, Mullis and Chrostowski, 2004) describe the design, development and implementation of the study, and the techniques used to analyse and report the results. To promote use of the results in participating countries, the *TIMSS 1999 User Gide for the International Database* (Gonzalez and Miles, 2001) provides educational researchers and policy analysts access to all of the TIMSS data in a convenient form, together with software to facilitate analysis and extensive documentation describing all data formats and variables. The TIMSS 2003 database will be documented and available to the public by June 2005. Since the data are extensive and complex, TIMSS provides training in processing and analysis to assist countries in using this valuable resource to address policy issues of relevance to their systems. The aim for TIMSS is maximum use of the data in each country to interpret the findings about differences among countries in the context of the uniqueness of their educational settings.

Notes

1. Originally named the Third International Mathematics and Science Study.
2. More specifically, a 95 percent confidence interval for a percentage of plus or minus 5 percent (see Foy and Joncas, 2000, for more detail).
3. More detailed information about the TIMSS 1999 questionnaires and their development may be found in Chapter 3 of this book, and also in Mullis, Martin and Stemler (2000).

References

Adams, R.J. and Gonzalez, E.J. (1996) The TIMSS test design. In M.O. Martin and D.L. Kelly (eds) *Third International Mathematics and Science Study Technical Report, Volume I: Design and Development*. Chestnut Hill, MA: Boston College.

Foy, P. and Joncas, M. (2000) TIMSS sample design. In M.O. Martin, K.D. Gregory and S.E. Stemler (eds) *TIMSS 1999 Technical Report*. Chestnut Hill, MA: Boston College.

Garden, R.A. and Smith, T.A. (2000) TIMSS test development. In M.O. Martin, K.D. Gregory and S.E. Stemler (eds) *TIMSS 1999 Technical Report*. Chestnut Hill, MA: Boston College.

Gonzalez, E.J. and Miles, A.M. (1999) *The TIMSS 1999 User Guide for the International Database*. Chestnut Hill, MA: Boston College.

Hiebert, J., Gallimore, R., Garnier, H., Givven, K.B., Hollingsworth, H., Jacobs, J., Chiu, A.M.-Y., Wearne, D., Smith, M., Kersting, N., Manaster, A., Tseng, E., Etterbeek, W., Manaster, C., Gonzales, P. and Stigler, J. (2003) *Teaching Mathematics in Seven Countries: Results from the TIMSS 1999 Video Study* (NCES 2003-013). US Department of Education, Washington, DC: National Center for Education Statistics.

Martin, M.O., Gregory, K.D. and Stemler, S.E. (eds) (2000) *TIMSS 1999 Technical Report*. Chestnut Hill, MA: Boston College.

Martin, M.O., Mullis, I.V.S., Gonzalez, E.J., Gregory, K.D., Smith, T.A., Chrostowski, S.J., Garden, R.A. and O'Connor, K.M. (2000) *TIMSS 1999 International Science Report: Findings from IEA's Repeat of the Third International Mathematics and Science Study at the Eighth Grade*. Chestnut Hill, MA: Boston College.

Martin, M.O., Mullis, I.V.S. and Chrostowski, S.J. (eds) (2004) *TIMSS 2003 Technical Report*. Chestnut Hill, MA: Boston College.

Martin, M.O., Mullis, I.V.S., Gonzalez, E.J. and Chrostowski, S.J. (2004) *TIMSS 2003 International Science Report: Findings from IEA's Trends in International Mathematics and Science Study at the Fourth and Eighth Grades*. Chestnut Hill, MA: Boston College.

Mullis, I.V.S., Martin, M.O. and Stemler, S.E. (2000) TIMSS questionnaire development. In M.O. Martin, K.D. Gregory and S.E. Stemler (eds) *TIMSS 1999 Technical Report*. Chestnut Hill, MA: Boston College.

Mullis, I.V.S., Martin, M.O., Gonzalez, E.J., Gregory, K.D., Garden, R.A., O'Connor, K.M., Chrostowski, S.J. and Smith, T.A. (2000) *TIMSS 1999 International Mathematics Report: Findings from IEA's Repeat of the Third International Mathematics and Science Study at the Eighth Grade*. Chestnut Hill, MA: Boston College.

Mullis, I.V.S., Martin, M.O., Smith, T.A., Garden, R.A., Gregory, K.D., Gonzalez, E.J., Chrostowski, S.J. and O'Connor, K.M. (2003) *TIMSS Assessment Frameworks and Specifications 2003, 2nd edition*. Chestnut Hill, MA: Boston College.

Mullis, I.V.S., Martin, M.O., Gonzalez, E.J. and Chrostowski, S.J. (2004) *TIMSS 2003 International Mathematics Report: Findings from IEA's Trends in International Mathematics and Science Study at the Fourth and Eighth Grades*. Chestnut Hill, MA: Boston College.

Robitaille, D.F., Beaton, A.E. and Plomp, T. (eds) (2000) *The Impact of TIMSS on the Teaching and Learning of Mathematics and Science*. Vancouver, British Columbia: Pacific Educational Press.

Robitaille, D.F., Schmidt, W.H., Raizen, S., McKnight, C., Britton, E. and Nicol, C. (1993) *Curriculum Frameworks for Mathematics and Science*. TIMSS Monograph No. 1. Vancouver, Canada: Pacific Educational Press.

Yamamoto, K. and Kulick, E. (2000) Scaling methodology and procedures for the TIMSS mathematics and science scales. In M.O. Martin, K.D. Gregory and S.E. Stemler (eds) *TIMSS 1999 Technical Report*. Chestnut Hill, MA: Boston College.

3

Using Indicators of Educational Contexts in TIMSS

Alka Arora, María José Ramírez and Sarah J. Howie

Introduction

The purpose of TIMSS, the Trends in International Mathematics and Science Study (formerly called the Third International Mathematics and Science Study), is to provide a base from which policymakers, curriculum specialists and researchers can better understand the performance of their educational systems. With this aim, TIMSS collects data on hundreds of contextual variables from nationally representative samples of students, their science and mathematics teachers, and their schools. TIMSS, like other IEA studies, aims to design and develop its questionnaires to maximize the validity and reliability of the data, while simultaneously keeping the instruments clear, simple, concise and manageable.

This chapter discusses how this background information was designed and developed (focusing on the TIMSS 1999 study) and a number of challenges that faced this process. Thereafter, it discusses how these data can be used and reported using an example from the 1999 assessment. This example illustrates the main concepts and steps involved in reporting background data, while presenting methods and statistical techniques that can be used for developing multiple-item indicators. Important criteria to evaluate the overall quality of the measures are introduced, paying special attention to their validity and reliability. The advantages of using these complex measures are discussed both from a conceptual and a measurement perspective.

Background Data in TIMSS

TIMSS administered a number of questionnaires in order to collect contextual information that would enhance the understanding of the test outcomes. First, there was the

Curriculum Questionnaire completed by the National Research Coordinators (NRCs) assisted by the curriculum specialists from the respective national departments and ministries of education. This questionnaire focused on information regarding the organization, phasing and content coverage of the mathematics and science curricula. Second, there was a School Questionnaire which addressed a number of issues at school level. These included staffing, resources, organization and administration and course offerings in mathematics and science (see Appendix 1). Third, questionnaires were designed for the mathematics and science teachers (Teacher Questionnaire) of those pupils who had been tested. The contents of these focused primarily on the classroom level and sought background information from the teachers about their training, experience, class preparation and beliefs about their subject and, furthermore, included information on their instructional practices in the classroom (see Appendix 2). Finally, a Students' Questionnaire was included to collect information about the pupils' home background, attitudes towards education and mathematics and science in particular, their academic self-concept, classroom activities and extra-mural activities (see Appendix 3).

The questionnaires for both the TIMSS 1995 and the 1999 studies were conceptualized based upon the Conceptual Framework for TIMSS (Robitaille, 1993). This in turn had been heavily influenced by the Second International Mathematics Study. The curriculum was considered the core and the single most significant explanatory variable in student achievement. Three aspects were identified, namely, the intended (what society intends children to learn), the implemented (what is actually taught in the classroom) and the attained curriculum (what children actually learn and experience). Through this framework, the study sought to identify factors that might play a role in affecting children's learning in mathematics and science.

The challenge of the 1999 study was that, in addition to being a cross-sectional study, it also aimed to analyze trends in the data collected between 1995 and 1999. Therefore the instruments were largely derived from the 1995 study, with some additional items being included or others excluded from the questionnaires based on the experiences of the 1995 study. The instruments were also modified as a consequence of the inclusion of a large number of developing countries in the 1999 study. The latter meant that some items were rendered unsuitable for the majority of these countries, given the very different contexts, while others needed some adaptation. Furthermore, the original 1995 questionnaires were considered too long by most NRCs. Examples of changes include the following: questions omitted in the teacher questionnaire included those about grades taught while questions about teacher education and preparation were added; a number of questions regarding teaching activities in a recent lesson were omitted and those on the coverage of topics were shortened and simplified; sections on opportunity to learn as well as the pedagogical approach were judged too lengthy and were removed; some additional questions were included regarding the use of computers and the Internet (Martin *et al.*, 2000: 75). However, all the changes posed a significant challenge to the study given the desire to study trends within the data.

The instruments were informed by a series of research questions. The main research questions of TIMSS were:

- What kinds of mathematics and science are children expected to learn?
- Who provides instruction?
- How is instruction organized?
- What have students learned?

In addition to these, there were also specific questions underpinning the development of each of the questionnaires.

The curriculum questionnaires were informed by:

- Is there a country-level curriculum? If so, how is implementation monitored?
- What is the nature of country-level assessments, if there are any?
- What content is emphasized in the national curriculum?

The following research questions were addressed by the school questionnaire:

- What staffing and resources are available at each school?
- What are the roles and responsibilities of the teachers and staff?
- How is the mathematics curriculum organized?
- How is the science curriculum organized?
- What is the school climate?

The teacher questionnaires were designed to address the following questions:

- What are the characteristics of mathematics and science teachers?
- What are the teachers' perceptions about mathematics and science?
- How do teachers spend their school-related time?
- How are mathematics and science classes organized?
- What activities do students do in their mathematics and science lessons?
- How are calculators and computers used?
- How much homework are students assigned?
- What assessment and evaluation procedures do teachers use?

The student questionnaires were underpinned by the following research questions:

- What educational resources do students have in their homes?
- What are the academic expectations of students, their families and their friends?
- How do students perceive success in mathematics and science?
- What are students' attitudes towards mathematics and science?

Due to the cultural and contextual diversity, countries were permitted to modify the items within certain parameters. In the TIMSS 1999 study, the instruments originally developed in English were also translated into 33 languages with at least 10 countries collecting data in two languages. An extensive process of verification was developed to ensure the accuracy of the modifications and translations, both nationally and inter-nationally, with every modification accounted for and argued. The verification ensured that the meaning of the question was not changed, that the reading level of the text was retained, and that the difficulty of the item was not altered.

Ultimately, the study had to ensure that all the instruments remained comparable across all the participating countries and that the questionnaires made provision for an enormously diverse population across 38 countries. In addition to this, the study had to retain its ability to measure trends from the 1995 study. One question that naturally arises is whether it is possible to develop questionnaire items which have cross-national validity or whether validity is more likely to be achieved regionally or nationally. Several examples of the difficulties facing the study arose from TIMSS 1995. In the following paragraphs four of those that were addressed in the 1999 study are presented.

The first example pertains to the questions asking about the educational levels of the parents in the students' questionnaire. In the 1995 version, no option was provided at the lowest level for 'did not attend schools or some years of primary school only'. With the inclusion of several Third World countries in 1999, it became obvious that this option was important to ensure the cross-national validity of a large number of countries participating in the study. One other challenge to asking children what education level their parents have attained is that they simply do not know. This is the case for both developed and developing nations. Another challenge to gaining this information is the issue of the comparability of education internationally. Once one has the data, how to interpret the different systems of post-secondary education becomes the next challenge.

The second example, namely the one of education resources in the home, presented the study with interesting options. Four common items were included (calculator, dictionary, study desk, computer) and the statistics provide one with the enormous inequities both within countries and between countries in terms of these. However, trying to gather data on other resources in the home so as to compare or reach a measure of social-economic status (SES) was too challenging in the end and an international option was provided to countries instead, which permitted countries to include additional items relevant for their context. For instance, in the South African Student Questionnaire, items such as running water, electricity and flushed toilets were included among others. Again, the difficulty of interpreting or equating resources is a complex issue for any international or comparative study as these are entirely contextually bound.

How to compare teacher training cross-nationally proved to be a very difficult item to analyze. With qualifications varying immensely in terms of content, level and resource allocation, this item demonstrated perhaps the worth of having more regional validity (for example, East Bloc countries in Europe where similar systems were in place) than cross-national validity. Obviously this is a key item for analysis and most countries are interested in how teachers are trained and to what level for both primary and secondary education. However, where many developing countries have traditionally educated their teachers in teacher training colleges, most teachers in industrialized countries have long since moved this responsibility to universities.

The last example is that of the type of community. Location has been shown to be an important variable in education. However, the difficulty in defining cross-nationally what is meant by rural or isolated is a great challenge. Rural in the Netherlands may mean that pupils in those schools are at most 30–40 minutes away from a large town or city with all the amenities. In contrast, in Africa this could be hours from a small town with very few amenities. Perhaps an index of isolation needs to be considered where the geographically isolated area means a lack of access to basic amenities such as electricity, water, libraries and teacher facilities.

While there are several such examples of the challenges of designing and developing items with cross-national validity, a more in-depth discussion goes beyond the scope of this chapter. TIMSS has merely scratched the surface when dealing with these and continues to be confronted with the delicate balance of satisfying many needs in the study.

In the following section there is a discussion of the indicators that can be designed based upon items in the questionnaires, which touches upon the discussion above.

How Are the TIMSS Background Indices Developed?

Once the background data are collected, one of the major challenges for TIMSS is reporting this vast array of information in a thorough and meaningful way. There is a need to focus on relevant educational contexts, inputs and processes, while avoiding overburdening the audiences with unmanageable amounts of information. Explaining the methodology used in reporting TIMSS data is important to help better understand the meaning of these indicators. It is also important for educational researchers and students interested in developing their own indices according to their local contexts, interests and research questions. The purpose of this section is to explain how the TIMSS data can be further analyzed to decide what information to report and how to report it.

The simplest and straightforward method to report TIMSS background data is by using the same response categories provided in the questionnaires, or by introducing only slight modifications in them. This method is appropriate when measuring something directly observable or a somewhat simple phenomenon (for example, teachers' age). While the direct reporting of data from the questionnaires presents the advantage of simplicity, it would be impossible to report the vast amount of information collected by TIMSS on a one-by-one basis. Some data reduction is required to summarize hundreds of contextual variables. Another weakness of the direct reporting method is its inability to measure more complex constructs not directly observable. Finally, single-item measures have more error, and therefore data are less reliable. Because of these weaknesses, the use of multiple-item indicators is strongly recommended in measuring students' contexts for learning.

A multiple-item indicator is a *derived variable* that combines data from several items in the TIMSS questionnaires. In TIMSS this kind of variable is often called an index, and is divided in three levels: high, medium and low. An index can be used as an indicator of more complex constructs not directly observable, such as students' attitudes or school climate. Because the items making up an index can target different facets of the same object, indices can provide a more global and thorough picture of the phenomenon being studied. Indices are also preferable to single-item indicators because they can provide more precise measures of the underlying construct. This is so because random measurement errors from different variables (items) are cancelled out when data are combined into a new derived variable. For detailed information about the increased validity and reliability of multiple-item indicators, see DeVellis (1991), Spector (1992) and Sullivan and Feldman (1979).

The development of indices involves seven main steps, which are described below. Real data from five countries were used to explain the main statistical analyses. The students' Positive Attitudes Toward Mathematics (PATM) index, reported in the TIMSS 1999 International Mathematics Report, was used as an example to explain each step. In this particular example, the index was based on an underlying continuous scale. There are other TIMSS indices that are based on a somewhat different methodology, which are not described in this chapter. For instance, response options from different questions may be directly combined into a new derived variable that does not assume an underlying scale. For more detailed information about how the TIMSS indices are developed, see Ramírez and Arora (2004). Despite these methodological differences, index development always involves common principles and approaches, which are emphasized in this chapter.

Step 1: Defining the Construct and Identifying Source Questions

The starting point to develop valid indicators of educational contexts is to clearly define the construct of interest, specifying its different facets. For instance, creating a positive attitude in students toward mathematics is an important goal of the curriculum in many countries. Accordingly, it was important for TIMSS to know if the students enjoy learning mathematics, place value on it and think this is an important subject for their future career aspirations (Mullis et al., 2001). The attitude construct is a complex psychological state involving feelings, pressure, behavior, and so forth. Using multiple-item indicators was more appropriate to capture the nature of students' attitudes. One or more items can be used to measure each facet of the construct, hence increasing the content validity of the overall scale. One index can condense information from two, five, or dozens of variables (questions), provided that these variables are all conceptually related to the same construct. For example, the students' PATM index was created based on five source items: 21a, 24a, 24b, 24d and 24e (see Figure 3.1). Item 24c was discarded because it focused on the perceived difficulty of mathematics rather than on how the students felt about this subject. Students from countries with very demanding mathematics curricula might perceive mathematics as very difficult, but still like the subject very much.

Step 2: Data Screening and Reverse Scoring

Once the source items have been pre-selected, the next step is to screen the data to ensure its quality. It is important to check that the data make sense conceptually, that they are within the range of possible response values, that there is enough variation in student responses, and that the response rate is adequate.

In the TIMSS questionnaires, most of the items have a four-point Likert scale format. For example, in the question 'How much do you like mathematics?' score points were distributed as follows: like a lot = 4 points; like = 3 points; dislike = 2 points;

21. How much do you like...

   Circle one letter, A, B, C or D, for each line.

	Like a lot	*Like*	*Dislike*	*Dislike a lot*
a) mathematics?..	A	B	C	D

24. What do you think about mathematics?

   Circle one letter, A, B, C or D, for each line.

	Strongly agree	*Agree*	*Disagree*	*Strongly disagree*
a) I enjoy learning mathematics.	A	B	C	D
b) Mathematics is boring.	A	B	C	D
c) Mathematics is an easy subject.	A	B	C	D
d) Mathematics is important to everyone's life.	A	B	C	D
e) I would like a job that involved using mathematics.	A	B	C	D

Figure 3.1 Source items for students' positive attitudes toward mathematics (PATM) index.

and dislike a lot = 1 point. In the statement 'I enjoy learning mathematics' score points were distributed as follows: strongly agree = 1 point; agree = 2 points; disagree = 3 points; and strongly disagree = 4 points.

Before data screening, it is recommended that all the items are scaled in the same direction. This means that the four-point options should always indicate the higher levels of the attribute being measured; that is, they should indicate more positive attitudes toward mathematics. When the items are not originally scaled in the same direction, it is necessary to *reverse score* for some of them. To build the students' PATM index, items 24a, 24d, and 24e were reverse scored so that: strongly agree = 4 points; agree = 3 points; disagree = 2 points; and strongly disagree = 1 point.

Table 3.1 presents the number of cases, minimum and maximum scores, mean and standard deviation for each item under analysis (after reverse scoring).[1] In all, 24,876 students (93.8 percent) provided a valid response to all five items. Consistent with the four-point Likert format, the minimum and maximum values for each item were 1 and 4, respectively. In item 21a, a mean of $M = 2.90$ indicates that, on average, students reported liking mathematics. In item 24d, $M = 3.42$ indicates that, on average, students were mid-way between agreeing and strongly agreeing with 'mathematics is important to everyone's life' (after reverse scoring).

Step 3: Looking for the Underlying Factor

The next phase was to test empirically if the selected items were in fact measuring the same latent factor. In this example, it was hypothesized that only one factor (attitudes

Table 3.1 Descriptive statistics for five items related to attitudes toward mathematics.

	N	Minimum	Maximum	Mean	Std. Deviation
21a_like	25,830	1	4	2.90	0.846
24a_enjoy[a]	25,675	1	4	2.96	0.845
24b_boring	25,539	1	4	2.67	0.901
24d_important[a]	25,640	1	4	3.42	0.735
24e_job[a]	25,488	1	4	2.65	0.937
Valid *N* (listwise)	24,876				

Note.   [a] Reverse scored variable.

toward mathematics) would explain most of the covariation among the items. For instance, a student with moderate PATM would be expected to *agree* with 'I enjoy learning mathematics' and *disagree* with 'mathematics is boring'. Principal component analysis (PCA) looks for underlying factors (components) that account for the common variance shared by the items.

The *correlation matrix* is the starting place for a PCA. As shown in Table 3.2, for the five items under analysis, all the Pearson's *r* were positive and significant at $p < 0.0005$. This was an expected outcome since all the items were scaled in the same direction and the large *N* made the standard errors to shrink. One correlation was high ($r = 0.72$), four moderate ($0.45 < r < 0.65$), and five low ($r < 0.35$). Items 21a 'how much do you like mathematics?' and 24a 'I enjoy learning mathematics' had the highest correlation ($r = 0.72$). This made sense considering how close the meaning of these two items was. Item 24d 'mathematics is important to everyone's life' had low correlation with all the other items. However, since conceptually it made sense to use this item as an indicator of attitudes toward mathematics, it was kept for further analyses.

Table 3.2 Inter-item correlation for attitude related items.

	21a	24a	24b	24d	24e
21a_like	—	0.720**	0.490**	0.260**	0.520**
24a_enjoy[a]		—	0.483**	0.309**	0.546**
24b_boring			—	0.183**	0.348**
24d_important[a]				—	0.304**
24e_job[a]					—

Notes.   ** p < 0.0005
          [a] Reverse scored variable.

Building upon the variance shared among all the items, Principal Components creates new variables (components) that account for most of the variance in the original items. As shown in Table 3.3, Principal Components identified one underlying component, with loadings (component-item correlations) ranging from $r = 0.87$ to $r = 0.49$. 'I like mathematics' (21a) and 'I enjoy learning mathematics' (24a) shared the most variance with the factor. Since these two items can be considered prototype indicators of students' attitudes toward mathematics, these results served as evidence that the correct construct was being targeted.

Step 4: Assessing the Reliability of the Scale

It is important to know how reliable a scale consisting of the pre-selected items would be. In Table 3.4, the item-total correlation indicates the relationship between each item and the rest of the items in the scale combined. If all the items are targeting the same underlying construct, high item-total correlations are expected. With the exception of item 24d ($r = 0.33$), correlations were of high and moderate sizes ($r > 0.5$), thus supporting the hypothesis that the items were measuring a common underlying construct.

Table 3.3 Component matrix.

Item	Component 1
21a_like	0.851
24a_enjoy[a]	0.868
24b_boring	0.751
24d_important[a]	0.486
24e_job[a]	0.685

Notes.   Extraction method: Principal Component Analysis.
         [a] Reverse scored variable.

Table 3.4 Item-total statistics.

	Corrected item-total correlation	Alpha if item deleted
21a_like	0.698	0.701
24a_enjoy[a]	0.728	0.691
24b_boring	0.500	0.768
24d_important[a]	0.332	0.810
24e_job[a]	0.579	0.742

Note.   [a] Reverse scored variable.

Cronbach's alpha (α) was used to measure the *internal consistency* of this five-item scale. An $\alpha = 0.79$ indicated that 79 percent of the total scale variance could be attributed to systematic variance in the latent variable (attitudes toward mathematics). This is a fairly respectable reliability considering the small number of items included in the analysis.

The last column in the table shows what alpha would be if an item were deleted from the scale. In this example, deleting item 24d would lead to a 0.02 increase in the reliability from $\alpha = 0.79$ to $\alpha = 0.81$. Deleting any other item would lower the reliability. However, considering that the reliability increased, though not that much, and that item 24d is conceptually related to the construct of interest, a decision was made to keep it for the following analyses.

Step 5: Computing Index Scores

An index score is a summated number derived from several source items. They are computed by averaging the score points associated with each student response. Because random errors are cancelled out when averaging scores from different sources, indices have the advantage of increasing the precision of the scale.

Following with the attitudes index, a student with highly positive attitudes could have answered: 21a = like at lot (4 points); 24b = strongly disagree (4 points); and 24a, 24d and 24e = strongly agree (4 points each after reverse scoring). This student would have a mean index score of $(4 \times 5) / 5 = 4$. Another student with somewhat negative attitudes may have responded: 21a = dislike (2 points); 24a = disagree (2 points); 24b = strongly agree (1 point); 24d = agree (3 points); and 24e = strongly disagree (1 point). The index score for this student would equal $(2 + 2 + 1 + 3 + 1) / 5 = 1.8$.

This procedure preserves the *original metric* of the item format, thus allowing for a straightforward interpretation of the index scores. For instance, a score of 4 means that the student consistently chose the options indicating the most positive attitude toward mathematics; it is equivalent to strongly agreeing to all the positive statements about mathematics. A score of 1.8, on the contrary, can be interpreted as on average disagreeing to positive statements about mathematics.

Index scores also present the advantage of *maximizing information* when dealing with *missing data*. If attitudes toward mathematics were solely based on item 21a 'how much do you like mathematics?' and if a student did not answer it, then she would be counted as missing in the final report. However, if item 21a is only one of the component items for an index, it is still possible to compute a derived score for that student by averaging her valid responses to the other four questions only. In computing index scores, TIMSS requires that at least two-thirds of the component items had valid responses. In a scale based on five items, this rule allows for one missing response only.

Step 6: Creating Index Levels

TIMSS sometimes classifies the students in three index levels: high, medium and low. The high level of an index is set so that it corresponds to conditions or activities generally associated with good educational practice or high academic achievement. The levels are also established so that there is a reasonable distribution of students across the three levels. TIMSS strongly emphasizes the *communication* aspect of the indices, which should be easy to understand and interpret by policymakers and school personnel. Efforts are made to create index levels that make sense in the educational practice, and that can be replicated easily.

To form three index levels, two cut-points must be established in the underlying continuous scale. The index levels have to make sense conceptually rather than just having a mathematical justification. It is important to note that these cut points have an *absolute* value and similar interpretation across all the countries. No matter from which country a student comes from, if he was classified in the medium level of the PATM index, it is possible to infer what his pattern of response could have been on the source questions. This shows how the countries differ in their distribution of students in the high, medium, and low levels of the PATM index. Readers can easily compare the percentage of students at each level of the index, both within and across the countries.

In the PATM scale, the cut points were established by assigning to the high level of the index all those students with an index score greater than 3, and to the low level those with index score less than or equal to 2. The remaining students (with scale score greater than 2 and less than or equal to 3) were assigned to the medium level. Conceptually, this corresponds to grouping in the high level those students who at least *agree* to the positive statements about mathematics, and in the low level those students who *disagree* to *strongly disagree* in their responses to the positive statements.

Step 7: Assessing the Index

Once the index scores have been computed, it is important to check if the data behave according to expectations. Failure to do so may suggest problems with the validity and/or reliability of the measures. An essential step in creating indices consists in providing evidence of *criterion related validity*. If the index is measuring the intended trait, and if the trait is supposedly related to an external criterion (e.g. mathematics achievement), then this relationship should be observed in the data. Hence, it is a prerequisite to formulate hypotheses about the relationship between the index levels and achievement scores in order to test them.

The discriminating power of an index indicates the strength of its association with the achievement outcome. Power varies from country to country, and it also varies when combining the information from all the countries together. In international studies such as TIMSS, it is a challenge to develop indices that discriminate well both within and across the countries.

Table 3.5 presents the percentage of students in the high, medium and low levels of the PATM index for five countries. According to expectations, students in the high level of PATM outperformed their peers in the medium and low levels. It is noteworthy that, across the five countries, the vast majority of the students were classified in the high and medium level of the index. This suggests that the countries have been relatively successful in developing positive attitudes toward mathematics in their students.

Table 3.5 Index of students' positive attitudes toward mathematics (PATM) and mathematics achievement.

	High PATM		Medium PATM		Low PATM	
	Percentage of students	Average achievement	Percentage of students	Average achievement	Percentage of students	Average achievement
Philippines	59	365	38	328	2	–
Chile	45	408	47	385	8	379
Israel	44	472	45	474	10	445
Australia	30	544	55	520	15	508
Chinese Taipei	23	643	59	582	18	529

Notes. Because results are rounded to the nearest whole number, some totals may appear inconsistent.
A dash (–) indicates insufficient data to report achievement.

Line graphs are useful for checking the hypothesized positive trend between index levels and achievement. Figure 3.2 shows the relationship between the PATM index and mathematics achievement for five countries separately, plus the international average for all the 38 countries participating in TIMSS 1999. The slope of the lines can be used as an indicator of how well the index discriminates among students with different achievement levels. The steeper the lines, the greater are the differences between the average mathematics scores of one index level and the next.

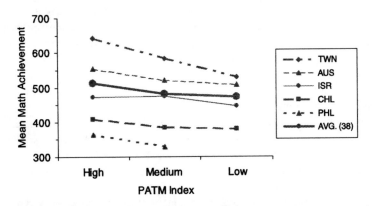

Note. AUS = Australia, CHL = Chile, ISR = Israel, PHL = Philippines, TWN = Chinese Taipei, AVG = International average for 38 countries participating in TIMSS 1999.

Figure 3.2 Relationship between the index of students' positive attitudes toward mathematics (PATM) and mathematics achievement.

The capacity of the PATM index to differentiate between students varied from country to country: it was higher for Chinese Taipei and lower for Israel. For Chile and Australia, the discrimination between high and medium levels was better than the discrimination between medium and low levels. For the Philippines, only 2 percent of the total cases lay in the low level of the index; achievement for this level was not reported due to insufficient data to produce a reliable estimate. The international average line shows that the index could discriminate well among students with different mathematics achievement.

Conclusion

In deciding what to report and how to report it, TIMSS evaluates the suitability of different indices for all countries together and separately for each country. Suitability is here understood as the conceptual and statistical adequacy of the measures. Indices that combine data from several background questions present the advantage of being more stable (reliable) measures, thus increasing the validity of the data. The indices reported by TIMSS are conceptually related to the traits or phenomenon being measured, and also are related to academic achievement, meaning that they can discriminate adequately both across and within the countries. Representatives from each country review the proposed measures and, by approving them, provide a seal of their utility and appropriateness to inform about their local educational contexts.

While there are many approaches for developing index scores (e.g. standardized measures, IRT scale scores) the ones used by TIMSS present the advantage of being valid and reliable measures across different contexts, while being simple enough to be understood by non-technical users. Parsimony in reporting background data allows for a better communication with the readers of the TIMSS reports. It also provides a base for replicating and developing new indicators of educational context. This is of special importance for those interested in carrying on secondary analyses using the TIMSS data.

Reporting background information from large-scale international surveys such as TIMSS is a challenging task. The unique characteristics of the educational systems of each country, language and cultural differences, must all be carefully taken into consideration to ensure the comparability of the data. In some countries, students may be more inclined to choose more extreme response options (strongly agree, strongly disagree), while in other places these options may be reserved for extreme opinions/feelings only. Standards for judging how prepared teachers are to teach, or how adequate the school facilities are, are also likely to vary from country to country. Survey data built on people's perceptions rather than on objective conditions and TIMSS data must be interpreted accordingly. Awareness of these cultural differences is a must to develop fair and meaningful indicators of educational contexts.

Appendices

Appendix 1. Contents of the school questionnaire.

Question number	Content	Description
1	Community	Situates the school within a community of a specific type.
2–4	Staff	Describes the school's professional full- and part-time staff and the percentage of teachers at the school for five or more years.
5	Years	Indicates the number of years students typically stay with the same teacher.
6	Collaboration policy	Identifies the existence of a school policy promoting teacher cooperation and collaboration.
7	Principal's time	Indicates the amount of time the school's lead administrator typically spends on particular roles and functions.
8	School decisions	Identifies who has the responsibility for various decisions for the school.
9	Curriculum decisions	Identifies the amount of influence various individuals and educational and community groups have on curriculum decisions.
10	Formal goals statement	Indicates the existence of school-level curriculum goals for mathematics and science.
11–12	Instructional resources	Provides a description of the material factors limiting the school's instructional activities.
13	Students in the school	Provides total school enrollment and attendance data.
14	Students in the target grade	Provides target grade enrollment and attendance data, student's enrollment in mathematics and science courses, and typical class sizes.
15	Number of computers	Provides the number of computers for use by students in the target grade, by teachers and in total.
16	Internet access	Identifies whether the school has Internet access as well as identifying whether the school actively posts any school information on the World Wide Web.
17	Student behaviors	Provides a description of the frequency with which schools encounter various unacceptable student behaviors.
18	Instructional time	Indicates the amount of instructional time scheduled for the target grade according to the school's academic calendar.
19	Instructional periods	Indicates the existence and length of weekly instructional periods for the target grade.
20	Organization of mathematics instruction (e.g., setting/streaming, tracking, and remedial/enrichment programs)	Describes the school's provision for students with different ability levels in mathematics.

21	Program decision factors in mathematics	Indicates how important various factors are in assigning students to different educational programs or tracks in mathematics.
22	Organization of science instruction (e.g., setting/ streaming, tracking, and remedial/enrichment programs)	Describes the school's provision for students with different ability levels in science.
23	Program decision factors in science	Indicates how important various factors are in assigning students to different educational programs or tracks in science.
24	Admissions	Describes the basis on which students are admitted to the school.
25	Parental involvement	Describes the kinds of activities in which parents are expected to participate (e.g., serve as teacher's aids, fundraising).

Source. Martin *et al*. (2000: 80).

Appendix 2. Contents of the teacher questionnaires.

Question number	Content	Description
1–2	Age and sex	Identifies teacher's sex and age range.
3	Teaching experience	Describes the teacher's number of years of teaching experience.
4–5	Instructional time	Identifies the number of hours per week the teacher devotes to teaching mathematics, science and other subjects.
6	Administrative tasks	Identifies the number of hours per week spent on administrative tasks such as student supervision and counseling.
7	Other teaching-related activities	Describes the amount of time teachers are involved in various professional responsibilities outside the formally-scheduled school day.
8	Teaching activities	Describes the total number of hours per week spent on teaching activities.
9	Meet with other teachers	Describes the frequency with which teachers collaborate and consult with their colleagues.
10	Teacher's influence	Describes the amount of influence that teachers perceive they have on various instructional decisions.
11	Being good at mathematics/ science	Describes teacher's beliefs about what skills are necessary for students to be good at mathematics/ science.
12	Ideas about mathematics/ science	Describes teacher's beliefs about the nature of mathematics/science and how the subject should be taught.
13	Document familiarity	Describes teacher's knowledge of curriculum guides, teaching guides, and examination prescriptions (country-specific options).

Appendix 2. Contents of the teacher questionnaires (continued).

14	Prepared to teach mathematics/science topics	Provides an indication of teacher's perceptions of their own preparedness to teach the TIMSS 1999 in-depth topic areas in mathematics or science.
15–18	Formal education and teacher training	Describes the highest level of formal education completed by the teacher, the number of years of teacher training completed, and the teacher's major area of study.
19–20	Career choices	Identifies whether teaching was a first choice and if the teacher would change careers if given the opportunity.
21	Social appreciation	Describes whether teachers believe society appreciates their work.
22	Student appreciation	Describes whether teachers believe students appreciate their work.
23	Books in home	Provides an indicator of teacher's cultural capital.

Source. Martin *et al.* (2000: 81).

Appendix 3. Contents of the student questionnaire (general version).

Question number	Content	Description
1–4	Student demographics	Provides basic demographic information such as age, sex, language of the home, whether born in country and if not how long he/she has lived in country.
5	Academic activities outside of school	Provides information on student activities that can affect their academic achievement (e.g. extra lessons, science club).
6	Time spent outside of school	Provides information about the amount of time student spends on homework and leisure activities on a normal school day.
7	Parents' education	Provides information about the educational level of the student's mother and father. Used as an indicator of the home environment and socio-economic status.
8	Student's future educational plans	Identifies the student's plans for further education
9	Parents' country of birth	Provides information regarding immigrant status.
10	Books in the home	Provides information about the number of books in the home. Used as an indicator of the home environment and socioeconomic status.
11	Possessions in the home	Provides information about possessions found in the home (e.g. calculator, computer, study desk, country-specific items). Used as an indicator of academic support in the home environment as well as an indicator of socio-economic status.

12	Mother's values	Provides information about the student's perception of the degree of importance his/her mother places on academics and other activities. Used as an indicator of the home environment and general academic press.
13	Student's behavior in mathematics class	Provides a description of typical student behavior during mathematics lessons.
14	Peers' values	Provides information about the student's perception of the degree of importance his/her peers place on academics and other activities. Used as an indicator of peers' values and student's social environment.
15	Student's values	Provides information about the degree of importance the student places on academics and other activities. Used as an indicator of student's values.
16	Competence in mathematics/ science	Provides an indication of student's self-description of academic competence in mathematics and science.[a]
17	Difficulty of mathematics	Provides a description of student's perception of the difficulty level of mathematics.
18	Doing well in mathematics	Identifies student's attributions for doing well in mathematics.
19	Difficulty of science	Provides a description of student's perception of the difficulty level of science.[a]
20	Doing well in science	Identifies student's attributions for doing well in science.
21	Liking mathematics/science	Identifies how much students like mathematics and science; a key component of student motivation.[a]
22	Liking computers for mathematics/science	Identifies how much students like using computers to learn mathematics and science.
23	Internet access	Identifies whether students are accessing the Internet and for what purposes they are using it.
24	Interest, importance, and value of mathematics	Provides a description of student's interest, importance rating, and value attributed to mathematics.
25	Reasons to do well in mathematics	Provides the extent to which students endorse certain reasons they need to do well in mathematics.
26	Classroom practices in mathematics	Provides a description of student's perceptions of classroom practices in mathematics instruction.
27	Beginning a new mathematics topic	Describes the frequency with which specific strategies are used in the classroom to introduce a new mathematics topic.
28	Taking science class(es)	Identifies whether or not the student is enrolled in science classes this year.[a]
29	Interest, importance, and value of science	Provides a description of student's interest, importance rating, and value attributed to science.[a]
30	Reasons to do well in science	Provides the extent to which students endorse certain reasons they need to do well in science.[a]
31	Classroom practices in science	Provides a description of student's perceptions of classroom practices in science instruction.[a]

Appendix 3. Contents of the student questionnaire (general version) (continued).

32	Beginning a new science topic	Describes the frequency with which specific strategies are used in the classroom to introduce a new science topic.[a]
33–34	People living in the home	Provides information about the home environment as an indicator of academic support and economic capital.
35–36	Cultural activities	Provides a description of student's involvement in cultural events or programming such as plays or concerts.
37	Report on student behaviors	Provides an indication of the student's perspective of the existence of specific problematic student behaviors at school.
38	Environmental issues	Provides an indication of student's beliefs about how much the application of science can help in addressing environmental issues.
39	Science use in a career	Identifies preference for sciences in careers.

Note. [a] Specialized version asks about biology, earth science, chemistry and physics separately.
Source. Martin *et al.* (2000: 83–84).

Note

1. Only data from five countries were included in the analyses presented in this chapter: Australia, Chile, Chinese Taipei, Israel and the Philippines.

References

DeVellis, R. (1991) *Scale Development, Theory and Applications. Applied Social Research Method Series Vol. 26.* Newbury Park, CA: SAGE.

Martin, M.O., Gregory, K.D. and Stemler, S.E. (2000) *TIMSS 1999 Technical Report. IEA's Repeat of the Third International Mathematics and Science Study at the Eighth Grade.* Chestnut Hill, MA: International Study Center, Lynch School of Education, Boston College.

Mullis, I.V.S., Martin, M.O., Smith, T.A., Garden, R.A., Gregory, K.D., Gonzalez, E. J., Chrostowski, S.J. and O'Connor, K.M. (2001) *TIMSS Assessment and Frameworks and Specifications 2003.* Chestnut Hill, MA: International Study Center, Lynch School of Education, Boston College.

Ramírez, M.J. and Arora, A. (2004) Reporting TIMSS 2003 questionnaire data. In M.O. Martin, I.V.S. Mullis and S. Chrostowski (eds) *TIMSS 2003 Technical Report* (pp. 309–24). Chestnut Hill, MA: TIMSS & PIRLS International Study Center, Lynch School of Education, Boston College.

Robitaille, D.F. (1993) *Curriculum Frameworks for Mathematics and Science.* TIMSS Monograph No. 1. Vancouver: Pacific Educational Press.

Spector, P. (1992) *Summated Rating Scale Construction – An Introduction. Sage University Papers Series on Quantitative Applications in the Social Sciences, Series No. 07-082.* Beverly Hills, CA: Sage.

Sullivan, J.L. and Feldman, S. (1979) *Multiple Indicators – An introduction. Sage University Papers Series on Quantitative Applications in the Social Sciences, Series No. 07-015.* Beverly Hills, CA: Sage.

4

Inconsistent Student Responses to Questions Related to Their Mathematics Lessons

Dirk Hastedt

Introduction

Several IEA studies questionnaires were administered to students. Most of them included questions related to students' school lessons. There is long standing concern as to the reliability of data gathered in this way (for example, Baumert *et al.*, 2000: 280; Gruehn, 2000: 93ff.; Lorsbach and Jinks, 1999). More precisely, can class means or modes based on information gathered from students' responses be deemed reliable information about their classes? Furthermore, how can reliability be verified? It is important that only questions that the respondent is able to answer and give reliable answers to should be asked (Martin *et al.*, 1999).

In some of the IEA studies a questionnaire was also administered to teachers of the students who had participated and also to their school principals. In these cases comparisons of student answers with those of the teachers and the principal can be made.

For example, in the TIMSS 1999 Grade 8 student questionnaire,[1] questions on the use of computers in mathematics lessons were asked in question SQ2-26g and SQ2-26t (Martin *et al.*, 1999). In the TIMSS 1999 mathematics teacher questionnaire,[2] the mathematics teachers of these students were asked if the students had computers available in their mathematics lessons (question TQM2B-8) and the school principals of the schools of these students completed a school questionnaire,[3] in which they were asked how many computers were available (SCQ2-15).

Those cases where principals and teachers reported that no computers were available, but students had answered that computers were used, were investigated. The result was that a number of inconsistencies were found. In countries where there were fewer computerized schools, the number of inconsistencies was higher than in other countries. A possible explanation is that the students did not know what a computer was, or could not imagine that a computer could be available in their school, or the students misinterpreted the computer to be a calculator when they were answering this question.

Now that it is acknowledged that inconsistencies exist in the student responses, an evaluation should be made of how reliable the student responses are if no alternative sources of information other than the student questionnaire are available.

This chapter evaluates the reliability of the information about classes and their activities gathered from student responses, and the possible reasons for discrepancies in what the students reported.

First, the problem itself is evaluated in the section 'Evaluation of Consistency within Classes'. In the following four sections, 'Examining the Reasons for Variance within Classes', 'Evaluating whether Students Understood the Questions Correctly', 'Evaluating the Relationship between Student Mathematics Achievement and Their Responses' and 'Evaluating the Grouping Effect in Classes', the possible reasons for discrepancies have been examined. In the last section, 'Obtaining More Reliable Responses on Mathematics Lessons', possible solutions are suggested.

Evaluation of Consistency within Classes

Question SQ2-26 was used to evaluate the TIMSS 1999 student questionnaire for all countries[4] that participated in TIMSS 1999 (Martin *et al.*, 1999) (see Figure 4.1). The data analyzed is that of TIMSS 1999 from all participating countries. The data included representative samples of eighth-grade students from 38 countries and resulting in a total of 180,000 students. For further details about the data, the reader is referred to the *TIMSS 1999 Technical Report* (Martin *et al.*, 2000).

I first analyzed the data to see whether the students in a class answered the subquestions in SQ2-26 similarly or whether differences existed. An analysis of variance (ANOVA) was performed both for each country and for each of the 20 parts of question 26 using an ANOVA.[5] This provided the within-class and between-class variance components. If the students in a class provided similar answers to the questions, then the variance should mostly be between the classes and not within the classes. All but 12 out of 678 classes had within-class variance above 50 percent. This means that for most of the variables and nearly all countries that there is more variance within the classes than between the classes. Although, at first glance, this result is quite surprising, it needs to be put in perspective.

If there is no or nearly no overall variance for a variable in a country then this analysis is quite useless because the variance is mostly caused by noise in the data. For example, for variable SQ2-26o (the teacher uses the board) in Japan only 4.6 percent of the variance was between the classes, but 95.4 percent within the classes. But when looking at the frequency distribution one sees that more than 94 percent of the students answered the question with 'Almost always'.[6] Only 4.4 percent of the Japanese students gave a different answer. Since these students were distributed quite evenly among the classes, the variance calculated was mainly within the classes.

Student General Science Questionnaire Main Survey **SQ2 [11]**

TIMSS-R Ref. No. 98-0035

26. How often does this happen in your mathematics lessons?
Circle one letter, A, B, C, or D, for each line.

			Almost always	Pretty often	Once in a while	Never
BS BM PROB	a)	The teacher shows us how to do mathematics problems.	A	B	C	D
BS BM NOTE	b)	We copy notes from the board.	A	B	C	D
BS BM TEST	c)	We have a quiz or test.	A	B	C	D
BS BM PROJ	d)	We work on mathematics projects.	A	B	C	D
BS BM WSHT	e)	We work from worksheets or textbooks on our own.	A	B	C	D
BS BM CALC	f)	We use calculators.	A	B	C	D
BS BM COMP	g)	We use computers.	A	B	C	D
BS BM EVLF	h)	We use things from everyday life in solving mathematics problems.	A	B	C	D
BS BM SG RP	i)	We work together in pairs or small groups.	A	B	C	D
BS BM HWGV	j)	The teacher gives us homework.	A	B	C	D
BS BM HWCL	k)	We can begin our homework in class.	A	B	C	D
BS BM HWTC	l)	The teacher checks homework.	A	B	C	D
BS BM HWFC	m)	We check each other's homework.	A	B	C	D
BS BM HWDS	n)	We discuss our completed homework.	A	B	C	D
BS BM USBT	o)	The teacher uses the board.	A	B	C	D
BS BM USOT	p)	The teacher uses an overhead projector.	A	B	C	D
BS BM USBS	q)	Students use the board.	A	B	C	D
BS BM USOS	r)	Students use the overhead projector.	A	B	C	D
BS BM RUPT	s)	The teacher gets interrupted by messages, visitors, etc.	A	B	C	D
BS BM IDEA	t)	The teacher uses a computer to demonstrate ideas in mathematics.	A	B	C	D

SQ2-11

Figure 4.1 Student general science questionnaire main survey SQ2-26.

This does not mean that all low between-class variance components were caused by a low overall variance or that there was a linear relation between the overall variance and the between-class variance component of a variable. For example, again in Japan, variable SQ2-26g (we use computers) had only half of the total variance of SQ2-26a (the teacher shows us how to do mathematics problems), but the between-class variance was 54.2 percent for SQ2-26g and only 9.9 percent for SQ2-26a.

If a variable has lower variance than the randomly appearing values that differ from the category chosen, most often it has more effect on the between-class variance.

Randomly appearing different values are usually caused by data entry mistakes or misunderstandings on the part of individual students. Variables with more than 90 percent of the responses in one response category should be excluded from the analysis. Across all countries there are 17 cases where a single response category accounted for more than 90 percent of given responses. The variables affected were: SQ2-26g (we use computers), which had more than 90 percent of the responses in the category 'never' in 7 countries; SQ2-26t (the teacher uses a computer to demonstrate ideas in mathematics), which had more than 90 percent of the responses in the category 'never' in 6 countries; the variable SQ2-26r (students use the overhead projector), which had more than 90 percent of the responses in the category 'never' in three countries; and SQ2-26o (the teacher uses the board), which had more than 90 percent of the responses in the category 'almost never' in one country. These variables were excluded from further analysis for these countries; they should only be used for descriptive purposes.

Examining the Reasons for Variance within Classes

There may well be different reasons why students in the same class answer questions on the lessons differently (see, for example, Gruehn, 2000). Some of these reasons might be:

1. The students do not understand the question correctly, or they interpret it differently.
2. The students have different views of the lessons, which might be affected by their own performance, their role/behavior in the class or other student related factors.
3. The students are grouped within their classes.

These suppositions have been evaluated in the following sections.

Evaluating whether Students Understood the Questions Correctly

When looking at the percentages of between-class variance across countries, it is readily apparent that some questions had a high percentage of between-class variance across all countries whereas for others the percentage was very small.

The three sub-questions in SQ2-26 with the smallest between-class variance were:

a. The teacher shows us how to do mathematics problems.
h. We use things from everyday life in solving mathematics problems.
s. The teacher gets interrupted by messages, visitors, etc.

These three questions had a very low percentage of between-class variance for all countries, with a mean across the countries of about 10 percent.

Conversely, the items with the highest percentages of between-class variance were:

f.   We use calculators.
j.   The teacher gives us homework.
p.   The teacher uses an overhead projector.

These questions had a mean percentage of between-class variance across the countries of 25–30 percent. So, these questions seem to have been answered more homogeneously within the classes.

When looking at the content of the questions, there were differences between the two categories of questions. Whereas questions (f), (j) and (p) could be answered without any further knowledge and could be measured quite accurately, questions (a) and (h) needed a deeper understanding of the situation and required some interpretation on the part of the students. Questions (a), (h) and (s) also contained terms that were quite vague with words and phrases such as 'to show', 'things', 'everyday life', 'messages, visitors, etc.' and the term 'mathematics problems'. Such terms are not often used by students and correspondingly were interpreted differently by individual students. 'Mathematics problems' might also have been one of the words that invoked a stereotypical reaction such as refusal. Students may not have wanted to be confronted with problems; especially if they experienced difficulty in mathematics which might have caused the term 'mathematical problems' to become what Oppenheim (1992) calls a 'loaded word'.

Without deeper investigation of the interpretation of these terms, it can be said that more precise, concrete and straightforward questions seemed to be answered more consistently by the students. It is important to formulate a question as precisely as possible without vague terms such as 'things' or 'etc.' and also without including technical terms (without explaining them) or words with a negative connotation. The importance of this is explained by Foddy (1993) discussing question construction and Oppenheim (1992) considering questionnaire wording.

Evaluating the Relationship between the Student Mathematics Achievement and Their Responses

There are some theories that suggest that students' perceptions of their lessons are correlated with their achievement (Clausen, 2002). This effect must also be taken into account when looking at the variables describing the mathematics lessons. Therefore, variable correlations were not limited to question SQ2-26 with which the mathematics achievement score were calculated. The following procedure was also used.

Class means were calculated for each variable derived from question SQ2-26. Then, the class means were subtracted from students' responses to eliminate the between-class variances. For these 20 newly derived variables, correlation with the mathematics achievement score were computed.

Correlations of the variables in SQ2-26 with mathematics achievement ranged from –0.27 (question SQ2-26j [teacher gives homework] in Cyprus) to 0.31 (question SQ2-26d [work on math projects] in Cyprus). The variables attracting the most negative correlations (lower than –0.1) across the countries were SQ2-26j (teacher gives homework), with 19 negative correlations out of 35 and SQ2-26o (teacher uses the board) with 20 negative correlations out of 35. The variables attracting positive correlation (above 0.1) for most countries were SQ2-26d (work on mathematics projects; 19 out of 35), p (teacher uses overhead projector; 18 out of 35), r (students use overhead projector; 26 out of 35) and t (teacher uses computer to demonstrate; 24 out of 35).[7]

This means that in most countries:

- Students with *lower* mathematics achievement reported that:
  - They were given homework less frequently.
  - They worked more frequently on mathematics projects.
  - Their teacher more frequently used a computer to demonstrate ideas in mathematics.
- Students with *higher* mathematics achievement reported that:
  - Their teacher used the board more often.
  - Overhead projectors were used less frequently.

However, not only did the variables present different levels of correlation but also between countries there was a big difference in terms of the number of correlations that indicated at least a weak effect.[8] In Japan there were no positive correlations and weak negative correlations for only two variables: SQ2-26a (the teacher shows how to do mathematics problems) and SQ2-26o (teacher uses the board). In Latvia there were no negative correlations and only SQ2-26t (teacher uses computer to demonstrate) had a weak positive correlation. In contrast, Macedonia exhibited 12 positive correlations (up to 0.26) and one negative correlation, the Philippines had six negative and seven positive correlations (from –0.25 to 0.29). Countries with more than half their correlations below –0.1 or above 0.1 were: Cyprus (11), Israel (11), Jordan (12) and Chinese Taipei (12).

Altogether, there were 89 negative correlations and 199 positive correlations. This was an indication that the effect of different mathematics lesson attributes, or the perception of the lessons that the students report, had a much closer connection to the students' mathematics achievement score in some countries than in others. A connection between these effects and the national mathematics achievement could not be observed – the number of correlations was not related to the mathematics achievement measured at the national level.

Correlation of deviation from the class mean of the students' responses with the 20 variables in question SQ2-26 with the mathematics score ranged between –0.26 (question SQ2-26d [we work on mathematics problems] in Cyprus) and 0.25 (question SQ2-26j [the teacher gives us homework] in Cyprus).

The variables with the most negative correlations were: SQ2-26t (the teacher uses a computer to demonstrate ideas in mathematics) having negative correlations in 15 countries; and SQ2-26r (students use the overhead projector) having negative correlations in 16 countries, which were exactly the variables that had the most positive correlations for the variable itself. The next most frequently found negative correlations were: SQ2-26p (teacher uses the overhead projector) with 12 negative correlations; and SQ2-26g (we use computers) with 10 negative correlations, which also had a high number of positive correlations for the original variables (20 and 15, respectively).

An interpretation of these figures is that, in Cyprus, students with a higher mathematics achievement reported having less homework than their classmates. Also taking into account the previously reported negative correlation of the variable SQ2-26j (teacher gives homework) at the class- and student-level, one can say that the students who performed better in mathematics came from classes where more homework was given, but they reported that less homework was given to them than to their classmates.

For variable SQ2-26d (we work on mathematics problems) the converse held. The students who performed better in mathematics tended to be in classes in which the students reported working less frequently on mathematics projects, but they reported that they worked more often on mathematics projects than their classmates had reported.

In general, the questions that had the most positive correlations with mathematics achievement had negatively correlated differences from class means.

This phenomenon is also described as the 'ecological fallacy' (Freedman, 1999). Nevertheless, the percentage of between-class variance not only differed over the different questions, but there were also huge differences between the countries.

Countries with a mean percentage of between-class variance of more than 25 percent were the USA (33.7 percent), Canada (29 percent), Slovenia (26 percent), Japan (26 percent), Australia (25.4 percent), and the Netherlands (25 percent). Countries with a low between-class variance of these variables were Tunisia (9.5 percent), Iran (10.2 percent), Cyprus (10.3 percent), Indonesia (10.7 percent), Turkey (10.9 percent) and South Africa (10.9 percent). These differences in the between- and within-class variances did not correspond to the within- and between-class variances in terms of the mathematics and science scores.

Evaluating the Grouping Effect in Classes

The extent to which students were grouped within classes and therefore reported differently about the mathematics lessons was examined. For this purpose, the class means for each of the 20 sub-questions in SQ2-26 were calculated and for each student the difference from the class mean was also calculated. For each class, the variances were calculated for each of the questions. In the mathematics teacher questionnaire, question TQM2-11c asked: 'In mathematics lessons, how often do students work together as a class with the teacher teaching the whole class?' The response categories were: 'never or almost never', 'some lessons', 'most lessons' and 'every lesson'.

The correlation between this variable and the class variance for each of the sub-questions in SQ2-26 was calculated for each country. Although the results were very different from country to country and between the questions, some quite strong correlations were found that indicated that some of the variation in the student responses was a result of grouping taking place in the mathematics lessons.

As expected, several negative correlations were found that indicated that, in classes where the teacher answered that the students worked together with the teacher teaching the whole class, there were fewer differences in the students' responses about the lessons. The sub-question SQ2-26e (we work from worksheets or textbooks on our own) was one with fewer and weaker negative correlations because the concept of working on their own was obviously not captured by the teachers' response about teaching the whole class together.

The variable with most correlations of –0.1 or less (12 out of 35) was SQ2-26g (we use computers) and the variable with the highest negative correlation (–0.45) is SQ2-26t (the teacher uses a computer to demonstrate). This can be interpreted as follows: computer use is an activity for which the students are split into groups and that not all students in the class get the same amount of computer instruction. Taking into account the correlations reported before, it can be said that the students who performed better in mathematics were separated more frequently in class for the use of computers.

When examining country by country across all variables, the country with most negative correlations was Belgium. Belgium had negative correlations for all variables except SQ2-26e (we work from worksheets on our own). The weakest correlations were found for SQ2-26n (we discuss our completed homework) (–0.07), SQ2-26m (we check each other's homework) (-0.11) and SQ2-26f (we use calculators) (–0.12). All other correlations in Belgium were between –0.20 and –0.45. The highest (negative) correlations were observed for SQ2-26t (the teacher uses a computer to demonstrate ideas in mathematics) (–0.45), SQ2-26r (students use the overhead projector) (–0.42) and SQ2-26q (students use the board) (–0.36). This leads to the interpretation that in Belgium students in different mathematic groups within a class were treated quite differently from those in other countries.

Obtaining More Reliable Responses on Mathematics Lessons

The next obvious step is to examine how to go about obtaining more reliable statements about the situation in the classes from the student responses. One possibility is to summarize information from different questions, for example, by calculating factor scores with a principal component factor analysis. This procedure was applied to some countries. The countries selected were the USA, where the highest between-class variance component for the question SQ2-26 was found; Tunisia, where the lowest between-class variance component for question SQ2-26 was found; and Japan, in order to demonstrate an interesting example of the ecological fallacy.

USA
Table 4.1 presents the US pattern for the between-class variances. Examination of the results of a principal components analysis with a VARIMAX rotation resulted in a number of factors[9] being identified.

Table 4.1 Between-class variance components and correlations with the mathematics score.

Question number	Description of question	Variance between-classes (%)	Student correlation with the math score[a]	Class correlation with the math score[b]
26a	Teacher shows how to do math problems	20.52	–0.15	–0.30
26b	We copy notes from the board	42.39	0.13	0.15
26c	We have a quiz or test	19.20	0.05	0.18
26d	We work on math projects	25.03	–0.01	0.03
26e	We work from worksheets on our own	13.40	–0.13	–0.33
26f	We use calculators	54.55	0.13	0.17
26g	We use computers	33.86	–0.21	–0.33
26h	Everyday life things for math problems	13.20	0.06	0.06
26i	We work together in pairs or small groups	41.51	0.23	0.33
26j	The teacher gives us homework	36.79	0.09	0.11
26k	We begin our homework in class	40.52	–0.09	–0.11
26l	The teacher checks homework	26.22	0.21	0.35
26m	We check each other's homework	41.27	0.20	0.47
26n	We discuss our completed homework	26.74	0.04	0.01
26o	The teacher uses the board	48.77	0.04	0.01
26p	The teacher uses an overhead projector	74.27	0.05	0.05
26q	Students use the board	43.53	–0.02	0.03
26r	Students use the overhead projector	33.78	0.10	0.11
26s	The teacher gets interrupted by messages	13.99	–0.04	–0.04
26t	The teacher uses a computer	24.08	–0.10	–0.30

Notes. [a] The correlations were calculated at the student level using the students' mathematics score and the students' response for the question.
[b] These correlations were calculated at the class level based on the class means of the student scores and the class means for each of the variables from question SQ2-26.

US factor 1: the board is used but not overhead projectors and calculators

This factor indicates schools that are not well equipped, as well as teachers and students who do not use 'modern technology' during the lessons. This factor was mainly driven by sub-questions SQ2-26f (use calculators), o (teacher uses the board), p (teacher uses an overhead projector), q (students use the board) and r (students use the overhead projector).

The negative factor loadings for SQ2-26o (teacher uses the board) and q (students use the board) mean that higher values for these variables resulted in a lower factor score for the students.[10] The percentage of between-class variance of the factor scores was 76.0 percent; these between-class variances can be seen in Table 4.2. The correlation between this factor score and the students' mathematics score at the student level was -0.08, which was not even a weak effect.

The class means of the factor scores and the students' mathematics achievement were calculated. The correlation between the individual student scores and the class mean of the factor score was –0.10. The correlation between-class means of the factor score and the mathematics score was –0.13.

When interpreting this factor, its loadings and correlations, the following must be taken into account. The coding scheme of the variables in question SQ2-26 was:

Almost always	1
Pretty often	2
Once in a while	3
Never	4

This means that a higher variable factor score that loads positively on the factor indicates that the student reports that something happens less often. Consequently, a positive correlation of a factor score with the mathematics score indicates that students who reported a less frequent occurrence of the variables with positive loadings tended to have a higher mathematics score. Looking at the 'overhead and calculators are used but not the board' factor, the interpretation is that students tend to have a lower mathematics achievement score if they are in classes where the students report that the board is used more frequently and an overhead projector and calculators less frequently. This result must be interpreted very carefully because it does not necessarily mean that bringing more overhead projectors and calculators into the classes would result in higher student mathematics achievement. However, the use (and therefore the availability) of an overhead projector and calculators as an attribute of the mathematics class indicates a higher mathematics achievement of the students. This might be caused by the fact that better equipped classes are located in areas where the students have a better mathematics achievement or better achieving students. Maybe also better mathematics teachers are attracted by better equipped schools.

Table 4.2 Factor loadings for US factor 1.

Question	SQ2-26f	SQ2-26o	SQ2-26p	SQ2-26q	SQ2-26r
Factor loadings	0.31	–0.76	0.79	–0.36	0.38
Between-class variance (%)	54.5	48.8	74.3	43.5	33.8

US factor 2: students do not work on their own

This factor was mainly generated by the high factor loading of sub-questions SQ2-26e (work from worksheets on our own), h (everyday life things for math problems), i (work in pairs or small groups), k (begin homework in class) and m (check each other's homework) (see Table 4.3).

The between-class variance for this factor was 47 percent and correlation with the mathematics score was –0.05. The correlation of the mathematics score and the class mean of the factor score was –0.07 and the correlation between both class means was –0.09.

The common conception of these variables is that there are student activities that are carried out without the direct involvement of the teachers. The students work from worksheets on their own, they use things from every day life (no teacher-prepared materials), they work in pairs or small groups, begin their homework in class and check each other's homework. The variable SQ2-26b (students copy notes from the board) had the next highest loading on that factor with –0.21. Although this variable was not included in the description of the factor above, students not copying notes from the board might also indicate that the teacher is less present in the teaching process. The fairly high between-class variance proportions showed that this was recognized very uniquely within the classes. Again, taking the coding scheme into account, the interpretation of a high factor score is that students do not work that much on their own. The correlation of –0.05 at the student level shows that students in classes in which students worked more independently from the teacher performed slightly better. However, not as significantly different from other classes as was true for the factor 'overhead and calculators are used but not the board' described earlier. This result must also be interpreted very carefully and does not mean that enabling students to work on their own increases their mathematics achievement, but rather, in classes with higher mathematics achievement, the students do work on their own.

Tunisia

Tunisia is one of the countries for which lower between-class variance fractions were found for all parts of question SQ2-26 as can be seen in Table 4.4. When examining the rotated factors, the following factors and their properties were found.

Tunisian factor 1: no homework or tests

This factor had loadings on the variables SQ2-26c (have a quiz or test), j (teacher gives homework), l (work together in pairs or small groups), m (check each other's

Table 4.3 Factor loadings for US factor 2.

Question	SQ2-26e	SQ2-26h	SQ2-26i	SQ2-26k	SQ2-26m
Factor loadings	0.34	0.24	0.28	0.73	0.62
Between-class variance (%)	13.4	13.2	41.5	40.5	41.3

Table 4.4 Between-class variance components and correlations with the mathematics score for Tunisia.

Question number	Description of question	Variance between-classes (%)	Student correlation with the math score	Class correlation with the math score
26a	Teacher shows how to do math problems	9.12	0.13	0.33
26b	We copy notes from the board	7.79	0.06	0.16
26c	We have a quiz or test	7.81	0.01	0.06
26d	We work on math projects	5.44	0.09	0.16
26e	We work from worksheets on our own	5.93	0.04	0.17
26f	We use calculators	10.53	0.04	0.26
26g	We use computers	7.01	0.05	0.17
26h	Everyday life things for math problems	6.98	0.02	0.18
26i	We work together in pairs or small groups	13.18	0.06	0.10
26j	The teacher gives us homework	12.68	−0.04	−0.23
26k	We begin our homework in class	9.02	−0.03	−0.10
26l	The teacher checks homework	19.18	−0.04	−0.15
26m	We check each other's homework	9.50	0.08	0.25
26n	We discuss our completed homework	8.21	0.14	0.45
26o	The teacher uses the board	7.17	−0.02	−0.06
26p	The teacher uses an overhead projector	12.90	−0.09	−0.34
26q	Students use the board	7.20	−0.05	−0.20
26r	Students use the overhead projector	12.54	0.10	0.25
26s	The teacher gets interrupted by messages	7.68	0.08	0.24
26t	The teacher uses a computer	9.13	−0.04	0.13

homework) and n (discuss completed homework). The loadings have been presented in Table 4.5.

The between-class variance of the factor scores was 20.8 percent. The correlation with the mathematics achievement on student level was 0.05. The correlation between the class mean of the factor score and the individual mathematics achievement scores was 0.08 and the correlation between the class mean mathematics achievement and the class mean of the factor score was 0.18.

Table 4.5 Factor loadings for Tunisian factor 1.

Question	SQ2-26c	SQ2-26j	SQ2-26l	SQ2-26m	SQ2-26n
Factor loadings	0.37	0.70	0.76	0.26	0.56
Between-class variance (%)	7.8	12.7	19.2	9.5	8.2

This increase in correlation was quite surprising. Whereas, for example, in the USA the increase was also visible for the factors presented,[11] it was not as high as in Tunisia. Interpretation of the factor is that the higher the factor score of a student is, the less frequently the student received homework (checked by the teacher and among the students and discussed) and less frequent testing and working in small groups took place. Students performed slightly better if they were in classes where the students reported getting homework less frequently, than did students in classes where more homework was given. It might be better to present this the other way around: in classes where the students' mathematics achievement was not that good, students reported that their teacher gave more homework (and checked and discussed it) and gave tests more often.

To investigate this further, a factor was generated from only the variables exhibiting high loading (SQ2-26c, j, l, m, n). This factor had a Cronbach's α of 0.72. The between-class variance component was 17.6 percent. Correlation with the math score was 0.04 at the student level and 0.22 at the class level. So, the effect observed at the class level was even stronger than for the factor based on all variables.

Tunisian factor 2: usage of board
The variables loading this factor were: SQ2-26b (copy notes from the board), o (teacher uses the board) and q (students use the board). The factor loadings of these variables have been given in Table 4.6.

The factor scores had a between-class variance of 9.4 percent, which was not very high, but higher than for the separate variables. The correlation between the factor score and the mathematics score was −0.09 at the student level, which was also higher than for the separate variables. The correlation between the class mean of the factor score and the mathematics achievement of the students was −0.15, the correlation of the class means was even −0.33. Here, the increase of the correlations when aggregating at the class level was even higher than for the factor presented earlier.

The interpretation of the factor is that the higher the factor score the less frequent the students and the teacher used the board and the students copied notes from the

Table 4.6 Factor loadings for Tunisian factor 2.

Question	SQ2-26b	SQ2-26o	SQ2-26q
Factor loadings	0.58	0.77	0.75
Between-class variance (%)	7.8	7.2	7.2

board. The more frequently these activities happened in the class, the higher the mathematics achievement of the students was in the class.

Japan

Japan is a very interesting case in which not only class effects on factors can be observed, but also strong student effects. The pattern of the between-class variance for Japan has been given in Table 4.7. The following factors were extracted.

Table 4.7 Between-class variance components and correlations with the mathematics score for Japan.

Question number	Description of question	Variance between-classes (%)	Correlation with the math score	Correlation of the deviation from class mean with math score[a]
26a	Teacher shows how to do math problems	9.91	0.02	0.13
26b	We copy notes from the board	22.96	0.02	0.02
26c	We have a quiz or test	33.47	−0.01	−0.01
26d	We work on math projects	13.45	0.02	−0.10
26e	We work from worksheets on our own	14.88	0.03	0.02
26f	We use calculators	29.19	0.01	−0.10
26g	We use computers	54.18	0.02	−0.05
26h	Everyday life things for math problems	8.17	0.03	0.10
26i	We work together in pairs or small groups	40.97	0.02	−0.01
26j	The teacher gives us homework	49.88	0.00	−0.01
26k	We begin our homework in class	20.85	−0.01	0.00
26l	The teacher checks homework	40.97	0.01	−0.04
26m	We check each other's homework	14.51	0.03	−0.06
26n	We discuss our completed homework	14.38	−0.00	−0.04
26o	The teacher uses the board	4.55	0.00	0.13
26p	The teacher uses an overhead projector	38.74	0.00	−0.08
26q	Students use the board	–	−0.02	0.01
26r	Students use the overhead projector	24.18	−0.00	−0.09
26s	The teacher gets interrupted by messages	12.08	0.00	−0.06
26t	The teacher uses a computer	46.42	0.03	−0.02

Note. [a] These correlation coefficients were obtained by calculating the class mean for each of the variables, calculating the difference of the student response from the class mean for each student and calculating the correlation between these differences and the students' mathematics achievement.

Japanese factor 1: no homework

This factor had loadings on the variables SQ2-26j (teacher gives homework), k (begin homework in class) and l (teacher checks homework). They have been presented in Table 4.8.

The between-class variance of the factor scores was 48.3 percent. This high value indicates that the students within the classes observe essentially the same thing. The correlation between the factor score and the mathematics score was –0.05 at the student level. The correlation between the student mathematics score and the factor class mean was –0.09. The correlation between factor mean and class mean at the class level was –0.24. So, by aggregating the information to the class level, the correlation increased significantly, although it must be considered that part of this might be an effect of aggregation bias.

The interpretation of the factor is that students who were rarely given homework, where homework cannot be started during the class and the teacher checked the homework less frequently, had a higher factor score and a lower mathematics score. If most of the students in a class reported this uniformly then the class mean of the mathematics score was significantly lower.

Japanese factor 2: not using calculator or computer

This factor had loadings on the variables SQ2-26f (use calculators), g (use computers) and t (teacher uses computer). The factor loadings of these variables can be seen in Table 4.9. The between-class variance of the factor score was 59.1 percent; a high value indicating very homogeneous responses within the classes.

The correlation between the factor score and mathematics achievement was at the student level 0.05, with a factor class mean of 0.05 and between the factor and the mathematics class means at the class level it was 0.13.

An interpretation could be that, if computers and calculators are used less frequently in the class, the factor score and also the mathematics achievement could be higher. The last statement is very unusual compared to results in other countries and should be investigated further. Maybe advanced mathematics is taught in the good

Table 4.8 Factor loadings for Japanese factor 1.

Question	SQ2-26j	SQ2-26k	SQ2-26l
Factor loadings	0.77	0.73	0.47
Between-class variance (%)	49.9	20.9	41.0

Table 4.9 Factor loadings for Japanese factor 2.

Question	SQ2-26f	SQ2-26g	SQ2-26t
Factor loadings	0.29	0.87	0.85
Between-class variance (%)	29.2	54.2	46.4

classes more often in Japan and for the more abstract mathematics no calculators and computers are necessary. Maybe the students in classes with higher mathematics achievement learn the basic mathematics elsewhere, for example, in private learning institutions like the *juku*.

Japanese factor 3: teacher explains on the board and students copy it
This factor is created by loadings on the variables SQ2-26a (teacher shows how to do math problems), b (copy notes from the board) and o (teacher uses the board). The factor loadings of these variables on the factor can be seen in Table 4.10. The between-class variance of the factor was 10.6 percent, which was quite low.

The correlation of the factor scores with the mathematics score differed quite markedly dependent on the level of aggregation. The correlations can be seen in Table 4.11.

This is a very interesting result that shows that students who reported that during mathematics lessons the teacher often showed the students how to do mathematics problems, used the board and where students often copied the notes from the board (indicated by a low factor score) had a higher mathematics achievement score. When aggregating the factor score at the class level and calculating the correlation with the mathematics score the same tendency can be observed, but the correlation was much lower, only –0.05. Obviously, this factor includes more than just the joint perception of the students. And indeed when calculating the differences of the student factor scores from the class mean, as done before for the single variables to eliminate the between-class variance, and then calculating the correlation between these differences and the mathematics achievement score of the students, a correlation of +0.19 was obtained. This means that the better performing students in a class had the perception that the teacher showed students less frequently how to do mathematics problems on the board and had the students copy it down, or that the better performing students did not have to do this as often as the other students in the class but did other things instead. Here again the phenomenon of the ecological fallacy can be observed.

Table 4.10 Factor loadings for Japanese factor 3.

Question	SQ2-26a	SQ2-26b	SQ2-26o
Factor loadings	0.67	0.54	0.68
Between-class variance (%)	9.9	23.0	4.6

Table 4.11 Correlations for Japanese factor 3 with mathematics achievement.

Correlation	Individual student's mathematics achievement	Class mean mathematics achievement
Individual student's factor scores	–0.19	–0.15
Class means of factor scores	–0.05	–0.14

This is a very extreme example of students' individual perceptions of lessons being driven by the fact that also the variables had a high perception effect. Correlation of the differences from the class mean for the high loading variables with mathematics achievement were: 0.13, 0.02 and 0.13, for SQ2-26a, b and o, respectively.

Conclusion

In some countries sub-questions of SQ2-26 contained data that provided a very good description of the activities being carried out during lessons. However, other sub-questions and also the sub-questions that worked in some countries proved problematic in others. This may have different causes.

Students' responses about their mathematics lessons were perceptions of classroom activities that depended also on the student's role in the class. For example, students with different levels of mathematics achievement reported differently about some aspects of their mathematics lessons. Although the fact that this information may be related to students' individual perceptions is very interesting, it clouds our ability to obtain a clear and reliable picture of the lessons.

Precise questions on measurable topics were answered more consistently than questions with vague terms, or that required some conceptual understanding. This is also true for the consistency across countries. It is therefore important to formulate precise questions with appropriate wording and in concrete terms.

Grouping of students within mathematics classes takes place in some countries. This grouping is connected with specific educational activities, for example, computer use whereby better performing students get more frequent computer access. Analysis of students' reports on their lessons should always take into account whether all students are in the same course or not, and whether they are grouped in their course or not.

Combining variables by deriving factor scores improves the consistency of the students' reports within classes, but they are not free of the students' individual perception and cannot solve all data-related problems. It remains difficult to get a clear picture about classroom activities only by asking students questions such as this.

Other sources of information are desirable, for example, questions administered to the teacher or reports from external observers.

Notes

1. The student questionnaire is abbreviated to SQ2 and questions in this questionnaire to SQ2. All questionnaires can be found in the *TIMSS 1999 User Guide for the International Database* (Gonzalez and Miles, 2001).
2. The mathematics teacher questionnaire is abbreviated to TQ2M and questions from the class-related part of this questionnaire to TQM2B.
3. This questionnaire answered by the school principal is abbreviated to SCQ2 and questions in this questionnaire to SCQ2.

4. Data for Israel and England were not analyzed because these countries did not sample mathematics classes within schools. Data for Lithuania were also not considered because the question was not administered.
5. This was done by using the GLM procedure of SAS.
6. The distribution is: 1.5 percent missing codes, 94.4 percent 'almost always', 2.7 percent 'pretty often', 1.1 percent 'once in a while' and 0.4 percent 'never'.
7. All the correlations have also been calculated with aggregated data on class level to ensure that there is no effect like the ecological fallacy observed. But the correlations were about the same also on class level.
8. Correlations between 0.10 and 0.29 are defined here as indicating a weak effect, correlations between 0.30 and 0.49 as a medium effect and correlations from 0.50 on as a strong effect
9. It must be noted that none of the factors presented met the criterion of the Cronbach's coefficient α above 0.6. This was not expected because factors constructed out of more than about ten variables usually do not meet that criterion. Future work should evaluate more clearly defined factors created only from variables which have a high loading on the factor.
10. To be precise the factor scores of the students were calculated not by using factor loadings but by using standardized scoring coefficients. But since these are just linear transformations of the factor loadings, the statement still holds.
11. There were also factors in the USA (which were not presented because they were not interpretable) for which the correlation of the mathematics scores with the factors scores went down when aggregating the factor scores on class level (for example, from 0.04 to 0.00).

References

Baumert, J., Bos, W. and Lehmann, R. Hrsg. (2000) *TIMSS/III Dritte Internationale Mathematik- und Naturwissenschaftsstudie*. Opladen: Leske + Budrich.

Clausen, M. (2002) Unterrichtsqualität: eine Frage der Perspektive. *Pädagogische Psychologie und Entwicklungspsychologie*. Bd. 29. Münster: Waxmann Verlag.

Foddy, W. (1993) *Constructing Questions for Interviews and Questionnaires*. Cambridge: Cambridge University Press.

Freedman, F.A. (1999) Ecological inference and the ecological fallacy. *International Encyclopedia of the Social & Behavioral Sciences* 6: 4027–30. [Technical report No. 549. University of California, Berkeley.]

Gonzalez, E.J. and Miles, A.M. (2001) *TIMSS 1999 User Guide for the International Database*. Chestnut Hill, MA: International Study Center, Boston College.

Gruehn, S. (2000) Unterricht und schulisches Lernen. *Pädagogische Psychologie und Entwicklungspsychologie*. Bd. 12. Münster: Waxmann Verlag.

Lorsbach, A.W. and Jinks, J.L. (1999) Self-efficacy theory and learning environment research. *Learning Environment Research* 2: 157–67.

Martin, M.O., Rust, K. and Adams, R.J. (1999) *Technical Standards for IEA Studies*. Delft: IEA.

Martin, M.O., Gregory, K.D. and Stemler, S.E. (2000) *TIMSS 1999 Technical Report*. Chestnut Hill, MA: International Study Center, Boston College.

Oppenheim, A.N. (1992) *Questionnaire Design, Interviewing and Attitude*. London: Cassel.

5

Raising Achievement Levels in Mathematics and Science Internationally: What Really Matters

Sarah J. Howie and Tjeerd Plomp

Introduction

It is within the spirit of the 'forefathers' of the IEA that this book was written and it is fitting that one chapter is dedicated towards reflecting upon treating the 'world as a laboratory'. While vast sums are spent in typically collecting data, less is spent on the secondary analysis of such data that seeks to probe the heart of the questions that cause policy-makers in education such sleepless nights. It therefore seems appropriate that this book seeks to stand on the shoulders of those of Robitaille and Beaton (2002) and others before them and to continue to pursue the IEA mission of conducting and disseminating data from its studies by this volume of secondary analyses.

TIMSS, like several of the IEA studies, was designed with the purpose of both seeking and providing answers to questions that matter and these typically include investigating the possible reasons behind how children perform in key subjects such as mathematics and science. Given the nature of the data collected, these answers can never include the causes for such performance but rather hypothesize, probe and explore the possibilities that the data permit.

This chapter highlights and reflects upon the work conducted by the authors of this book, most of whom were intimately involved in TIMSS and gave significantly of their time in order to explain children's performance in mathematics and science. Forty authors from 17 countries investigated data from altogether more than 40 countries in compiling the research for this book. In keeping with the original TIMSS projects, it was a substantial task to analyze the findings given the richness and complexity of their analyses. While this chapter seeks to identify and tease out some of the main findings across all the 24 chapters, we in no way suggest that this process is exhaustive, nor fully representative of the considerable analyses that were undertaken by all the authors. At best we hope to illustrate that, first, school really does matter and, second, to contribute to the debate on raising the standard of education through identifying factors that make a difference to students' achievement.

The TIMSS data, starting with the 1995 study, were unique in that within one database a huge amount of data was available on three different education levels, namely the student-, class- and school-levels. This has allowed a large number of secondary analyses to occur where vast numbers of interrelationships could be explored. The advent of multi-level analyses has also permitted researchers to explore the data between different levels taking into account the nested nature of the data. There are several examples of this approach within this book.

Given the investment into TIMSS 1995 and 1999, it is crucial to extract the most out of the data both nationally and cross-nationally. Important questions posed by policy-makers internationally are 'what can be done to raise student achievement in mathematics and science?' and 'what really matters?'

In the following sections, we reflect upon the main findings of the contributions to this book and upon those questions. The analyses are presented on school-level within countries for mathematics, then for science and followed by observations about the cross-national studies first for mathematics and then for science. This pattern is also followed for the student-level analyses and thereafter the class- and teacher-level analyses. Finally the conclusions are presented and some reflections are made.

Factors that Matter at School Level

A vast amount of literature exists on factors that predict pupil achievement at school level. In particular, the strategic importance of mathematics and science, as well as the perceived difficulty of these subjects at school, has increased the interest in ascertaining the most important predictors of achievement. Among this vast literature are key studies that have helped to shape the conceptual basis for much of this research. One of the first was that by Shavelson et al. (1987), based on more than 2000 studies which drafted a conceptual framework for the elements and relationship between these that would affect science achievement (see Chapter 11, in which Howie uses a conceptual framework adapted from the Shavelson et al. model). A second example is the educational productivity model by Walberg (1992), based on the results of about 8000 studies that demonstrates the effects of the nine key factors identified in the original model related to students' affective, behavioral, and cognitive development and grouped into three general groups, namely personal, instructional and environmental variables.

Both of these models (among others) have been instrumental in conceptualizing various large-scale studies of achievement in mathematics and science.

One of the first reflections relates to the contribution made by schools and to what extent schools matter. While schools are essential components of a civilized society, conveying and embodying the necessary cultural and societal attributes of broader society, their effectiveness and contribution in terms of learning, has been extensively researched in the Western world and questioned as well. From the contributions within this book, some interesting patterns emerged.

First, the extent to which the variance in mathematics achievement could be explained at school level differed vastly. For instance, the European experience reflected by Van den Broeck *et al.* (Belgium, Flanders) and Kupari (Finland) revealed that only 14 per cent and 10 per cent, respectively, could be explained by the differences between schools. The Asian experience revealed by Park and Park showed that, in Korea, as far as mathematics was concerned only 4 per cent of the variance in performance (science was 5 per cent) could be explained at school level. In significant contrast, the South African story exposes that 55 per cent of the difference in performance is explained at the school level. To what extent these national studies are truly representative of their respective continents is unknown, but it does reveal that, while it is not so significant which school the Belgian, Finnish or Korean child attends, it appears to be a key factor with regard to the South African child.

Second, the only study within the book that could statistically measure the effect of class level was that of Van den Broeck, Opdenakker and Van Damme (Chapter 6) given the design of their TIMS 1999 sample – as a national option, they sampled two classes per school. Some 28 per cent of the variance in achievement was explained by the class-level. Hence in Flanders, Belgium, it was most significant which class you were in rather than which school.

Third, the variance in achievement explained by the student level once again varied greatly from 45 per cent in South Africa to 95 per cent in Korea. Most of the variance in the Belgian Flemish performance was explained at student level (58 per cent) as was that of the Finnish (90 per cent). Ramirez also found that the biggest differences were found between the Chilean students as opposed to between Chilean schools.

Although the literature suggests that school-level variables have a lesser effect on achievement than student- and class-level variables, nonetheless these are important factors when ascertaining what matters. As already mentioned the variance explained by school factors varied hugely between countries, from 4 per cent between schools in South Korea to 55 per cent between schools in South Africa for mathematics. Two school-level variables were found to be significant among the studies. First, Bos and Meelissen found that *students' perceptions of their safety* at school were associated with achievement in Flanders, Germany and the Netherlands but to varying degrees. Second, the *location of the school* was found to be a significant factor in the South African context and Howie revealed that students attending schools in more rural areas were more likely to achieve lower scores in mathematics than those attending schools in urban areas.

Factors that Matter at Student Level

Factors at student level were extensively explored within nine countries and a significant number coincided in their findings. While policy-makers are most interested in variables that are alterable, it is nonetheless important to continue to monitor variables that are unalterable such as socio-economic status (SES) or social class. This variable was found significantly affecting the mathematics performance of pupils from

Belgium, Chile and South Africa while *home educational resources* was found to be significant in the Philippines. A dominant variable across the studies was that of *students' attitudes towards mathematics* and eight studies (Ramírez [Chapter 7], Papanastasiou and Papanastasiou [Chapter 8], Kupari [Chapter 9], Kiamenesh [Chapter 10], Park and Park [Chapter 12], Bos and Meelissen [Chapter 13], Talisayon, Balbin and de Guzman [Chapter15] and Ogura [Chapter 20] – especially intrinsic motivation) found this to be significant in nine countries (Chile, Cyprus, Finland, Germany, Iran, Japan, Korea, the Netherlands and the Philippines) in mathematics and one in science (Japan). In particular, Park and Park found student motivation to be the most significant factor explaining achievement in Korea, as did Kiamenesh (Iran) and Bos and Meelissen (Flanders, the Netherlands and Germany).

Another important variable was that of *self-concept* in mathematics and was significant in the performance of Finnish, Iranian and South African pupils and, in fact, in Finland and Iran it was the most significant predictor of achievement. However, to what extent these personal variables could be said to be causal or consequential cannot be ascertained and this should be borne in mind.

There were also individual factors that were significant within certain studies. For instance, Van den Broeck *et al.* included an *intelligence* test as a national option which was administered simultaneously with the mathematics and science tests of TIMSS. This was the single most significant predictor of Belgium (Flanders) pupils' performance in mathematics.

Ogura focused on the issue of out-of-school *extra lessons* (in the so-called *juku*) in Japan, which is famous world wide with many believing that it explains the superior Japanese performance in TIMSS. Ogura revealed the difference within the Japanese urban and rural societies as extra lessons within an urban context had a positive effect upon performance, possibly attributed to the competitive environment and the drive for higher performance and getting into the 'right schools'. This is in stark contrast to the negative correlation of achievement with extra lessons for rural children, most of whom would be enrolled for remedial purposes, and therefore dispels the popular international interpretation and perception of Japanese extra-mural education.

Papanastatiou and Papanastatiou discovered that in Cyprus, the *educational background* of the parents was a key factor in their children's performance in addition to student attitudes. This is naturally a variable which is also linked to resources and wealth.

In South Africa, most of the pupils had to write the TIMSS tests in a second language. Howie (Chapter 11) uncovers the strong relationship between *language* proficiency and mathematics achievement as the South African children had to write an English Language Proficiency test in addition to the TIMSS tests. This proved to be the most significant determinant of achievement in the South African sample. Furthermore, there was a strong relationship between home language and the language that pupils listened to on radio (often the only exposure to English outside school).

In the cross-national research that is reported in the book, a number of student-level factors were found to be significant. Shen (Chapter 24) revealed that, in his cross-national study of the USA, Japan, South Korea, Hong Kong, Singapore and Taiwan, that the *self-perceived easiness of mathematics* had a negative effect on achievement, while *literacy* in the home had a positive effect (with country as unit of analysis). Among the same group of countries pupils' *absenteeism* also had a negative effect. In the cross-national study within Europe, Bos and Meelissen found three significant variables for Flanders, Germany and the Netherlands namely, *gender*, the *home educational background* and friends *academic expectations* of which the former two had a positive effect on achievement and the latter a negative effect. They found that boys outperformed girls, and the higher the expectations of friends to do well in mathematics, the lower the test score of the students. What is interesting in this particular case is that, in three neighbouring countries, the outcomes of the exploration (resulting in very few common variables having an effect on the results of the three countries) appeared to be so different, some of which can be attributed to the very different education systems, culture and context.

While few studies concentrated on science especially, there were some interesting results. Ogura's study uncovered the strength of Japanese student's intrinsic *motivation* in their science performance. Cross-national research by Ericikan, McCreith and Lapointe (Chapter 14) revealed that SES mattered in the USA, although not in Canada and Norway overall (however, an effect was found for Canadian female students). Students' *attitudes* towards science mattered only with regard to Canadian males and not to students from the USA or Norway. Finally, students' *expectations* about science performance only mattered as far as Norwegian students are concerned and not those from Canada and the USA.

Shen, in his cross-national research, found that countries doing well in mathematics also performed well in science and he uncovered two variables that had effects on the pupils' performance in science. First, students who *perceived science to be easy* were generally from the poor performing countries and, second, *absenteeism* had a significant negative effect on science achievement.

Factors that Matter at Class and Teacher Levels

Within countries, a number of authors explored a large variety of classroom and teacher variables in relation to mathematics achievement. Ramírez found in her study of Chilean students that the *school curriculum* and its implementation in the classroom accounted for the largest differences in student performance. She found that *content coverage, instructional resources* and *teacher preparation* also had a significant effect on achievement. The latter was also found to be significant in Howie's research within the South African context, with higher achievement being associated with more *time spent by mathematics teachers on lesson planning*. Furthermore, students of South African teachers who spent more *time at school* also seemed to achieve higher results.

In Papanastasiou and Papanastasiou's study of Cypriot students' performance, the strongest direct influence on mathematics achievement was the teachers' *teaching practices*, and a more in-depth look (by video) at teaching practices by Kunter and Baumert (Chapter 21) revealed that no direct effects of constructivist teaching elements could be found on interest development nor on achievement in mathematics.

Furthermore, they found that, whereas mathematics achievement seemed to be influenced mainly by class-level predictors, these predictors did not influence students' interest development – students' individual perception of challenge was the best predictor. The latter was shown to be important in their study as the students' sense of challenge was found to be higher in classes with more effective *classroom management*, meaning in turn an indirect effect on mathematics achievement.

The medium of instruction at class level was found to have a significant effect on the performance of South African students in mathematics. Howie found that there appeared to be a strong association between the frequency of English used as a medium of instruction in the class and higher performance in mathematics.

One classroom variable that had mattered in previous IEA studies, but was not much explored in the research in this book, was *Opportunity to Learn*. However, Gregory and Bankov's (Chapter 16) research on Bulgarian student mathematics performance revealed that Opportunity to Learn measures had an effect on the Bulgarian performance. They attempted to explain the drop in Bulgarian performance (since 1995) in 1999 and discovered that much of the curriculum covered in the TIMSS assessment was in fact taught in Bulgaria between one up to four years before grade and they felt that this might explain the drop in performance.

While the practice of *tracking* or streaming children into a variety of ability groupings was commonplace within the countries participating in TIMSS, one study investigated this phenomenon in some depth. Zuzovsky (Chapter 23), in her study of the effect of ability grouping on Arab and Jewish children in Israel, found that whilst ability grouping had a positive effect on children's achievement in mathematics, this was not the case in science.

Other individual teacher characteristics found to have a significant effect on country-level were *teachers' attitudes* towards teaching and their *beliefs* about mathematics. Both of these were explored in the South African data where Howie found that, while teachers' perceptions of self-worth regarding their profession had a counter-intuitive negative effect on student achievement, the more complex construct of beliefs about mathematics itself had a positive effect.

In the cross-national studies exploring the classroom variables, some interesting findings were made. Bos and Meelissen revealed that *time on task* (that is, the number of minutes scheduled for mathematics) and the proportion of the *teachers' time spent on mathematics* assigned as a percentage of their total workload both had positive although varying degrees of effects on mathematics achievement across Flanders, Germany and the Netherlands. When aggregated to the class level Bos and Meelissen also found that the parents' education background and the students' attitude towards mathematics had a significant effect across the same countries.

At class level, Shen's research revealed that both *student repetition* and *student body mobility* had a negative effect on mathematics achievement. Likewise, students in classes where teachers were frequently *interrupted* also tended to perform poorer than those in classes where this did not happen.

From a curriculum perspective, Prawat and Schmidt (Chapter 17) made a fascinating revelation with the observation that, in their analysis of data from the USA, Japan, South Korea, Hong Kong, Singapore, Belgium and the Czech Republic, more child-centered and thus more process-oriented countries such as the USA (and Norway) opt for a 'content in service of process' orientation to curriculum organization, while countries whose students perform at the highest level on the TIMSS mathematics test employ the converse of this: a 'process in service of content' approach. They observed that while the curriculum profile in mathematics in the USA is flat and it is assumed that students can revisit the same content in a profitable manner from one grade to the next due to their increasing process skills, top performing countries have a more hierarchically structured curriculum that does not make the same assumption.

An often debated variable is that of *class size* and Wößmann (Chapter 22) conducted a comprehensive analysis of the TIMSS data from 18 countries with regard to the effects of the number of children in the class and the effect of this on their mathematics achievement. He concluded that at least in 11 countries he analyzed, a large causal effect could be ruled out within meaningful estimates. However, he did find significant effects in Iceland and Greece.

With regard to science, Angell, Kjærnsli and Lie (Chapter 18) found that *Opportunity to Learn* and the *teacher coverage* measure were significant factors in their study of nine selected countries. Their findings indicated that curricular effects (measured as teacher coverage of the 'topic') seem to be able to explain between 25 and 50 per cent of the variance between countries for individual items. Also related to instruction, Li, Ruiz-Primo and Shavelson's (Chapter 19) research revealed the importance of *instructional experiences* and *knowledge types* with regard to science achievement by distinguishing between declarative (as found mostly in TIMSS), procedural, schematic and strategic forms of knowledge. They concluded that the science achievement measured by the TIMSS 1999 test could be distinguished as different types of knowledge. They also suggest the importance of including frameworks for science achievement developed on the basis of cognitive research and related fields in order to provide accurate interpretations of students' achievement scores and also meaningful insights into the relationship between instructional experiences and students' learning.

Conclusion

The analysis above serves to show the richness of the data available in the TIMSS database and simultaneously the futility of over-simplistic answers to the question of what really matters with regard to achievement in mathematics and science. Certainly the authors within this book have identified and discussed a number of variables having an effect at school, student and class levels. The findings also reveal the fascinating world of cross-national research, where certain variables may be significant within a single country, but may lose their effect in other contexts given the effects of different variables. Few variables were found to be significant across several countries. Those that did were mostly on student level and included SES (in three countries), students attitudes towards the subject (in eight countries), student motivation (in four countries) and self-concept (in three countries). However, the concern here is that several of these are personal variables and are difficult to distinguish between being causal or consequential and furthermore are not easy to intervene on. Cross-nationally, absenteeism, gender, home educational background and academic expectations were significant.

While many variables at class level were investigated, few were found to have an effect in more than one country. The exceptions were time on task, teachers' load in mathematics, student repetition and mobility, opportunity to learn, teacher coverage of topics and class size. The latter is the source of much debate and Wößmann concluded in a meaningful contribution to the class size debate that '*it may be better policy to devote the limited resources available for education to employing more capable teachers rather than to reducing class size thus moving more to the quality side of the quantity-quality tradeoff in the hiring of teachers.*' However, while this may be an appropriate suggestion in terms of the developed nations, this 'luxury' of choice would probably not be feasible in those developing nations where the notion of quality is a lot less nuanced and 'quality teachers' are less available.

What was interesting to observe is that, in recent times, secondary analyses of the kind included in this book are driving various forms of analyses forward through the advances made methodologically. There are a variety and multitude of statistical techniques for secondary analysis, as well as those laying the foundation for collecting survey and achievement data such as TIMSS does, but also compiling valid and reliable indicators, scores and scaling. However, stepping back from this cutting-edge methodological advancement, it would seem that we still have some work to do conceptually with regard to such large-scale cross-national research. This is highlighted in the suggestions from Li, Ruiz-Primo and Shavelson and those of Bos and Meelissen. These contributions are noteworthy in their value of the future orientation of studies such as TIMSS and provide one with not only food for thought but a significant starting point for conceptualizing such studies in the future. The secondary analyses demonstrate the worth and importance of this research which lends itself to being complemented by national longitudinal studies and a combination of both quantitative and qualitative research.

Although TIMSS may seem, for many, to be a horse-race of sorts and with the media being over-infatuated with league tables of achievement, the real depth of the data is beginning to emerge through secondary analyses such as these. One perspective often overlooked is the utilization of the data in terms of monitoring different subpopulations within countries where large disparities lie, thus allowing policy-makers the opportunity to monitor educational issues related to equity. Another perspective revealed through the data in terms of equity is the increasing effects of globalization and the resulting increase in the divide between rich and poorer nations. This cross-national research provides a number of indicators whereby this may be monitored in terms of the conditions of schooling across many countries. Clearly the gap is not closing, but by continuing to monitor this disparity, the opportunity for intervention exists. What this does highlight is Fuller's (1987) finding that in developing countries schools have a great influence on students' achievement once the effect of home background has been accounted for. Students from disadvantaged families simply do not have access to an appropriate learning environment and therefore schools are the main source of providing learning opportunities and are often a poor child's only opportunity for upward mobility out of a sentence of squalor. Given the extent of poverty experienced by developing nations, this perspective should not be overlooked among the many priorities of policy-makers internationally and certainly not at the expense of short-term elevations of achievement.

References

Fuller, B. (1987) What school factors raise achievement in the Third World? *Review of Educational Research*, 47(3), 255–92.

Robitaille, D.F. and Beaton, A.E. (2002) *Secondary Analysis of the TIMSS Data*. Dordrecht: Kluwer Academic Publishers

Shavelson, R.J., McDonnell, L.M. and Oakes, J. (1987) *Indicators for Monitoring Mathematics and Science Education: a Sourcebook*. Santa Monica, CA: The RAND Corporation.

Walberg, H.J. (1992) The knowledge base for educational productivity. *International Journal for Educational Reform* 1(1): 5–15.

Part 2

Background Variables and Achievement: Single Country Studies

6

The Effects of Student, Class and School Characteristics on TIMSS 1999 Mathematics Achievement in Flanders

Ann Van den Broeck, Marie-Christine Opdenakker and Jan Van Damme

Introduction

Students differ with respect to their academic achievement. International research indicates that educational practices and characteristics of schools are partly responsible for these differences. However, research also reveals that, at the start of secondary education, children differ with regard to other characteristics that are relevant for their achievement. The main objective of this chapter is to identify such characteristics, or – in other words – to examine which explanatory variables at the student, class and school level reduce the variance in achievement scores after controlling for background characteristics of the students.

Literature on the Importance of Student, Class and School Characteristics

The importance of student background and family characteristics for academic achievement has already been stressed in the research of Coleman *et al.* (1966) on equal educational opportunity in the USA. General ability, intelligence, prior achievement, self-concept and motivation are often mentioned as important student characteristics (see Grisay, 1996; Opdenakker *et al.*, 2002), while the socio-economic status (SES) of the family and ethnicity are often mentioned as important family characteristics (see e.g., Dekkers *et al.*, 2000; Opdenakker *et al.*, 2002) for the explanation of differences in academic achievement and attainment. Parental education, which can be interpreted as an indicator of SES, is found to correlate positively with mathematics achievement (Beaton *et al.*, 1996; Opdenakker and Van Damme, 2001).

Also the learning environment and, in particular, the way the student perceives the learning environment in the class, his/her attitudes towards the subject matter and the courses the student has chosen, are mentioned in the literature to have potential in explaining student achievement.

Emotions, attitudes and beliefs strongly influence the way in which problems are treated, the amount of time and the intensity of the effort one spends on learning activities in a specific domain and eventually the learning result itself. The opposite is also true: good learning results stimulate a positive development of these affective components (Vermeer and Seegers, 1998). Emotions, attitudes and beliefs are not only influential with respect to cognitive learning processes and learning results. They are also a (by)product of learning processes inside and outside the school (see the research of Schoenfeld, 1988).

Constructivism, as a development within cognitive psychology, emphasizes that human learning is active and constructive and stresses the importance of the contextual character of human cognition. Constructivism and self-regulation of learning processes underline the responsibility of the learner for his own learning processes (Boekaerts, 1999). In the constructivist view of learning, knowledge and skills are constructed by students themselves during the learning process. The constructivist approach to teaching is often viewed as an alternative for the direct instruction approach, sometimes called 'the structured approach'. The last approach has received quite a lot of empirical support (e.g. Scheerens and Creemers, 1996); the constructivist approach has also received empirical support, although to a lesser extent (de Jager *et al.*, 2002). In this chapter, the degree to which the student (and also the teacher) perceives the learning environment as constructivist is measured. Research indicates that student perceptions can act as intermediate variables between instruction variables and learning processes and outcomes (see McDowell, 1995). It is mentioned that student perceptions of the learning environment are sometimes even more relevant than teacher perceptions of their own teaching in explaining learning outcomes, which was confirmed by research of Brekelmans (1989).

With respect to the curriculum, research concerning the first two grades of secondary education in Flanders provides some evidence that students' choices from the available optional programmes can explain differences in achievement between students (Van Damme *et al.*, 2000). All students have, in principle, the same mathematics programme. But a limited number of hours per week is dedicated to either Latin and/or Greek or additional hours covering general subjects such as Dutch and French, or a programme of technical subjects. In contrast to the limited number of hours dedicated to the optional programme, students choosing the same optional programme are very often grouped into the same class and their optional choice often reflects their intellectual level and SES, with the brightest students choosing Latin and/or Greek and the less able ones choosing technical subjects. Belonging to a class in which all the students take Latin and/or Greek (this is a class with a high intellectual level) appears to have a positive effect on achievement scores.

With respect to potential effectiveness-enhancing class, teacher and school characteristics several meta-analyses (Wang *et al.*, 1993) show that the impact of school characteristics such as school demographics, culture, climate policies and practices is lower than the influence of factors closer to the actual learning process such as classroom practices.

Furthermore, the importance of characteristics reflecting the inputs in schools such as the teacher–pupil ratio, class size, per pupil expenditure and school resources is very low in Western countries (compared to non-industrialized countries and compared to other characteristics of Western schools and classes). This is the case, perhaps, because the minimum input is available in almost all Western schools or because Western schools do not differ significantly from each other with respect to their input. A meta-analysis of Hedges *et al.* (1994) indicates some evidence for a small positive effect of teacher experience. Also Scheerens and Bosker (1997) conclude that teacher experience is a potential achievement-enhancing factor but they also state that 'this interpretation is still contested'.

Evidence of the effects of group composition (of classes or schools), or to state it differently, of the effects of peer-group background characteristics on achievement (and other outcomes) is often found and its existence has been recognized for a long time. Evidence is found for a positive effect of the intellectual composition (Dar and Resh, 1986; Leiter, 1983; Opdenakker and Van Damme, 2001; Opdenakker *et al.*, 2002) and the composition with respect to the SES of students' families (Caldas and Bankston, 1997; Opdenakker *et al.*, 2002; Sammons *et al.*, 1997).

With respect to classroom instructional practices evidence is often found for the importance of structured instruction emphasizing testing and feedback (see Scheerens and Creemers, 1996; Slavin, 1996; Wang *et al.*, 1993).

The climate of a classroom or school – a concept that is sometimes criticized because of its vague nature – seems to matter with respect to achievement and other student outcomes. The classroom climate is often described in terms of the orderliness in the class, good work attitudes, quietness, study-orientedness, the learning climate and the quality of the relationships between students and teacher. A positive school climate is often specified as 'a safe and orderly climate' or is operationally defined by means of questions about frequency and severity of problematic and disruptive behaviour and questions about the orderliness of the environment. The relevance of the class and school climate has received empirical support (Anderson, 1982; Opdenakker, 2003; Opdenakker and Van Damme, 2000; Scheerens and Bosker, 1997).

Next to the already mentioned class and school characteristics, consensus and cooperation among the staff and the evaluative potential of schools (monitoring at school) are often mentioned as potential effectiveness-enhancing factors. With respect to the consensus factor, some evidence has been found in qualitative reviews (Levine and Lezotte, 1990; Sammons *et al.*, 1995). However, the effect of this factor is not always positive and Scheerens and Bosker (1997) are convinced that consensus and cohesion among staff may not be sufficient in themselves to enhance effectiveness, but need an underlying shared achievement-oriented effort. Their meta-analysis indicates that cooperation between staff, which leads to consensus and cohesion, seems to be important for student achievement only in secondary education. With respect to monitoring at school and class level their meta-analysis indicates significant effects of monitoring at school level (for all subjects), while monitoring at class level seems to matter only with respect to mathematics and language achievement.

In many educational effectiveness studies the effects of the above-mentioned characteristics of schools, teachers and classes on student outcomes are studied without taking into account the relationships between these school, teacher and class characteristics. However, research indicates the existence of relationships between characteristics of group composition, and school or class process characteristics (Opdenakker, 2003; Opdenakker and Van Damme, 2001; Weinert *et al.*, 1989). A high intelligence group composition or a group composition implying positive family characteristics seems to be related to more favorable school and class process characteristics, which implies at least a joint effect of composition and process characteristics on student outcomes because both characteristics are also related to student outcomes. Kreft (1993) also indicates such a relationship and finds that a disadvantaged school composition gives rise to lower (teacher) expectancies than an advantaged school composition and that lower expectancies seem to result in lower achievement.

Sample Design, Data, Method and Variables

In the general TIMSS 1999 sample design only one class per school was selected. In Flanders,[1] the sample design differed from the general design: instead of selecting one class in each school, two classes per school were selected. This makes it possible to perform multi-level analyses with three differentiated levels. In the analyses, the dependent variable is the Rasch-score for mathematics.[2] The total dataset contains data of 5,119 students, 269 classes and 136 schools.[3] To answer the research questions several datasets were created. The size of each dataset will be presented together with the results. The multi-level analyses were conducted with the programme MLwiN (Rasbash *et al.*, 2000).

In Flanders, the TIMSS data collection was considerably extended. The questionnaires of the principals, the mathematics teachers and the students were extended, and a questionnaire for the parents was also administered. The data collection was also extended with an intelligence test for the students.

A number of variables used in this contribution were derived from these extensions. Table 6.1 gives an overview of the variables of this study. More information about the intake and student characteristics can be found in Van den Broeck *et al.* (2005).

Table 6.1 Overview of variables.

Intake characteristics
- Numerical and spatial intelligence[a] (two numerical subtests and two spatial subtests with each 20 items) (80 items, $\alpha = 0.91$) (Demeyer, 1999).
- Educational level of the parents[a] (this variable is based on the answers of the students and on the answers of the parents).
- Language at home (whether = 1 or not = 0 the student speaks Dutch at home).
- Possessions at home (students had to indicate whether they have each of 17 items, e.g. a microwave oven, a computer, Internet access, etc.).

Student characteristics
- Attitude towards mathematics[a] (e.g. 'I like to study mathematics', 'mathematics is an easy subject', 'mathematics is boring') (39 items, $\alpha = 0.93$).
- Constructivist learning environment as perceived by the student[a] (e.g. 'I get the chance to discuss my methods to solve a problem with other students', 'the teacher asks me how much time I need to solve a mathematical problem') (six items, $\alpha = 0.76$).
- Optional programme[a] (there are three categories: the student attends a course in classical languages (Latin and/or Greek), general subjects or technical subjects).

Class characteristics
- Constructivist learning environment as perceived by the teacher[a] (e.g. 'I give the students the chance to discuss their methods to solve a problem with other students', 'I take the remarks of the students into consideration while looking for suitable materials') (six items, $\alpha = 0.74$).
- Experience of the teacher (number of years teaching) (TIMSS Study Center, 1998a).
- Disruptive students (e.g. 'to what extent do uninterested students/students with special needs limit how you teach your mathematics class') (TIMSS Study Center, 1998) (six items, $\alpha = 0.81$).
- Average intelligence score of the class.[a]
- Class size (TIMSS Study Center, 1998).
- Age of the teacher (TIMSS Study Center, 1998).
- Quiet class[a] (e.g. 'the class is noisy (–)', 'the students of this class are quiet') (seven items).
- Study-oriented class[a] (e.g. 'to learn something is very important for the students of this class?' 'the students of this class prepare their tests seriously') (four items).
- Structured teaching and discussing errors[a] (e.g. 'at the end of a chapter, I distinguish between main issues and side-issues', 'do you discuss the most frequent errors in this class?') (six items, $\alpha = 0.72$).
- Consultation with teachers about students[a] (e.g. 'I discuss the results of students with other teachers') (three items, $\alpha = 0.79$).
- Consultation with teachers concerning teaching methods[a] (e.g. 'I discuss the teaching methods with other teachers') (five items, $\alpha = 0.82$).

School characteristics
- Frequency of problematic behavior of the students in the school (e.g. arriving late at school, vandalism, skipping class) (TIMSS Study Center, 1998b) (18 items, $\alpha = 0.96$).
- General shortcomings in the school (e.g. instructional materials, budget for supplies) (TIMSS Study Center, 1998b) (six items, $\alpha = 0.71$).
- Shortcomings for mathematics instruction (e.g. computers for mathematics instruction) (TIMSS Study Center, 1998b) (six items, $\alpha = 0.72$).
- Seriousness of problematic behaviour of the students in the school (e.g. arriving late at school, vandalism, skipping class) (TIMSS Study Center, 1998b) (18 items, $\alpha = 0.97$).
- Percentage of absent students (on a typical school day) (TIMSS Study Center, 1998b) (one item).
- Percentage of migrant students[a] (one item).
- (Frequency of) registration of students' progress[a] (one item).
- (Frequency of) school report[a] (one item).
- (Frequency of) discussions with the parents[a] (one item).

Note. [a] The variable is based on the extensions of the data collection in Flanders.

Results

Decomposition of the Variance of the Mathematics Scores

The empty model based on a dataset of 4,168 students, 261 classes and 133 schools reveals that almost 58 per cent of the total variance in mathematics scores is due to differences between students within a class, 28 per cent is due to differences between classes and 14 per cent is due to the school. Thus, more than 42 per cent of the variance is situated at the higher levels and the percentage variance at the class level appears to be greater than the percentage variance at the school level.

Relevance of Each of the Variables Separately as Predictors of Mathematics Achievement

To determine the total effect of each variable in itself and to avoid non-significant coefficients of potentially relevant variables due to problems of multicollinearity, multi-level models were built with only one independent variable added to the empty model.

Intake and Student Characteristics

The results of the analyses regarding the intake and student characteristics are discussed in detail in Van den Broeck *et al.* (2005). The most important results are summarized here. There is a strong correlation of 0.61 between 'mathematics score' and 'intelligence score'. The reduction of the variance in mathematics achievement by the variable 'intelligence score' is noticeable: 22 per cent at the student level; 60 per cent at the class level; and 34 per cent at the school level. This is a reduction of the total variance by 35 per cent. The variable 'educational level of the parents' reduces the variance at the class level by 9 per cent and the variance at the school level by 10 per cent. The reduction of the variance at the student, the class and the school levels by the variable 'attitude towards mathematics' is 12, 30 and 8 per cent, respectively. This is a reduction of the total variance by 16.2 per cent. The variable 'optional programme – technical subjects' (as compared with the other optional programmes) reduces the variance to 41 and 49 per cent at the class and the school levels, respectively. With regard to the total variance this is 18 per cent.

Class Characteristics

Correlations among the class variables themselves vary between –0.45 and 0.91. A moderate correlation is found between the variables 'average intelligence score' and 'disruptive students' ($r = -0.38$); 'quiet class' and 'disruptive students' ($r = -0.36$); 'study-oriented class' and 'constructivist learning environment as perceived by the teacher' ($r = 0.34$); 'study-oriented class' and 'disruptive students' ($r = -0.45$); and 'consultation with teachers about students' and 'constructivist learning environment as perceived by the teacher' ($r = 0.37$). A (rather) strong correlation is found between 'age of the teacher' and 'experience of the teacher' ($r = 0.91$) and 'study-oriented class' and 'average intelligence score' ($r = 0.50$). The correlations between the class variables and the mean mathematics score range from –0.46 to 0.82. These variables correlate to a considerable extent with the mean mathematics score: 'disruptive students' ($r = -0.46$); 'average intelligence score' ($r = 0.82$); 'study-oriented class' ($r = 0.51$) and (to a much smaller extent) 'quiet class' ($r = 0.28$).

The results of the multi-level models (based on 4,328 students, 244 classes and 129 schools) with one class characteristic as the only independent variable are presented in Table 6.2.

The reduction of the variance by the variable 'average intelligence score' is striking. At the class level 87 per cent of the variance is removed and at the school level 34 per cent of the variance is removed. This means a reduction of 30 per cent of the total variance. The variable 'disruptive students' is also important at both the class and the school level: respectively, 21 and 29 per cent reduction of the variance at both levels. This is a reduction of almost 10 per cent of the total variance. With regard to the class level, the variables 'constructivist learning environment as perceived by the teacher' and 'study-oriented class' reduce the variance by a considerable amount (11

Table 6.2 Variance reduction (%) – models with one variable (class characteristic).

Variable	Level					Difference in deviance[a]	Coefficient
	Student (%)	Class (%)	School (%)	Total (%)	Deviance		
Constructivist learning environment – (teacher)	–	10.8	0.0	2.0	29379.87	12.62***	1.67***
Experience of the teacher	–	5.2	0.0	0.9	29387.05	5.44*	0.09
Disruptive students	–	20.5	29.4	9.8	29340.62	51.87***	–0.81***
Average intelligence score	–	86.7	33.7	30.3	29125.75	266.74***	0.76***
Class size	–	0.0	5.4	0.6	29389.99	2.50	0.15
Age of the teacher	–	5.8	0.0	0.5	29388.30	4.19*	0.79*
Quiet class	–	6.1	13.7	3.6	29375.55	16.94***	1.70***
Study-oriented class	–	47.8	0.0	11.8	29311.64	80.85***	3.47***
Structured teaching and discussing errors	–	0.0	0.7	0.0	29392.47	0.02	0.08
Consultation with teachers about students	–	0.0	0.7	0.1	29392.22	0.27	0.25
Consultation with teachers about didactical matters	–	1.9	0.3	0.6	29389.35	3.14	–0.77*

Notes. In some cases, the variance at a certain level can increase when an independent variable is included into the model. In that case, the variance reduction is negative. In the table, a negative percentage is reported as 0.0%.
[a] The deviance of the empty model is 29392.49.
Level of significance: * = < 0.05, ** = < 0.01 and *** = < 0.001.

and 48 per cent, respectively), which implies a reduction of the total variance with 2 and 12 per cent, respectively. Three variables reduce the variance at the class level with 5 to 6 per cent: 'experience of the teacher'; 'age of the teacher'; and 'quiet class'. Regarding the school level variance, the variable 'class size' reduces it by 5 per cent and the variable 'quiet class' reduces it by 14 per cent.

School Characteristics

The correlations between the school variables vary between –0.14 and 0.41. Most correlations are very low. A moderate correlation is found between 'seriousness of problematic behaviour' and 'frequency of problematic behaviour' ($r = 0.41$), between 'percentage of absent students' and 'frequency of problematic behaviour' ($r = 0.32$), and between 'school report' and 'registration of students' progress' ($r = 0.38$).

The results of the multi-level analyses (based on 3,704 students, 193 classes and 98 schools) with one explanatory variable are presented in Table 6.3.

Table 6.3 Variance reduction (%) – models with one variable (school characteristic).

Variable	Level					Difference in deviance[a]	Coefficient
	Student (%)	Class (%)	School (%)	Total (%)	Deviance		
Frequency of problematic behaviour	–	–	22.1	3.1	25081.86	11.18***	–0.16***
General shortcomings in the school	–	–	1.9	0.3	25092.12	0.92	–0.17
Shortcomings for mathematics instruction	–	–	0.1	0.0	25092.98	0.06	–0.04
Seriousness of problematic behaviour	–	–	13.9	1.9	25086.32	6.72**	–0.10**
Percentage of absent students	–	–	40.9	5.7	25071.26	21.78***	–1.08***
Percentage of migrant students	–	–	8.6	1.1	25089.35	3.69	–0.17*
Registration of students' progress	–	–	2.0	0.2	25092.24	0.80	0.44
School report	–	–	1.8	0.2	25092.28	0.76	0.56
Discussion with the parents	–	–	0.3	0.0	25092.93	0.11	–0.40

Notes.   In some cases, the variance at a certain level can increase when an independent variable is included into the model. In that case, the variance reduction is negative. In the table, a negative percentage is reported as 0.0%.
[a] The deviance of the empty model is 25093.04.
Level of significance: * = < 0.05, ** = < 0.01 and *** = < 0.001.

There are four variables that reduce the variance at the school level to a considerable extent: 'frequency of problematic behaviour' (22 per cent); 'seriousness of problematic behaviour' (14 per cent); 'percentage of absent students' (41 per cent); and 'percentage of migrant students' (9 per cent). With regard to the total variance, these reductions are rather small.

Multiple Characteristics as Independent Variables

Introducing explanatory variables into the empty model one by one will gradually reduce the variance at the different levels. The intake characteristics are added first, next the student variables, then the class variables and finally the school variables. Within each category the variable that reduced the deviance the most in the model with one explanatory variable (see previous sections) is introduced first in this stepwise procedure. Table 6.4 gives an overview of the results of the models with several variables.

Intake and Student Characteristics

The results of the models with intake and student characteristics are discussed in detail in Van den Broeck *et al.* (2005). It was found that the intake characteristics together reduce the total variance by 36 per cent. With respect to the separate levels, the four intake characteristics together reduce the variance at the student level by 23 per cent, at the class level by 62 per cent and at the school level by 41 per cent. All the intake and student characteristics together reduce the total variance by 47 per cent. This means that the total variance in mathematics scores is almost halved by the introduction of these variables at the student level. With regard to the separate levels, 27 per cent of the variance at the student level is removed, 93 per cent at the class level and 46 per cent at the school level. A remarkable result is that the variance situated at the class level is already reduced by 93 per cent by the introduction of characteristics of the students in the class. A possible explanation is the way in which some variables are operationalized. For example, the variables about the optional programme students take were considered as student variables but a class can be composed of stu-

Table 6.4 Variance reduction (%) – models with several variables.

Characteristics	Level				Deviance	Difference in
	Student (%)	Class (%)	School (%)	Total (%)		
Intake	22.7	61.5	40.7	35.6	22620.41	994.79**
Intake and student	26.5	92.5	46.0	47.2	22311.59	308.82**
Intake, student and class	26.5	95.6	46.0	47.8	22281.04	30.55**
Intake, student, class and school	26.5	95.6	46.5	47.9	22280.76	0.28

Note. [a] The difference between the deviance of the previous model and the model with the additional explanatory variables.

dents with the same optional programme, and this is a common practice in secondary education in Flanders. (For more details: see Van den Broeck *et al.*, 2005.) So, these variables are to a certain extent class characteristics that are probably related to class-room practice. The same applies probably to the attitude towards mathematics.

Class Variables

In previous analyses (Van den Broeck *et al.*, 2004) models were built with all the class variables (but without the student variables). The results showed that the deviance of the model did not decrease any further once the following class variables were introduced: 'average intelligence score of the class'; 'study-oriented class'; and 'disruptive students'. These are the variables considered in the next model. In this model the class variables are introduced together with the intake and the student characteristics. The additional reduction of the variance is limited to 3 per cent of the variance at the class level. This means that once the student-level variables are taken into account, the class characteristics can not reduce the variance (much) further. However, the effects of classes are underestimated in models with intake, student and class variables when the optional programme students take is characterized by specific educational arrangements (for example, when more experienced teachers are assigned to classes with students taking Latin and/or Greek) and when the attitude towards mathematics of students is influenced by the classroom practice, the class group or the teacher. Comparing the results of several models (models with combinations of the intake characteristics, optional pro-gramme and attitude, and class characteristics) gives evidence that this is the case. Comparing the results of the model with intake characteristics and the model with intake and class characteristics indicates that the three class characteristics explain an additional fraction of almost 8 per cent of the variance in mathematics achievement.

School Variables

When models (Van den Broeck *et al.*, 2004) were built in which only the school vari-ables were successively added to the empty model, it appeared that the deviance does not reduce anymore once the variables 'percentage absent students' and 'frequency of problematic behaviour' are taken into consideration. In the next model, only these two variables are considered. In the final model, the school variables are added after the intake, the student and the class variables. The additional reduction of the variance at the school level is only 0.1 per cent. Apparently the considered school variables can not explain the residual differences between schools if the student and class charac-teristics are taken into account.

Conclusion

Concerning the differences in TIMSS 1999 mathematics achievement between stu-dents, classes and schools in Flanders, 58 per cent is attributable to differences between students within classes and 42 per cent of the total variance in mathematics scores is situated at the class and school levels together; 28 per cent is due to differences between classes within a school and 14 per cent is due to differences between schools. Compared to educational effectiveness research conducted in other countries the percentage of the variance situated at the class and school level is rather high. However, the results are in accordance with other Flemish effectiveness studies on mathematics and mother language achievement in the second grade of secondary education, in which curriculum based achievement tests were used (De Fraine *et al.*, 2003; Opdenakker *et al.*, 2002). In accordance with those Flemish studies, the present study also revealed that a lot of variance situated at the class and school level is related to intake characteristics of students, primarily intelligence. Despite this explanation of differences between schools and between classes within schools in terms of intake characteristics, one fifth of the total variance[4] in mathematics achievement is still related to other characteristics of classes and schools, which is quite a large amount of variance compared to international effectiveness studies.

With respect to relevant student background characteristics, the numerical and spatial intelligence appears to be most important – as could be expected from the research literature – and reduces the variance considerably at all levels. The reduction is remarkable especially with respect to the class level. This reduction is most probably a consequence of the fact that, in Flanders, the composition of classes is mainly based on the achievement level of the students. Other intake characteristics (mainly SES) reduce the (remaining) variance only to a small extent.

Once the intake characteristics are taken into account, the other explanatory student characteristics reduce the variance with about 12 per cent. Regarding the student variables related to their education, the importance of attitudes is high, but this is in accordance with the existing literature. The optional programme students take is also important and reduces the variance at the class and at the school level, even after the background characteristics of the students have been taken into account. Perhaps different classroom practices in classes with a different composition are responsible for this. Evidence for this explanation is found in a study of Opdenakker (2003) about mathematics achievement in Flanders in the second grade of secondary education. In our TIMSS study, a constructivist learning environment for mathematics achievement (as perceived by the student) did not prove important. However, one has to bear in mind that mathematics classes and mathematics instruction in Flanders are seldom (completely) organized by the principles of a constructivist learning environment.

With respect to relevant class characteristics, we found evidence for the importance of group composition (intellectual level and disruptive behaviour) and climate characteristics. Although the perception of the learning environment as constructivist (by the teacher) and the experience of the teacher could also explain some variance at the class level, their effect disappears once group composition characteristics are taken into account. The effects of group composition and climate characteristics are additional to those of student characteristics.

With respect to school characteristics, we also found that problematic behaviour of students and percentage of absent students affect mathematics achievement. However, after taking into account student and class characteristics, the effects of the investigated school characteristics are not significant anymore. These results could be attributed to methodological limitations of the TIMSS study and especially on the kind of school level data that were collected. A suggestion may be to include a broader range of relevant school characteristics in a next TIMSS study.

Acknowledgements

The authors would like to thank the editors, Sarah Howie and Tjeerd Plomp, for their support, Dirk Hermans and Georges Van Landeghem for their contribution and the Flemish government (Department of Education) for its financial support.

Notes

1. In Belgium, only the schools of the Flemish Community participated.
2. Because of the test design (not all students solved all the mathematics items), IRT scaling was used (Yamamoto and Kulick, 2000).
3. The dataset contains the data of the students in the academic stream (which is taken by more than 80 per cent of the students and offers programmes with Latin and/or Greek, programmes with general subjects and programmes with technical subjects). The students in the vocational stream of secondary education are not taken into consideration in this study.
4. The exact number is 19.08 and is calculated as follows (see also Table 6.4 and the section 'Decomposition of the Variance of the Mathematics Scores'). At the class level: $28 - (28 \times [61.5 / 100]) = 10.78$; at the school level: $14 - (14 \times [40.7 / 100]) = 8.30$. The sum of 10.78 and 8.30 is 19.08.

References

Anderson, C.S. (1982) The search for school climate: a review of the research. *Review of Educational Research*, 52, 368–420.

Beaton, A., Mullis, I., Martin, M., Gonzalez, E., Kelly, D. and Smith, T. (1996) *Mathematics Achievement in the Middle School Years: IEA's Third International Mathematics and Science Study* (TIMSS). Chestnut Hill, MA: Boston College.

Boekaerts, M. (1999) Self-regulated learning: Where are we today? *International Journal of Educational Research*, 31, 445–56.

Brekelmans, M. (1989) *Interpersoonlijk gedrag van docenten in de klas* [Interpersonal behaviour of teachers in the class]. Utrecht: WCC.

Caldas, S. and Bankston, C., III (1997) Effect of school population socioeconomic status on individual academic achievement. *Journal of Educational Research*, 90, 269–77.

Coleman, J., Campbell, E., Hobson, C., McPartland, J., Mood, A., Weinfall, F. and York, R. (1966) *Equality of Educational Opportunity*. Washington, DC: Department of Health, Education and Welfare.

Dar, Y. and Resh, N. (1986) *Classroom Composition and Pupils' Achievement.* London: Gordon and Breach.

De Fraine, B., Van Damme, J., Van Landeghem, G., Opdenakker, M.-C. and Onghena, P. (2003) The effect of schools and classes on language achievement. *British Educational Research Journal,* 29, 841–59.

de Jager, B., Creemers, B.P.M. and Reezigt, G. (2002) Constructivism and direct instruction: competing or complementary models? Paper presented at the International Congress of School Effectiveness and Improvement, Copenhagen.

Dekkers, H.P.J.M., Bosker, R.J. and Driessen, G.W.J.M. (2000) Complex inequalities of educational opportunities. A large-scale longitudinal study on the relation between gender, social class, ethnicity, and school success. *Educational Research and Evaluation,* 6(1), 59–82.

Demeyer, W. (1999) *Niet-verbale testbatterij voor het lager secundair onderwijs. De gestandaardiseerde aanpassing en bewerking in het Nederlands van de BCR (Batterie de tests du Centre de Recherches) [Non-verbal Test for Lower Secondary Education. The Standardized Adaptation in Dutch of the BCR (Battery of Tests of the Centre for Research)].* Brussel: EDITEST.

Grisay, A. (1996) *Evolution des acquis cognitifs et socio-affectifs des élèves au cours des années de collège [Evolution of cognitive and affective development in lower secondary education].* Liège: Université de Liège.

Hedges, L.V., Laine, R.D. and Greenwald, R. (1994) Does money matter? A meta-analysis of studies of the effects of differential school inputs on student outcomes. *Educational Researcher,* 23(3), 5–14.

Kreft, I.G.G. (1993) Using multilevel analysis to assess school effectiveness: a study of Dutch secondary schools. *Sociology of Education,* 66, 104–29.

Leiter, J. (1983) Classroom composition and achievement gains. *Sociology of Education,* 56, 126–32.

Levine, D. and Lezotte, L. (1990) *Unusually Effective Schools: A Review and Analysis of Research and Practice.* Madison, WI: National Center for Effective Schools Research and Development.

McDowell, L. (1995) The impact of innovative assessment on student learning. *Innovations in Education and Training International,* 32, 302–13.

Opdenakker, M.-C. (2003) Leerling in Wonderland? Een onderzoek naar het effect van leerling-, lesgroep-, leerkracht- en schoolkenmerken op prestaties voor wiskunde in het secundair onderwijs [Student in Wonderland? A study on the effect of student, class, teacher, and school characteristics on mathematics achievement in secondary education]. PhD thesis, K.U. Leuven, Leuven.

Opdenakker, M.-C. and Van Damme, J. (2000) Effects of schools, teaching staff and classes on achievement and well-being in secondary education: similarities and differences between school outcomes. *School Effectiveness and School Improvement,* 11, 165–96.

Opdenakker, M.-C. and Van Damme, J. (2001) Relationship between school composition and characteristics of school process and their effect on mathematics achievement. *British Educational Research Journal,* 27, 407–32.

Opdenakker, M.-C., Van Damme, J., De Fraine, B., Van Landeghem, G. and Onghena, P. (2002) The effect of schools and classes on mathematics achievement. *School Effectiveness and School Improvement*, *13*, 399–427.

Rasbash, J., Browne, W., Goldstein, H., Yang, M., Plewis, I., Healy, M., Woodhouse, G., Draper, D., Langford, I. and Lewis, T. (2000) *A User's Guide to MLwiN*. London: Institute of Education.

Sammons, P., Hillman, J. and Mortimore, P. (1995) *Key Characteristics of Effective Schools: A Review of School Effectiveness Research*. London: OfSTED.

Sammons, P., Thomas, S. and Mortimore, P. (1997) *Forging links: effective schools and effective departments*. London: Paul Chapman.

Scheerens, J. and Bosker, R. (1997) *The Foundations of Educational Effectiveness*. Oxford: Pergamon Press.

Scheerens, J. and Creemers, B.P.M. (1996) School effectiveness in the Netherlands: the modest influence of a research programme. *School Effectiveness and School Improvement*, *7*, 181–95.

Schoenfeld, A. (1988) When good teaching leads to bad results. The disasters of well-taught mathematics courses. *Educational Psychologist*, *23*, 145–66.

Slavin, R.E. (1996) *Education for All*. Lisse: Swets & Zeitlinger.

TIMSS Study Center (1998a) *Mathematics Teacher Questionnaire Main Survey*. TIMSS-R Ref. No. 98-0037. Boston, MA: Boston College.

TIMSS Study Center (1998b) *School Questionnaire Main Survey*. TIMSS-R Ref. No. 98-0039. Boston, MA: Boston College.

Van Damme, J., Van Landeghem, G., De Fraine, B., Opdenakker, M.-C. and Onghena, P. (2004) *Maakt de school het verschil? Effectiviteit van scholen, leraren en klassen in de eerste graad van het middelbaar onderwijs [Does the school make the difference? Effectiveness of schools, teachers, and classes in the first cycle of secondary education]*. Leuven: Acco.

Van den Broeck, A., Van Damme, J. and Opdenakker, M.C. (2004) The effects of student, class and school characteristics on mathematics achievement: explaining the variance in Flemish TIMSS-R data. Paper presented at the IEA International Research Conference, Lefkosia.

Van den Broeck, A., Opdenakker, M.-C. and Van Damme, J. (2005) The effects of student characteristics on mathematics achievement in Flemish TIMSS 1999 data. *Educational Research and Evaluation*, *11*, 107–21.

Vermeer, H. and Seegers, G. (1998) Affectieve en motivationele elementen binnen het leerproces [Affective and emotional elements of the learning process]. In L. Verschaffel and J. Vermunt (eds.), *Het leren van leerlingen. Onderwijskundig lexicon*, 3rd ed. (pp. 99–114). Alphen aan den Rijn: Samsom.

Wang, M.C., Haertel, G.D. and Walberg, H.J. (1993) Towards a knowledge base for school learning. *Review of Educational Research*, *63*, 249–94.

Weinert, F., Schrader, F.-W. and Helmke, A. (1989) Quality of instruction and achievement outcomes. *International Journal of Educational Research*, *13*, 895–932.

Yamamoto, K. and Kulick, E. (2000) Scaling methods and procedures for the TIMSS mathematics and science scales. In M.O. Martin, K.D. Gregory and S.E. Stemler (eds.), *TIMSS 1999 Technical Report*. Chestnut Hill, MA: Boston College.

7

Factors Related to Mathematics Achievement in Chile

María José Ramírez

Introduction

TIMSS 1999 brought 'bad news' to the educational community in Chile. During the 1990s, expenditure on education more than doubled and an ambitious reform was launched with the goal of improving the quality and equity of education. When, following TIMSS 1999, Chile sat near the bottom of the country league table (35th out of 38 countries), and only half of its students provided evidence that they could do basic arithmetic; the impression was that the efforts of a decade had been fruitless.

Since the political costs of the 'bad news' of the TIMSS results are high, it is valid to wonder what motivates Chile to take part in this major assessment project. One good reason is that TIMSS facilitates analyses of hundreds of background variables relevant to an understanding of the likely causes of poor performance. The purpose of this chapter is to further contribute to understanding mathematics achievement by analyzing its relationship with some important characteristics of the Chilean school system.

The pertinent literature refers to several factors that may help in understanding why some schools attain higher performance levels than others. Schools operating in more socially advantaged contexts systematically attain higher achievement levels than schools serving more disadvantaged communities (McEwan and Carnoy, 2000; Mizala and Romaguera, 2000; Ministerio de Educación [MINEDUC], 2001). The importance of socio-economic status (SES) in Chile could be even greater considering that the school system is strongly segregated by social class: public schools serve the poorest families; private-subsidized (voucher type) schools serve the middle class; and the elite-paid schools serve the richest (Mella, 2003). The urban–rural location of the schools is closely related to social class. More affluent communities are usually urban, while poorer communities are usually concentrated in rural areas.

Several studies have shown how schools' assets are related to SES and to academic achievement (Secada 1992; Cueto et al., 2003). Students enrolled in schools serving more affluent communities are more likely to be taught more content in their classes, and more likely to take advanced mathematics courses. Gau (1997) stated that less affluent students are in a doubly disadvantaged situation. The scarce educational support they get at home is further compounded by the few learning opportunities they have at school. Are schools' assets unevenly distributed across Chilean schools? Are these differences related to school performance?

Other important school characteristics may be related to school performance. Variations in grade retention policies and class size may affect the learning experiences of students, which in turn may affect students' mathematics performance. In Chile, little is known about how these variables relate to school performance.

The official curriculum may have played its part in producing low mathematics performance and in structuring unequal learning opportunities. At the time of the TIMSS 1999 data collection, the Chilean curriculum (or program of study) was more of a framework in which schools developed their own curriculum than a structured curriculum in itself. Schools and teachers were left to interpret and implement this pre-reform curriculum. Better-resourced schools (mainly subsidized and elite-paid) developed their own curriculum or programs of study to support mathematics instruction. However, most of the schools (especially public ones) did not have the necessary resources to do so. In this chapter the effect of schools having their own curriculum on students' mathematics performance is explored.

In Chile, school principals are mainly focused on administrative tasks. There is a need to increase their capacity to provide pedagogical support (OECD, 2004). Ethnographic studies have stressed the importance of this problem. McMeekin (2003) provided evidence of how an atmosphere of cooperation and trust in school, with a principal focused on pedagogical support, positively affected school performance in Chile.

Concern exists about how well prepared Chilean teachers are to enable their students to obtain high achievement levels. Teacher training programs are among the least selective university programs (Ávalos, 1999). TIMSS 1999 showed that Chilean teachers have relatively low confidence levels in their preparation to teach mathematics (Mullis et al., 2000: 192). In Chile, there is also concern about the few opportunities teachers have to meet with other colleagues to discuss their teaching experiences.

The school effectiveness literature has suggested that schools with more parental involvement are in a better position to produce higher achievement levels than schools with less parental involvement (Reynolds and Teddlie, 2000). Hence, a further research consideration is the relationship between parental involvement and school performance.

International evidence shows that, within the same schools and classrooms, students vary widely in their academic performance (Beaton and O'Dwyer, 2002). This also seems to be the case in Chile (Ramírez, 2003). Investigation of what factors account for achievement differences among classmates is important, regardless of the school they attend. Gender differences favoring boys in mathematics (McEwan, 2003) may arise as a consequence of the differential feedback boys and girls receive from significant others (parents, teachers). Variations based on student age may result from students repeating grades, and may also relate to mathematics performance.

Benham (1995) and Stemler (2001) reported that students with positive attitudes toward learning and who believe that success depends more on their own effort out-perform their peers with more negative attitudes and those who weight more heavily the influence of external forces (for example, good luck or innate talent). Among parents and educators, there is usually concern about how students spend their out-of-school time. Some claim that students spend too many hours watching poor quality television, at the cost of playing sports or studying harder. Martin *et al.* (2000) reported that students that spent some time each day studying or doing homework in mathematics, science or other subjects had higher achievement than students who did not.

This study examines the distribution of mathematics achievement in Chile, and estimates the effect of several background factors on students' performance. Three research questions guided the analyses:

1   How is mathematics achievement distributed in the eighth grade?
2   What factors account for differences in schools' mathematics achievement?
3   What factors account for differences in achievement among students from the
     same classes and schools?

Method

A crucial step in this study was the selection of background variables that would be used to better understand mathematics performance. Bearing in mind the research questions of this study, the TIMSS 1999 questionnaires were reviewed extensively for questions that could be used as indicators of the background variables of interest. An extensive exploratory analysis of the relationship between the background variables and achievement was carried out. Eleven areas of interest were identified: (1) schools' social contexts; (2) school characteristics; (3) school climate; (4) principals' pedagog-ical involvement; (5) implemented mathematics curriculum; (6) teacher quality; (7) parental involvement; (8) students' social backgrounds; (9) students' demographics; (10) students' beliefs and perceptions; and (11) students' out-of-school time. Each area was measured by one or more variables. Tables 7.1 and 7.2 present detailed infor-mation about the background variables used in this study.

Single-item indicators were used to measure simpler constructs (for example, school size) whereas multiple-item indicators (indices) were used to measure more complex constructs (for example, topic coverage). The number of items used to create an index ranged between 2 and 18. When the source items of an index meas-ured the same facet of a construct, the strength of the association between the vari-ables was measured using *ad hoc* statistics (chi-square, correlations, Cronbach's

Table 7.1 Predictor variables at the school/class levels.

Block	Variable description (questionnaire and sequence number[a]).
BLOCK 1 Social context	Aggregated socio-economic index. Derived from averaging and aggregating mean parents' education (SQ_7a,b), number of books in the home (SQ_10), and having 12 possessions at home (SQ_11a–l). Dummy variable for type of community (SCQ_1): 1 = urban; 0 = other.
BLOCK 2 School structure	School size (SCQ_13a). Class size (Derived from achievement file). Number of students repeating eighth grade (SCQ_14b). Index of school resources for mathematics instruction. Derived from the average responses to 10 questions about shortages that affect instruction (SCQ_12a–e, g–k): instructional materials, budget, buildings and grounds, heating/cooling and lighting, instructional space, computers, software, calculators, library, audiovisual resources.
BLOCK 3 School climate	Index of student behavioral problems. Derived from the average responses to 18 questions (SCQ_17Aa–r). Index of schools' reports on school and class attendance. Derived from direct combination of three questions (SCQ_17Ba–c) about the severity of students' behaviors: tardiness, absenteeism, skipping classes.
BLOCK 4 Principals' activities	Time principal spends on instructional leadership activities. Derived from the sum of the responses to four questions (SCQ_7g,h,l,m). Influence principal has in determining curriculum (SCQ_9e).
BLOCK 5 Curriculum	Dummy for same content at different difficulty levels (SCQ_20a): 0 = no, 1 = yes. Dummy for students grouped by ability within mathematics classes (SCQ_20b): 0 = no, 1 = yes. Dummy for offering of enrichment mathematics (SCQ_20c): 0 = no, 1 = yes. Dummy for offering of remedial mathematics (SCQ_20d): 0 = no, 1 = yes. Dummy for written statement of curriculum (SCQ_10a): 0 = no, 1 = yes. Topic coverage index (TQMB_13a1–f34). Derived from the average responses to 34 TIMSS topics. Dummy for subject matter emphasis (TQMB_2): 0 = mainly numbers, 1 = others.
BLOCK 6 Teachers' background	Index of confidence in teaching mathematics. Derived from the average response to nine questions (TQMA_14a–i). Dummy for having a bachelors' or master's degree in mathematics (TQMA_17a,f and TQMA_18a,f): 0 = no, 1 = yes. Dummy for meetings with other teachers (TQMA_9): 0 = never; 1 = at least once a year.
BLOCK 7 Parents' involvement	Dummy for parental expectations of notification if their child is having problems (SCQ_25a): 0 = no, 1 = yes. Dummy for parents' checking if homework has been completed (SCQ_25e): 0 = no, 1 = yes.

Note. [a] SQ = students' questionnaire general version, SCQ = schools' questionnaire, TQM = mathematics teachers' questionnaire (part A or B).

Table 7.2 Predictor variables at the student level.

Block	Variable description (questionnaire and sequence number[a]).
BLOCK 1 Family background	Socio-economic index. Derived from the average responses to three questions: mean parents' education (SQ_7a,b), number of books at home (SQ_10), and having 12 possessions at home (SQ_11a–l).
BLOCK 2 Demographics	Dummy for sex of students (SQ_2): 0 = boy, 1 = girl. Age of students (SQ_1).
BLOCK 3 Attitudes and beliefs	Index of locus of control. Derived from the average responses to four questions (SQ_17c,d, SQ_18a,b): I am not talented in mathematics, I will never understand, I need natural talent to do well in mathematics, I need good luck to do well in mathematics. Index of pressure to do well. Derived from the average responses to two questions: My mother thinks it is important to do well in mathematics (SQ_12b), I think it is important to do well in mathematics (SQ_15b). Index of perceived difficulty of mathematics. Derived from the average responses to two questions (SQ_17a,b): I would like mathematics more if it were not so difficult, mathematics is more difficult for me than for classmates. Student's educational expectations (SQ_8). Index of positive attitudes toward mathematics. Derived from the average responses to five questions (SQ_21a, SQ_24a,b,d,e): How much do you like mathematics, I enjoy learning mathematics, mathematics is boring, mathematics is important to everyone's life, I would like a job with mathematics.
BLOCK 4 Use of time	Time spent watching television (SQ_6a). Time spent watching television squared (SQ_6a). Time spent playing sports (SQ_6e). Time spent playing sports squared (SQ_6e). Dummy for daily study in mathematics, science, and others (SQ_6g–i): 0 = no, 1 = yes

Note. [a] SQ = students' questionnaire general version.

alpha). In some other cases, the source items feeding an index targeted somewhat different aspects of a construct (for example, coverage of basic versus advanced topics), reason why the items were not expected to covary necessarily. In this latter case, it was checked that the pattern of responses makes sense conceptually.

In the TIMSS sampling design, students were nested in classes and classes were nested in schools. Since only one eighth-grade class was sampled within each school, the schools and the classes are further referred as the school/class unit. Because of this design, the final sampling units (students) were not independent of each other, and so hierarchical linear models (HLM) were statistically appropriate to model mathematics scores. At the student level, student variables served to model students' mathematics scores. At the school/class level, school and teacher variables served to model the schools/classes' mean performance. The only exception was the schools' socio-economic index, which was based on the average index value of all the students within the same school/class.

At each level of analysis, several models were tested, and each model tested the effect of different combinations of predictor variables. At the school/class level, the models were built giving priority to variables external to the schools, to move then to the school- and teacher-level variables that may have a more indirect effect on students' achievement, finally to school and teacher variables that might have a more direct effect on achievement. At the student level, the aim was to understand the achievement performance of students within the same class, regardless of the schools they attended. Predictor variables at this level were centered on the school/class mean (that is, they measure individual deviations from the school/class mean, $x_{ij}-\bar{x}_{.j}$). The predictors were entered in blocks that represented different conceptual categories. Model 1 used predictors from block 1, model 2 from blocks 1 and 2, and so on. In all of the models, mathematics scores were used as the outcome variable.

As a consequence of missing data, the analyses were run using a restricted sample: 66 percent ($n = 123$) of the sampled schools/classes and 49 percent ($n = 2,898$) of the sampled students. Since it is unclear to what extent these reductions may have affected the estimates, caution should be used in interpreting the results.

Results and Discussion

In this section, the results are presented in response to the three research questions. First, the distribution of the achievement scores is mapped; then, the mean differences among the schools/classes are modeled; and finally, the achievement differences among classmates are modeled. Results are interpreted and discussed considering the main characteristics of the Chilean educational system, and in light of the relevant literature.

Where Were the Differences in Mathematics Achievement at the Eighth Grade?

A preliminary (null) model showed that 35 percent of the differences in mathematics achievement corresponded to differences among schools/classes; the other 65 percent corresponded to differences among the students within these classes. Within the classes, the average achievement range was 232 points, equivalent to almost three standard deviations in the Chilean distribution of mathematics scores (M = 394, SD = 82). These findings indicate that most of the mathematics achievement spread is concentrated within the eighth-grade classes. This implies that fellow classmates cover a wide range of mathematics knowledge and skills. This situation poses a tremendous challenge for teachers and the implementation of effective instructional methods.

This is not to mean that Chilean schools were similar in their mathematics performance: they varied widely in their average achievement, too. The lowest performing school/class had a mean of 308 points in the TIMSS 1999 mathematics test while the highest performing had a mean of 565 points (similar to the average score of Belgium Flemish). The differences were so large that the achievement distributions of these two schools/classes did not overlap. The maximum score in the bottom performing school was max = 444 points, while the minimum score in the top performing one was min = 462 points.

What Factors Accounted for the Differences in Mean Mathematic Achievement among the Schools/Classes?

Table 7.3 shows the results of nine models that try to account for the achievement spread that lay among schools/classes. Model 1 used two variables: a socio-economic index of the communities served by the schools and a dummy variable for urban/rural location. This model accounted for 65 percent of the achievement differences among schools/classes. The socio-economic index was highly significant ($p < 0.0005$), while the school location was not. These results confirm the strong positive relationship existing between the cultural and economic capital of the students' families and schools' performance (Beaton and O'Dwyer, 2002; MINEDUC, 2001; Martin et al., 2000). The lack of significance of school location should be interpreted as a consequence of its correlation with socio-economic index ($r = 0.50$).

Model 2 expanded the previous model by including four new variables: school size; class size; number of students repeating the eighth grade; and school resources for mathematics instruction. This model accounted for 69 percent of the between schools/classes variance – an increase of 4 percentage points as compared to the previous model. School size had a positive and significant relationship with achievement ($p = 0.014$), thus indicating that once the other variables in the model were held constant, the bigger the school, the higher its achievement. A plausible explanation for this is that bigger schools are concentrated in urban settings, and accordingly may have access to better technical support from the Ministry of Education, universities or other private organizations. Class size did not make a significant unique contribution to predict the outcome. Since larger classes are more frequently found in larger schools, the effect of the former variable may have been weakened by that of the latter.

School resources did not make a significant unique contribution to predict achievement. While school resources were a significant predictor when entered alone in the equation ($p = 0.001$; not shown in the table), its effect was neutralized by the schools' socio-economic index. This is an indication of the positive relationship existing between school resources and the socio-economic status of the communities served by the schools ($r = 0.40$). This finding indicates that school resources are unequally distributed across social classes in Chile.

The number of students repeating the grade was not a significant predictor of achievement ($b = -5.92, p < 0.066$). While this variable failed the standard criteria for significance ($p < 0.05$), it is worth noting that its partial association with the outcome was negative in this and subsequent models.

Model 3 extended the previous models by including two school climate variables: an index of behavioral problems; and an index of schools' reports of good school and class attendance. Neither of these new predictors had a significant partial correlation with the outcome, and the model did not account for additional achievement variance. Their effect on the outcome was neutralized by the school socio-economic level.

Table 7.3 Predictors of schools/classes' mean achievement.

	MODEL (BLOCKS)								
	M1 (B1)	M2 (B1–2)	M3 (B1–3)	M4 (B1–4)	M5 (B1–5)	M6 (B1–6)	M7 (B1–7)	M8 Restricted	M9 Restricted and unadjusted by SEI
Between schools/classes variance accounted for by model (R^2) (%)	65	69	69	69	75	75	76	77	64
	Betas	Betas	Betas	Betas	Betas	Betas	Betas	Betas	Betas
Mathematics score (intercept)	394	394	394	394	394	394	394	394	394
Block 1: school social context									
Mean socio-economic index (SEI)	38.87*	34.55*	33.80*	33.33*	24.61*	25.86*	25.66*	26.80*	
Type of community	1.07	−0.53	−0.13	0.54	2.72	1.91	0.69		
Block 2: school structure									
School size		13.75**	12.95**	13.30*	10.80*	9.78**	10.20**	9.32*	17.76*
Class size		−0.31	0.25	0.15	−0.16	−0.15	−0.22		
Number of students repeating grade		−5.92***	−4.71	−4.87	−2.06	−1.96	−2.64		
Index of mathematics resources		2.10	1.06	0.70	1.11	0.64	0.47		
Block 3: school climate									
Index of behavioral problems			−4.94	−4.74	−3.18	−3.37	−4.12		
Index of good school and class attendance			0.88	1.14	0.28	−0.35	−1.21		

	(1)	(2)	(3)	(4)	(5)	(6)
Block 4: principal activities						
Time spent on instructional activities	-2.71	-1.34	-1.11	-1.15		
Influence in determining curriculum	3.73	1.73	1.36	0.93		
Block 5: curriculum						
Same content at different difficulty levels		-3.85	-3.67	-3.60		
Students grouped by ability within mathematics classes		-3.42	-3.19	-2.42		
Enrichment mathematics is offered		3.19	3.01	3.05		
Remedial mathematics is offered		-2.61	-2.25	-2.39		
Written statement of curriculum		11.78*	11.67*	11.38*	12.41*	25.82*
Topic coverage index (TCI)		4.82***	4.45***	3.87	4.28**	2.00
Subject matter emphasis		-0.13	0.35	0.90		
Block 6: teacher background						
Index of confidence to teach mathematics			-3.09	-2.48		
Have a bachelor or master in mathematics			-0.13	0.64		
Meeting with other teachers			2.16	2.39		
Block 7: parent involvement						
Expect parents to notify if child is having problems				-2.29		
Expect parents to check if homework complete				6.63*	5.86*	7.13*

Note. HLM random intercepts models weighted with HOUWGT. All predictors standardized and grand-mean centered. Method: Restricted maximum likelihood with robust standard errors. Significant alpha levels: * = <0.01; ** = <0.05; *** = <0.10.

Model 4 assessed the effect of the variables included in the previous models, plus two variables: the time principals had spent on instructional leadership activities, and the influence he/she had on determining the curriculum. This model did not account for additional variance, and none of the new predictors made a significant partial contribution to explain the outcome. More information about the kinds of pedagogical support the principals provided is needed to interpret these findings.

Model 5 tested the effect of all previous variables plus a block of curriculum implementation variables. This model accounted for an additional 6 percent of the total between schools/classes variance, which increased from 69 to 75 percent. A school having its own curriculum was a strong predictor of mathematics achievement ($b = 11.78, p < 0.0005$). This effect was significant even after controlling statistically for the socio-economic index. This sounds reasonable considering the little guidance the national curriculum provided to the Chilean teachers.[1] Around 15 percent of the schools reported having their own written statement of the curriculum; almost all of these schools were private (elite-paid or private-subsidized). Lack of technical and material resources may have precluded public schools from developing their own curriculum. This is another example of how school assets are unequally distributed among the Chilean schools.

The results of the fifth model also show that teachers serving similar populations of students produced higher achievement when they covered more mathematics topics in classes ($b = 4.82, p = 0.052$). While topic coverage failed the standard criteria for significance ($p < 0.05$), its effect showed up strongly in subsequent models. Chilean teachers reported substantial variations in the coverage of the topics included in the TIMSS test. They mainly focused on teaching fractions and number sense, giving the students fewer opportunities to learn more advanced topics (for example, geometry, algebra). The unique contribution of subject matter emphasis (numbers versus geometry, algebra, and so forth), while significant when analyzed alone, was overridden by the effect of topic coverage.

In order to accommodate students with different needs and interests, more than three-fourths of the schools provided remedial mathematics, and two-thirds used the same curriculum for the students but at different difficulty levels. Neither these nor other curriculum differentiation strategies (that is, streaming, ability grouping) made a unique contribution to explaining the outcome variance. More information about the nature of these courses is needed to interpret these findings.

Model 6 expanded the previous models by including a block of teacher background variables: having a major in mathematics, confidence in teaching mathematics and meeting with other teachers. This model did not account for additional variance above and beyond that already accounted for by the previous models. The effect of teacher variables was confounded with SES. It is noteworthy that only three-fourths of the sampled teachers of mathematics reported that they majored in mathematics.

Model 7 created a new model that included all previous predictors plus two parental involvement variables: if the school expected the 'parents to notify the school about any problems their child may be having at home or with classmates', and whether the school expected the 'parents to be sure that their child completes his/her homework'. The latter variable made a unique significant contribution to the outcome

(b = 6.63, p = 0.004), and the percentage of between schools/classes variance accounted for increased slightly, from 75 to 76 percent.

Model 8 showed that an equivalent amount of between school/class variance could be accounted for by entering the predictors with a significant effect on previous models. The socio-economic index, school size, schools' having their own written statement of the curriculum, topic coverage and expectations that parents check homework, accounted for 77 percent of the schools' mean performance. Model 9 showed that if the socio-economic index were deleted, the other four predictors could still account for 64 percent of the differences among schools/classes. It is interesting to note that models 1 and 9 accounted for similar amounts of achievement variance (65 and 64 percent, respectively). This result vividly illustrates how school and teacher characteristics were related to the SES of the communities served by the schools.

What Factors Account for the Differences in Mathematics Achievement among Classmates?

Table 7.4 presents six models that try to explain the achievement differences among classmates, regardless of the schools they attend. In Model 1, the socio-economic index was entered alone. While a significant effect was detected (b = 10.95, p = 0.012), this index accounted for only 1 percent of the differences in achievement among classmates.

Model 2 tested the effect of the socio-economic index, sex and age of the students. This model accounted for 6 percent of the differences among classmates (5 percentage points more than the previous model). Girls obtained lower scores than boys (b = −8.49; p = 0.003), net the effect of the other variables in the model. One could speculate that this gender gap is due, at least in part, to the differential reinforcement boys and girls receive from relevant others (teachers, peers, parents).

The results also show that older students obtained lower scores than their younger peers (b = −12.57; p < 0.0005), net the effect of the other variables in the model. This result may suggest that students repeating grades are not benefiting from being recycled through the curriculum. This result is also consistent with the finding that school achievement tended to be lower when there were more students repeating the grade.

Model 3 extended the previous models by including five new variables measuring students' beliefs and perceptions. Overall, this model accounted for 22 percent of the differences in achievement among classmates – an additional 16 percent over the previous model. Regardless of school characteristics, students who (1) expected to graduate from university, (2) thought that doing mathematics was not so difficult, and (3) thought that their academic performance did not depend on good luck or innate talent, attained significantly higher mathematics achievement (p < 0.02), after controlling for the effect of the other predictors in the model. Pressure to do well failed to show a significant partial association with the outcome. Its effect was blurred because of its relationship with other variables in the model.

Model 4 expanded the previous model by including a block of variables related to the students' use of their out-of-school time: (1) students spent some time each day

Table 7.4 Predictors of mathematics achievement among classmates.

| | MODEL (BLOCKS) | | | | | |
	M1 (B1)	M2 (B1-2)	M3 (B1-3)	M4 (B1-4)	M5 Restricted	M6 Restricted and unadjusted by SEI
Within-schools variance accounted for by model (R^2) (%)	1	6	22	24	22	22
	Betas	Betas	Betas	Betas	Betas	Betas
Mathematics score (intercept)	394	394	394	394	394	394
Block 1: social context						
Socio-economic index (SEI)	10.95**	9.66**	3.13	3.66	3.08	
Block 2: students' demographics						
Sex		−8.49*	−6.65*	−9.05*	−6.93*	−7.07*
Age		−12.57*	−9.03*	−8.51*	−9.09*	−9.13*
Block 3: students' beliefs and perceptions						
Locus of control			9.46**	9.26**	9.93*	10.10*
Pressure to do well			−0.01	0.08		
Perceived difficulty of mathematics			19.52*	19.39*	20.21*	20.29*
Students' education expectations			6.78*	6.48**	6.87*	7.39*
Positives attitudes toward mathematics			3.03	3.78		
Block 4: students' use of time						
Time spent watching television				28.61		
Time spent watching television squared				−25.54		
Time spent playing sports				−0.55		
Time spent playing sports squared				−4.91		
Time spent each day studying mathematics, science, others				−1.11		

Note. HLM random intercepts models weighted with HOUWGT. All predictors standardized and group-mean centered. Method: Restricted maximum likelihood with robust standard errors. Significant alpha levels: * = < .01; ** = < 0.05; *** = < 0.10.

studying mathematics, science, or others; (2) number of hours watching television and videos; and (3) number of hours playing sports. Exploratory data analysis suggested a curvilinear (inverted U-shape) relationship between number of hours watching television/videos and mathematics achievement, and between hours playing sports and achievement. Accordingly, two squared variables were included in the equation. Model 4 increased the amount of accounted variance by two percentage points. Nevertheless, none of the new variables made a unique significant contribution to explain the outcome. A plausible explanation for this is that students do their homework and play sports in school; accordingly, they may not have reported these activities as 'out-of-school time'.

Two restricted models that included only the significant predictors – sex, age, importance of luck and innate talent, perceived difficulty of mathematics, and educational expectations – were run. Model 5 adjusted the outcome by the socio-economic index while Model 6 did not. Both models accounted for 22 percent of the achievement variance – that is, two percentage points less than Model 4. In both models, the demographics variables and the three variables from the students' beliefs and perceptions' block had a significant effect on students' achievement. The socio-economic index remained non-significant in the fifth model.

Conclusion

While overall mathematics achievement is low in Chile, there are important differences in school performance. These differences are strongly related to social class: schools serving more affluent communities have higher achievement levels than schools serving more disadvantaged communities. Less affluent students are in a doubly disadvantaged situation. On the one hand, their parents have fewer years of schooling, and have fewer educational resources at home; on the other hand, their schools have fewer resources to provide quality instruction, and their teachers are less prepared to teach mathematics. This situation poses a tremendous challenge: fostering equity in the school system.

It seems that the official (pre-reform) curriculum did not provide enough guidance for its effective implementation. Schools needed to have their own curriculum to reach higher performance. However, most of the schools (especially the poorest ones) never had the necessary resources to do so. Currently, a new updated and more clearly specified curriculum is in use. Educational authorities should closely monitor its implementation at the classroom level to ensure its effective implementation.

In Chile, the debate about how to improve achievement levels usually points to structural factors as explanations of low performance. Less attention has been given to students' beliefs and perceptions toward schooling and mathematics. This study shows that students' beliefs and perceptions are closely related to achievement, regardless of the school they attend. It may be desirable to pay more attention to students' perceptions in order to ensure the effective implementation of educational reform policies.

Note

1. The curriculum in use at the time data were collected for this study was *Planes y Programas de la Educación General Básica* (MINEDUC, 1980). This curriculum was replaced by an updated and more detailed national curriculum in 2002.

References

Ávalos, B. (1999) Mejoramiento de la formación inicial docente [Improvement in teacher pre-service training]. In J.E. García-Huidobro (ed.), *La reforma educacional chilena* (pp. 195–214). Madrid: Editorial Popular.

Beaton, A.E. & O'Dwyer, L.M. (2002). Separating school, classroom, and student variances and their relationship to socio-economic status. In D.F. Robitaille and A.E. Beaton (eds.), *Secondary analysis of the TIMSS data* (pp. 211–31). Boston, MA: Kluwer Academic Publishers.

Benham, J.M. (1995) *Fostering self-motivated behavior, personal responsibility, and internal locus of control in the school setting.* University of Southern Maine. (ERIC Document Reproduction Service No. ED386621)

Cueto, S., Ramírez, C., León, J. and Pain, O. (2003). *Opportunities to learn and achievement in mathematics in a sample of sixth grade students in Lima, Peru.* Global Development Network. Retrieved 9 September 2004, from http://www.gdnet.org/pdf/2002AwardsMedalsWinners/EducationKnowledge Technology/santiago_cueto_paper.pdf

Gau, S. (1997) *The distribution and the effects of opportunity to learn on mathematics achievement.* Paper presented at the annual meeting of the American Educational Research Association, Chicago, IL, March. (ERIC Document Reproduction Service No. ED407231)

Martin, M.O., Mullis, I.V.S., Gregory, K.D., Hoyle, C. and Shen, C. (2000) *Effective schools in science and mathematics IEA's third international mathematics and science study.* Chestnut Hill, MA: International Study Center, Lynch School of Education, Boston College.

McEwan, P.J. (2003) Peer effects on student achievement: evidence from Chile. *Economics of Education Review*, 22, 131–41.

McEwan, P.J. and Carnoy, M. (2000). The effectiveness and efficiency of private schools in Chile's voucher system. *Educational Evaluation and Policy Analysis*, 22(3), 213–39.

McMeekin, R. (2003) Networks of schools. *Education Policy Analysis Archives*, 11(16). Retrieved 14 May 2003 from http://epaa.asu.edu/epaa/v11n16/

Mella, O. (2003) 12 años de reforma educacional en Chile: algunas consideraciones en torno a sus efectos para reducir la inequidad. *Revista Electrónica Iberoamericana sobre Calidad, Eficacia y Cambio en Educación.* Retrieved 30 December 2003, from http://www.ice.deusto.es/rinace/reice/vol1n1/Mella.pdf

Ministerio de Educación(MINEDUC) (1980) *Planes y programmas de la educación general básica, Decreto No. 4002* [*National curriculum for primary school, Decree No. 4002*]. Santiago: MINEDUC.

Ministerio de Educación (MINEDUC) (2001) *Informe de resultados 8vo básico SIMCE 2000* [*SIMCE 2000 report for the 8th grade*]. Retrieved 20 April 2004, from http://www.simce.cl/doc/informe_2000_8basico.pdf

Mizala, A. and Romaguera, P. (2000) School performance and choice. *Journal of Human Resources*, 35(2), 392–417.

Mullis, I.V.S., Martin, M.O., Gonzalez, E.J., Gregory, K.D., Garden, R.A., O'Connor, K.A., Chrostowski, S.J. and Smith, T.A. (2000) *TIMSS 1999 international mathematics report: Findings from IEA's repeat of the third international mathematics and science study at the eighth grade*. Chestnut Hill, MA: International Study Center, Lynch School of Education, Boston College.

Organization for Economic Co-Operation and Development (OECD) (2004) *Review of national policies for education: Chile*. Paris: OECD.

Ramírez, M.J. (2003) The distribution of mathematics knowledge among Chilean fourth graders and related explanatory factors. *EducarChile*. Retrieved 20 November 2004, from http://200.68.0.6/medios/20030604142903.pdf

Reynolds, D. and Teddlie, C. (2000) The processes of school effectiveness. In C. Teddlie and D. Reynolds (eds.), *The international handbook of school effectiveness research* (pp. 134–59). London: Falmer Press.

Secada, W. (1992) Race, ethnicity, social class, language, and achievement in mathematics. In D.A. Grouws (ed.), *Handbook of research on mathematics teaching and learning: A project of the National Council of Teachers of Mathematics* (pp. 623–60). New York: Macmillan Publishing Company.

Stemler, S.E. (2001) Examining school effectiveness at the fourth grade: a hierarchical analysis of the Third International Mathematics and Science Study (TIMSS). (Doctoral dissertation, Boston College). *Dissertation Abstracts International*, 62 (03A), 213–919.

8

Modeling Mathematics Achievement in Cyprus

Constantinos Papanastasiou and
Elena C. Papanastasiou

Introduction

The purpose of this study was to identify the background factors that influence mathematics achievement and to estimate the strength of their effects on students in Cyprus. Participating students were in the eighth grade in the academic year 1998/9.

The educational system of Cyprus is highly centralized due to the small size of the country. The Ministry of Education and Culture is responsible for the enforcement of educational laws. The structure of the public educational system is based on four sectors: the preprimary education; the primary education (providing six years of compulsory schooling); the lower secondary education (gymnasium) (offering a three-year program); and the upper secondary education (lyceum, technical-vocational) (offering three-year programs).

The TIMSS study was undertaken with the support of the Ministry of Education for the primary purpose of comparing the Cyprus educational system to the educational systems of other countries. Before the participation of Cyprus in the IEA studies, the common belief was that the Cyprus educational system was in good health. This belief encompassed the teachers, the students and their behavior, the students' study habits, as well as the cooperation between the school and the community. In general the parents regarded the value of education as high. So overall, people in Cyprus thought that the educational system in Cyprus worked very effectively. However, the TIMSS as well as other IEA studies came to drastically change these kinds of beliefs.

For example, in the TIMSS 1999, the average student mathematics achievement of grade 8 students in Cyprus ($\bar{X} = 476$) was below the international average. In addition, the rank order of Cyprus was 24th out of the 38 countries that took part in the study, while the achievement of the female students was statistically higher than that of males.

In light of such research results, the Ministry of Education and Culture in Cyprus is now struggling to find ways to improve the country's educational system. In addition, societal confidence in education is also starting to weaken. A lot of discussion is taking place about the important issues and problems that the Cypriot educational system faces. These issues include the vast curriculum that the teachers have to teach, the way in which teachers are hired, which is more often based on the teacher's age than on any meritocracy, as well as an invalid evaluation system for teachers.

Literature Review

Attitudes toward mathematics have been of interest to researchers for many years (Cooper *et al.*, 1998; Ma and Kishor, 1997; Ross and Cousins, 1995; Vanayan *et al.*, 1997). Positive attitudes toward mathematics are highly valued since it is understood that attitudes are related to achievement, and achieving a positive attitude toward school subjects should be an important school goal. Therefore, it is likely that a student who feels very positive about mathematics will achieve at a higher level than a student who has negative attitudes toward mathematics. It is also likely that a high achiever will enjoy mathematics more than a student who does poorly in mathematics (Reyes, 1984). According to Aiken (1986), attitudes toward mathematics begin to develop whenever the children are exposed to the subject, but are particularly important for students aged 11–13-years old. This is the time when negative attitudes towards mathematics become especially noticeable (Cheung, 1988). Research studies on attitudes have led mathematics educators to study affective differences in comparison with student achievement (Fennena, 1980; Leder, 1990). Such studies have found that attitudes play significant roles in both learning mathematics (Lester *et al.*, 1989; Shaughnessy *et al.*, 1983) and in maintaining a continued interest in the subject (Eccless *et al.*, 1985).

The belief that positive affects might lead to positive achievement is fairly widespread (McLeod, 1992). However, in contrast to this view, Eisenhardt's (1977) research indicated that achievement influences attitudes more than attitudes influences achievement in mathematics. Previous research on the relation between achievement and attitudes toward school mathematics has shown a modest correlation between them. Fraser and Butts (1982) reviewed a meta-analysis by Willson (1981) and concluded that the empirical evidence is insufficient to support the claim that attitudes and achievement are highly related. Such inconclusive evidence has also been reached for the subject area of science, where in some countries it was attitudes that affected achievement, while in other countries achievement affected the students' attitudes (Papanastasiou and Zembylas, 2004). What remains to be seen is the direction of the relationship in this study.

There is much empirical work that suggests that the home environment plays an important role in learning. A number of studies have found evidence of strong relationships between home environment and cognitive skills (Crane, 1996). There are at least two ways in which parents can affect their children's mathematics scores. One way is to make opportunities to learn available by exposing the child to particular social and cultural environments or by spending money in a particular way. There is plenty of evidence that students from homes with more educational resources have higher achievement in mathematics and in other subjects than students from less advantaged backgrounds (Mullis *et al.*, 2000; Papanastasiou, 2003; Papanastasiou and Ferdig, 2003). This has been documented in the TIMSS study of the eighth-grade results in 1999 (Martin *et al.*, 2000). The TIMSS report showed that students with more books at home, or with more highly educated parents, also had higher mathematics achievement (Beaton *et al.*, 1996). The number of books in the home is a very useful indicator of home literacy support, and is one of the variables that correlates positively with student achievement in mathematics and science in all TIMSS countries (Martin *et al.*, 2000). In general students with a wide range of reading materials at home can strengthen and deepen their understanding of concepts covered in class through the use of the books at home and can foster academic interest that encourages learning. According to Martin *et al.* (2000), students whose parents have attained a high level of education (at least one parent had completed a university degree) are more likely to place high value on academic achievement.

Relationships between school climate and achievement and between affect and achievement have received more research attention than has the relationship between learning environments and attitudes to mathematics (Forgasz, 1995). Mathematics teachers organize the learning environments of their students and consequently are in a critical position to influence their attitudes. Learning environment variables are characteristics of a particular learning experience that can be affected by students, teachers or the school climate in general. Students' attitudes and expectations regarding mathematics have been considered to be very significant factors underlying their school achievement. Teachers establish the climate of their classrooms by setting the goal of orientation, degree of control, level of competition, system of rewards, pattern of expectations and degree of support of students. All these factors influence students' achievement (Newmah and Schwager, 1993).

The purpose of this study was to identify some background factors that influence mathematics achievement and to estimate the strength of their effects. The first question guiding this analysis is whether the exogenous factors of school climate and the family's educational background are statistically significant predictors of mathematics achievement. The second question is whether both exogenous predictors can influence the teaching of mathematics and the student's attitudes toward mathematics. A third question asked is whether the endogenous factors of teaching and attitudes are predictors of the mathematics outcome.

Data Source

This study used the TIMSS 1999 data collected from Cypriot eighth-grade students in the school year 1998/9. The average age of students tested was 13.8-years old. This age group was selected since it is believed that these are the students who are mostly influenced by their family, their close environment, as well as their school. The variables used in this study were all collected from the student questionnaires. All 61 gymnasia (the secondary junior schools) in Cyprus participated in this project, and two eighth-grade classes were selected from each school. Within each class all students were tested and both achievement and background data were obtained from the students. The participating students completed questionnaires on home and school experiences related to learning mathematics. Data were obtained from 3,116 students, which represented about 31.8 percent of the entire population (9,786 students). However, among those students, only the ones who had completed all of the questions that would be used in the analysis were eventually used. This led to listwise deletion of the subjects from the data set, which narrowed the final sample to 2,447 eighth-grade students, which represented 78.5 percent of the original sample of students.

Student Variables

The factors and items used in this study, their respective descriptive statistics, and the number of valid cases are included in Table 8.1. The 17 questions used in this study were grouped into separate categories, related to the following latent variables: the family educational background; the disciplinary climate of the school; student attitudes on mathematics; and teachers' teaching practices in the mathematics class.

The observed variables of 'mother's education', 'father's education' and 'number of books in the home', are assumed to be indicators of the factor 'family educational background'. The observed variables of 'something was stolen', 'a student hurt me', 'friends skipped a class' and 'friends were hurt' are assumed to be indicators of the factor 'disciplinary climate'. Also, the variables of 'like mathematics', 'enjoy mathematics', 'mathematics is boring' and 'mathematics is easy' are related to attitudes, and are assumed to be indicators of the corresponding latent variable 'attitude'. Finally, the observed variables 'we work on projects', 'teacher uses things from everyday life', 'teacher checks homework', 'we discuss homework', 'we discuss practical problems' and 'relation to a new topic' are assumed to be indicators of the factor 'teachers' teaching practices'.

The measures used to define the variables or factors used in the analysis are presented next. The family educational background measures which included the highest education level of the parents was measured on a scale from 1 to 7 (1 = some primary school or did not go to school, 2 = finished primary school, 3 = some secondary school, 4 = finished secondary school, 5 = some vocational/technical education after secondary school, 6 = some university and 7 = finished university). The size of the home library which did not include student textbooks was measured on a scale from 1 to 5 (1 = 0–10 books, 2 = 11–25 books, 3 = 26–100 books, 4 = 101–200 books and 5 = more than 200 books).

Disciplinary climate measures included questions related to the school environment. For example, 'some of my friends skipped a class' or 'some of my friends were hurt by other students' were measured on Likert scales that ranged from 1 to 5 (1 = never, 2 = once or twice, 3 = 3–4 times, 4 = 5 or more). Attitudes measures that

Table 8.1 Items[a] that were used in the study and their respective descriptive statistics.

Items		Descriptive statistics		
		$\bar{X}$	SD	N
Family educational background				
Bsbgedmo:	How far in school did your mother go?	3.96	1.78	2926
Bsbgedfa:	How far in school did your father go?	4.09	1.90	2869
Bsbgbook:	About how many books are there in your home?	3.28	1.10	3074
Disciplinary climate				
Bsbgsstl:	Something of mine was stolen.	1.45	0.70	3025
Bsbgshrt:	I thought another student might hurt me.	1.51	0.79	3025
Bsbgfskp:	Some of my friends skipped a class.	2.29	1.09	3015
bsbgfhrt:	Some of my friends were hurt by other students.	2.01	0.95	3029
Attitudes				
Bsbmlike:	I like mathematics.	2.99	0.88	3072
Bsbmenjy:	I enjoy learning mathematics.	1.84	0.81	3058
Bsbmbore:	Mathematics is not boring.	1.16	0.90	3034
Bsbmeasy:	Mathematics is an easy subject.	2.59	0.86	3002
Teachers' instructional practices				
Bsbmproj:	We work on mathematics projects.	2.96	0.96	3043
Bsbmevlf:	We use things from everyday life in solving mathematics problems.	2.46	0.93	3059
Bsbmhwtc:	The teacher checks homework.	1.80	0.91	3066
Bsbmhwds:	We discuss our completed homework.	2.01	0.95	3032
Bsbmprac:	We discuss practical or story problems related to everyday life.	2.22	0.93	3065
Bsbmask:	The teacher asks us what we know related to the new topic.	2.09	0.92	3055

Note. [a]Items with negative direction within the factors have been changed in the analysis.

included questions to determine whether students like mathematics were measured on four-point Likert scales (1 = like a lot, 2 = like, 3 = dislike, 4 = dislike a lot). The same was the case for the questions of whether the students enjoy mathematics, do not find it boring and think it is an easy subject (1 = strongly agree, 2 = agree, 3 = disagree, 4 = strongly disagree). Teachers' teaching practices measures included questions on activities related to the mathematics lesson; that is, if students work on mathematics projects, if they use things from everyday life in solving mathematics problems, if their teachers check and discuss homework, if they begin the lesson discussing a practical problem, and if the teacher asks questions related to the new topic (1 = almost always, 2 = fairly often, 3 = once in a while, 4 = never).

According to Table 8.1, the average parent in Cyprus, according to the students' responses, has completed secondary school. In addition, the average amount of books that exist in students' houses are around 26–100 books. In terms of the school climate, students in Cyprus rarely have ever had something of theirs stolen, or thought that another student might hurt them. In addition, the average student indicated that their friends have skipped class about once or twice, which is about the same frequency with which they indicated that some of their friends were hurt by other students. In terms of attitudes, the majority of the students in Cyprus like and enjoy mathematics, and do not find the subject boring. These same students were somewhat neutral in their responses of whether they thought that mathematics was an easy subject. In terms of teaching activities, the students responded that the activity that was performed most often was that of the teacher checking the student's homework. The rarest activity, as indicated by the students, was that of working on mathematics projects, which was typically done every once in a while. As for the rest of the teaching activities such as discussing completed homework, discussing a story problem related to everyday life, and asking what the students know about a new topic, these were activities that were performed fairly often, as indicated by the students.

Hypothetical Model

Structural equation models are often used to analyze relationships among variables, and in many different fields, such as sociology (Alsup and Gillespie, 1997), psychology (Raykov, 1997), medicine (Papa *et al.*, 1997), economics (Kaplan and Elliot, 1997) and education (Dauphinee *et al.*, 1997). While structural equation modeling supposes that cross-product covariances or Pearson correlations have been derived from variables that are continuous and measured on an interval scale, this is rarely the case for survey data (Coenders *et al.*, 1997). Data collected through questionnaires or interviews are usually based on ordinal observed variables, that is, the responses are classified into different ordered categories, although they are conceptually continuous. An ordinal variable y_i may be regarded as a measurement of an underlying unobserved continuous variable y_i^*, and therefore y_i would be related to y_i^* through the step-function: $y_i = k$ when $\tau_{i,k-1} < y_i^* < = \tau_{i,k}$ for $k = 1,\ldots$, mi, where $\tau_{i,o} = -\infty$, $\tau_{i,k} < \tau_{i,k+1}$, $\tau_{i,mi} = \infty$. The parameters $\tau_{i,1},\ldots,\tau_{i,mi-1}$ are called thresholds of the ith variable. This method appears most suitable for the social sciences (Coenders *et al.*, 1997), where many variables are conceptually continuous, and measurement instruments may be discrete and have only ordinal properties. In this study, all 17 variables were ordinal.

The analysis of the data used in this study revealed statistical differences in achievement between students whose parents have high versus low educational background (Martin *et al.*, 2000). Thus, this factor was selected as one of the two exogenous constructs of the proposed model (see Figure 8.1), which assumed that students' mathematics outcomes were initially affected by students' family educational background and the disciplinary climate of the school. The family educational background factor is included in the model as a partial explanation of students' mathematics achievement, and represents variables brought into the school learning environment, that can influence attitudes, teachers' teaching practices and mathematical outcomes. In this model, mathematics achievement is seen as part of a greater context that includes family, school, teaching and attitudes. The model implies that the educational background of the family and the disciplinary school climate affect student attitudes and support the teaching quality and all these affect, either directly or indirectly, the mathematics achievement of the students. A hypothetical initial factor model is presented in Figure 8.1.

Actual Model

Throughout the process of building the model, certain latent variables which had been assumed as valid, proved not to fit. Therefore, a modified version of the structural model had been created, which is presented in Figure 8.2. Table 8.2 presents the factors, the items that were used in this study, the weighting least squares standardized and unstandardized LISREL estimates, the standard errors and the corresponding t values, which show that all lambdas are statistically significant. These results indicate the magnitude of the relationship between the various factors and variables.

Figure 8.2 presents the unstandardized solution of the path model. In this model family educational background has strong direct effects on mathematics outcomes, on teachers' teaching practices and attitudes toward mathematics. As Figure 8.2 shows, the paths from family educational background to attitudes $(0.75, (se = 0.05),$

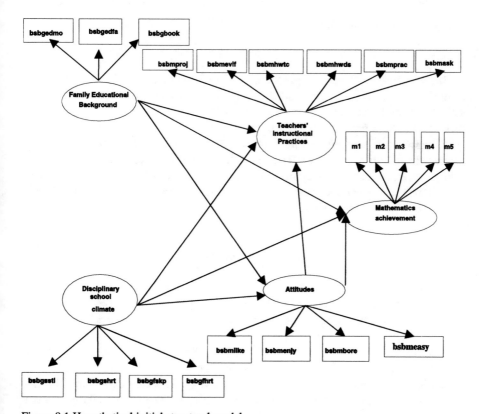

Figure 8.1 Hypothetical initial structural model.

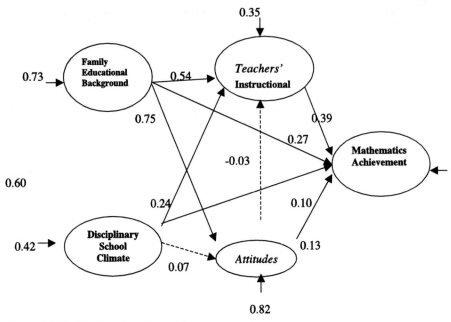

Figure 8.2 Model of mathematics achievement.

$t = 15.31$), to teachers' teaching practices (0.54, (se = 0.06), $t = 9.45$) and mathematics achievement (0.27, (se = 0.06), $t = 4.49$) were significant. The paths from disciplinary school climate to teachers' teaching practices (-0.24, (se = 0.03), $t = -7.32$) and to mathematics outcomes (−0.10, (se = 0.03), $t = -2.95$) were also significant, but not to attitudes toward mathematics (−0.07, (se = 0.04), $t = -1.86$). The path from attitudes to teachers' teaching practices (−0.03, (se = 0.03), $t = -0.96$) was also not significant. The paths from teachers' teaching practices to mathematics achievement (0.39, (se = 0.05), $t = 7.46$) and from attitudes to mathematics achievement (0.13, (se = 0.03), $t = 4.76$) were also significant.

Fit Statistics

The most commonly used model fit criteria are chi-square, goodness-of-fit index (GFI), adjusted goodness-of-fit index (AGFI) (Joreskog and Sorbom, 1989), comparative fit index (CFI) and root mean square error of approximation (RMSEA). All the above fit statistics were applied to assess the 'goodness of fit' of the hypothetical initial model. As we can see from Table 8.3, the chi-square is significant ($p = 0.0 < 0.05$) indicating that the model is significantly different from the data, which does not support the fit of the model. However, it is well known that the chi-square is sensitive to sample size and when the sample size increases the chi-square test has the tendency to indicate a significant probability level, even with trivial dif-

Table 8.2 LISREL estimates (weighted least squares).

Factors	Items	Unstandardized Lambda-X	se	Standardized Lambda-X	t^a
Family	How far in school did your mother go?	1.03	0.03	0.88	33.58
Educational	How far in school did your father go?	1.00	–	0.85	–
Background	About how many books are there in your home?	0.03	0.66	25.09	0.78
	Something of mine was stolen	0.85	0.03	0.55	25.13
Disciplinary	I thought another student might hurt me	1.00	–	0.65	–
Climate	Some of my friends skipped a class	0.78	0.03	0.512	3.38
	Some of my friends were hurt by other students	0.99	0.04	0.64	27.44

Factors	Items	Unstandardized Lambda-Y	se	Standardized Lambda-Y	t^a
	I like mathematics	1.00	–	1.10	–
Attitudes	I enjoy learning mathematics	0.73	0.02	0.80	34.25
	Mathematics is not boring	0.57	0.02	0.63	30.43
	Mathematics is an easy subject	0.51	0.02	0.57	27.26
	We work on mathematics projects	1.00	–	0.70	–
Teacher's Instructional Practices	We use things from everyday life in solving mathematics problems	0.44	0.03	0.32	14.53
	The teacher checks homework	0.23	0.03	0.17	8.85
	We discuss our completed homework	0.46	0.03	0.34	16.59
	We discuss practical or story problems related to everyday life	0.51	0.03	0.38	18.35
	The teacher ask us what we know related to the new topic	0.62	0.03	0.46	21.26

Note. [a] All *t*-values are significant at $p < 0.05$.

ferences between the original population matrix and the matrix produced by the model. Recalling that the number of cases is 2,447, and according to Marcoulides and Hershberger (1997) the fit of a model should be independent of the size of the sample used to test it. All other indices support the hypothesis that the models fit the sample data. More specifically, the three fit indices GFI and AGFI and CFI have large values and generally values above 0.9 are considered to indicate well-fitting models (Broome *et al.*, 1997; Marcoulides and Hershberger, 1997). Furthermore, the CFI is the least affected by sample size (Hu and Bentler, 1995). The RMSEA index is small, and, according to Browne and Cudeck (1993), RMSEA values between 0.0 and 0.05 reflect a close fit, less than 0.08 reflect reasonable fit, and greater than 0.08 reflect poor fit. In general it seems from the indices of Table 8.3 that there is a close fit between the matrix produced by the models and the original population matrix in the Cypriot data.

Table 8.3 Goodness-of-fit indices.

Fit statistics	Cyprus
Chi-square	1974.23
df	199
p	0.00
Goodness-of-fit index (GFI)	0.99
Adjusted goodness-of-fit index (AGFI)	0.99
Comparative fit index (CFI)	0.99
Root mean square error of approximation (RMSEA)	0.06

Discussion

This study explored how predictors related to family and schools affect mathematics outcomes. For the initial analyses we decided to use only the student questionnaire data. For this study we began by posing a simple question: How can we best explain student achievement (based on the TIMSS 1999 data) in relation to the family educational background, to disciplinary school climate, teachers' teaching practices and students' attitudes? The results of this study indicated that two exogenous factors – the educational background of the family and the disciplinary school climate – define a second order-factor structure, which includes the endogenous predictors, the students' attitudes toward mathematics and the teachers' teaching practices.

According to the proposed model, the strongest direct influence on mathematics achievement was the teachers' teaching practices. The next strongest influence was the educational background of the family, followed by attitudes. The weakest effect was exerted by the disciplinary school climate. The educational background of the family exerted the strongest direct effect on student attitudes. Teachers' teaching practices were most directly influenced by the family educational background, followed by the disciplinary school climate. Finally, teaching practices were not influenced by the student attitudes and attitudes were not influenced by the disciplinary school climate.

Among the four factors that influence mathematics achievement, the teachers' teaching practices is the factor with the strongest influence on achievement. Since the general achievement of Cypriot students is below average and is strongly related to the teachers' teaching practices, we can conclude that the teaching methods that are currently used in mathematics classes in Cyprus need to be changed. Part of this problem could be due to the fact that the curriculum that needs to be covered by the teachers is huge. As a result, such teachers might attempt to cover the course content superficially by teaching lower order thinking skills, without worrying about whether the students have actually grasped the mathematical concepts that are being taught. Therefore, the Ministry of Education and Culture should take the initiative to take the proper measures to reduce the heavy load of the curriculum, as well as to help learn how to select the best possible teaching methods based on each separate school subject, content area and grade.

Going back to the factor model, the path from family's educational background to student's attitudes toward mathematics is the strongest one in the model. This is a good sign for the society of Cyprus. It seems that the parents respect education and the subject of mathematics, which are important factors which can be used to improve the country's educational system. In addition, since attitudes influence students' achievement, it is up to the system to find ways to increase student attitudes toward mathematics beyond the factors that are school related.

Another important finding of this study is the influence of the school disciplinary climate on the teacher's teaching practices. This result is not surprising since the misbehavior of students has been significantly increasing recently. Therefore, the Ministry of Education should also take the initiative of examining this issue to help resolve such problems.

The findings of this study indicate that further in-depth research should be done to examine the influence of family educational background, disciplinary school climate, attitudes and teachers' teaching practices on the mathematics outcome, although academic performance can be attributed to a complex and dynamic interaction between cognitive, affective and motivational variables (Volet, 1997). The model presented here has implications for future research in the modeling of mathematics achievement. First of all, a generalization of the model with other populations and with other datasets would strengthen the claims on the variables that affect mathematics attitudes and achievement. Second, as with any modeling approach, cross-validation and replication are required (Bollen, 1989). Third, the elements of the present model may provide empirical measures for a broader conceptualization of mathematics outcomes in the developmental model (Papanastasiou, 2000).

References

Aiken, L.R. (1986) Attitudes toward mathematics. In T. Husén and T.N. Postlethwaite (eds), *The International Encyclopedia of Education* (pp. 4538–44). New York: Pergamon.

Alsup, R. and Gillespie, D.F. (1997) Stability of attitudes toward abortion and sex roles: A two factor measurement model at two points in time. *Structural Equation Modeling*, 4(4), 338–52.

Beaton, A.E., Mullis, I.V.S., Martin, M.O., Gonzalez, E.J., Kelly, D.L. and Smith, T.A. (1996) *Mathematics achievement in the middle school years*. Boston College, MA: IEA.

Ballen, K.A. (1989) *Structural equations with latent variables*. New York: Wiley.

Broomer, K.M., Knight, K., Joe, G.W., Simpson, D.D. and Cross, D. (1997) Structural models of antisocial behavior and during treatment performance for probationers in a substance abuse treatment program. *Structural Equation Modeling*, 4(1), 37–51.

Browne, M.W. and Cudeck, R. (1993). Alternative ways of assessing fit. In K.A. Bollen & J.S. Long (eds.), *Testing structural equation models*. Newbury Park, CA: SAGE.

Coenders, G., Satorra, A. and Saris, W.E. (1997) Alternative approaches to structural modeling of ordinal data: A Monte Carlo study. *Structural Equation Modeling*, 4(4), 261–82.

Cheung, K.C. (1988) Outcomes of schooling in mathematics achievement and attitudes towards mathematics learning in Hong Kong. *Educational Studies in Mathematics*, 19, 209–19.

Cooper, H., Lindsay, J.J. and Nye, B. (1998) Relationships among attitudes about homework, amount of homework assigned and completed and student achievement. *Journal of Educational Psychology*, 90(1), 70–83.

Crane, J. (1996) Effects of home environment, sex, and maternal test scores on mathematics achievement. *The Journal of Educational Research*, 89(5), 305–14.

Dauphinee, T.L., Schau, C. and Stevens, J.J. (1997) Survey of attitudes toward statistics: Factor structure and factorial invariance for women and men. *Structural Equation Modeling*, 4(2), 129–41.

Eccless, J., Adler, T., Futterman, R., Goff, S., Kaczala, C., Meece, J. and Midgley, C. (1985) Self-perceptions, task perceptions, socializing influences, and decision to enroll in mathematics. In S.F. Chipman, L.R. Brush and D.M. Wilson (eds.), *Women and mathematics: Balancing the equation*. Hillsdale, NJ: Lawrence Erlbaum.

Eisenhardt, W.B. (1977) A search for predominant causal sequence in the inter-relationship of interest in academic subjects and academic achievement. A cross-lagged panel correlation study. *Dissertation Abstract International*, 37, 4225A.

Fennena, E. (1980) Sex-related differences in mathematics achievement: Where and why? In L.H. Fox, L. Brody and D. Topin (eds.), *Women and the mathematical mystique*. Baltimore, MD: Johns Hopkins University Press.

Forgasz, H.J. (1995) Gender and the relationship between affective beliefs and perceptions of grade 7 mathematics classroom learning environments. *Educational Studies in Mathematics*, 28, 219–39.

Fraser, B, and Butts, W.L. (1982) Relationship between perceived levels of classroom individualization and science related attitudes. *Journal of Research in Science Teaching*, 19, 143–54.

Hu, L. and Bentler, P.M. (1995) Evaluating model fit. In R.H. Hoyle (ed.), *Structural equation modeling: Concepts, issues, and applications* (pp. 76–99). Thousand Oaks, CA: SAGE.

Joreskog, K. and Sorbom, D. (1989) Lisrel 7: *User's reference guide*. Mooresville, IN: Scientific Software, Inc.

Kaplan, D. and Elliot, P.R. (1997) A didactic example of multilevel structural equation modeling applicable to the study of organizations. *Structural Equation Modeling*, 4(1), 1–24.

Lester, F., Garofalo, J. and Kroll, D. (1989) Self-confidence, interest, beliefs and metacognition: Key influences on problem solving behavior. In D. McLeod and V. Adams (eds.), *Affect and mathematical problem solving*. New York: Springer-Verlag.

Ma, X. and Kishor, N. (1997) Assessing the relationship between attitude toward mathematics and achievement in mathematics: A meta-analysis. *Journal of Research in Mathematics Education*, 28(1), 26–47.

McLeod, D.B. (1992) Research on affect in mathematics education: A reconceptualization. In D.A. Grouws (ed.), *Handbook of research on mathematics teaching and learning* (pp. 575–96). New York: Macmillan.

Marcoulides, G.A. and Hershberger, S.L. (1997) *Multivariate statistical methods: A first course*. Mahwah, NJ: Lawrence Erlbaum.

Martin, M.O., Mullis, I.V.S., Gregory, K.D., Hoyle, C.D. and Shen, C. (2000) *Effective schools in science and mathematics: IEA's Third International Mathematics and Science Study*. Chestnut Hill, MA: Boston College.

Mullis, I.V.S., Martin, M.O., Gonzalez, E.J., Gregory, K.D., Garden, R.A., O'Connor, K.M., Chrostowski, S.J. and Smith, T.A. (2000) *TIMSS 1999 international mathematics report*. Boston College, MA: IEA.

Newmah, R.S. and Schwager, M.T. (1993) Students' perceptions of the teacher and classmates in relation to reported help seeking in math class. *The Elementary School Journal*, 94(1), 3–17.

Papa, F.J., Harasym, P.H. and Scumacher, R.E. (1997) Evidence of second-order factor structure in a diagnostic problem space: Implications for medical education. *Structural Equation Modeling*, 4(1), 25–36.

Papanastasiou, E.C. (2000) Effects of attitudes and beliefs on mathematics achievement. *Studies in Educational Evaluation*, 26(1), 27–42.

Papanastasiou, E.C. and Ferdig, R.E. (2003) Computer use and mathematical literacy. An analysis of existing and potential relationships. *Proceedings of the third Mediterranean conference on mathematical education* (pp. 335–42). Athens: Hellenic Mathematical Society.

Papanastasiou, E.C. and Zembylas, M. (2004) Differential effects of science attitudes and science achievement in Australia, Cyprus, and the USA. *International Journal of Science Education*, 26(3), 259–80.

Raykov, T. (1997) Growth curve analysis of ability means and variances in measures of fluid intelligence of older adults. *Structural Equation Modeling*, 4(4), 283–319.

Reyes, L.H. (1984) Affective variables and mathematics education. *The Elementary School Journal*, 84(5), 558–81.

Ross, J.A. and Cousins, J.B. (1995) Impact of explanation seeking on student achievement and attitudes. *The Journal of Educational Research*, 89(2), 109–17.

Shaughnessy, J., Haladyna, T. and Shaughnessy, J. (1983) Relations of student, teacher, and learning environmental variables to attitude toward mathematics. *School Science and Mathematics*, 83, 21–37.

Vanayan, M., White, N., Yuen, P. and Teper, M. (1997) Beliefs and attitudes toward mathematics among third- and fifth-grade students: A descriptive study. *School Science and Mathematics*, 97(7), 345–51.

Volet, S.E. (1997) Cognitive and affective variables in academic learning: The significance of direction and effort in students' goals. *Learning and Instruction*, 7(3), 235–54.

Willson, V.L. (1981) A meta-analysis of the relationship between science and attitude. Paper presented at the annual meeting of the National Association for Research in Science Teaching. New York.

9

Student and School Factors Affecting Finnish Mathematics Achievement: Results from TIMSS 1999 Data

Pekka Kupari

Introduction

One of the key challenges facing all school systems is working out what kind of curricula and instructional practices will yield the best learning experiences and outcomes. International comparative evaluation studies – such as the Third International Mathematics and Science Study Repeat (TIMSS 1999) – provide some information to help answer this question. These evaluation studies can be seen to serve both *a descriptive function and an understanding function* (Plomp, 2001; Bos, 2002). At the *descriptive* level, these evaluation studies provide the educational policy-makers and practitioners in the participating countries with information about the outcomes of teaching and learning within each system by comparing the respective performance data across all participating countries. In addition, such comparisons also concern educational investments, circumstances and processes through various indicators. Beside their descriptive function, evaluation studies make it possible to *understand* the differences and similarities detected in learning both within and between education systems. What kinds of background factors explain differences in student performance? Why do some countries perform better than others? How could students' learning be promoted and/or the differences be diminished by national education policies?

The results of the national and international assessments conducted in Finland have shown that the mathematics performance of our students is highly equal and of a relatively high standard. Performance differences between Finnish schools have been quite small and in the international studies (TIMSS 1999 and PISA 2000) the variation of our student performance has been among the smallest (Kupari *et al.*, 2001; Välijärvi *et al.*, 2002).

The study reported here has two goals. First, it examines patterns of mathematics achievement at grade 7 of the Finnish comprehensive school and partitions variance by means of multi-level modeling procedures to estimate the amount of variance that can be explained at the student and school (classroom) levels. The second aim is to analyze the relationships of some background variables – based on student, teacher and school questionnaires – to students' mathematics performance in order to identify factors that explain the variation in student performance.

Selection of Potential Background Factors

In view of this study it is important to know what are the relevant background factors concerning the school, mathematics instruction and the student that could potentially explain student performance in mathematics. The IEA research framework of three curriculum levels (Travers and Westbury, 1989) – also applied in the TIMSS 1999 study – establishes a basic guide for the classification of potentially effective factors on mathematics achievement in different education systems. In addition to this framework we drew on research knowledge about educational effectiveness and especially the educational effectiveness model introduced by Creemers (1994) in exploring the relevant and potentially effective factors.

The main assumption of the Creemers model is that student achievement is strongly influenced by student factors like social background, intelligence and motivation, but also by what the students do during lessons and how they use their opportunities to learn. The main purpose of the model is to give an overview of the classroom, school and context factors that influence achievement. Quality, time and opportunity are key concepts characterizing all levels above the student level (Reezigt *et al.*, 1999). In the present study the Creemers model is used as an underlying organizational frame. Furthermore, when selecting the potential explaining factors for mathematics performance, we have also drawn on the extensive empirical research available on the TIMSS 1995 and TIMSS 1999 data (Zabulionis, 1997; Bos and Kuiper, 1999; Köller *et al.*, 1999; Fullarton and Lamb, 2000; Lamb and Fullarton, 2000; Bos, 2002; Howie, 2002).

Background Factors Included in the Modeling

The set of background data collected in TIMSS 1999 by means of the student, teacher and school questionnaires consisted of a very large number of variables concerning the students themselves and their activities, their mathematics instruction as well as their school environment (altogether more than 600 separate background variables). These data were explored to determine the potentially effective educational factors at student, class and school levels, and related indicators from the TIMSS 1999 questionnaires were selected for multi-level modeling analyses. At this final stage of variable selection we also made use of previous correlative findings on the TIMSS 1999 data

(Gonzalez and Miles, 2001) as well as the results of explorative path analyses (Kupari, 2001).

As a result of the pre-existing analyses, altogether 24 factors were selected in the modeling and divided into five groups as suggested by Fullarton and Lamb (2000). Three groups were composed of student-level factors, while the rest consisted of school-level factors. Table 9.1 provides descriptive statistics on the factors included in the multi-level modeling.

Student's background factors

- *Gender of students*. A positive link between gender and achievement score means that boys 'do better' than girls.
- *Number of books in the home*. This is a proxy indicator. It reflects the social or cultural background of the student's home (Fullarton and Lamb, 2000; Bos,

Table 9.1 Descriptive statistics on the student-level variables included in the multi-level model analysis (TIMSS 1999/grade 7).

Variable	N	M	SD	Range	α^a	r^f
Achievement score[b]	2920	521.8	71.8	231–844	0.86[c]	–
Student's gender	2920	0.50[d]	–	1–2	–	ns
Number of books	2886	3.42	1.08	1–5	–	0.21
Educational aids	2890	1.70	0.46	1–2	–	0.18
Self-concept	2873	2.17	0.64	1–3	0.88	0.51
Positive attitude	2851	2.02	0.63	1–3	0.83	0.27
Perceived importance	2823	2.88	0.43	1–4	0.67	0.17
Success attribution	2848	2.78	0.30	1–4	0.65	0.06
Maternal expectations	2838	3.19	0.47	1–4	0.84	0.04
Peer expectations	2819	2.66	0.59	1–4	0.88	−0.05
Self expectations	2864	3.17	0.56	1–4	0.84	0.08
Lesson climate	2781	2.55	0.57	1–4	0.71	ns
Teaching style[e]	2747	2.04	0.66	1–3	0.52/0.70	0.12
Studying at home	2877	0.57	0.45	0–7	–	−0.11
Extra lessons	2849	0.14	0.54	0–7	–	−0.19
Out-of-school activities	2769	2.05	0.94	0–7	–	−0.28
Teacher's gender	153	0.60[d]	–	1–2	–	ns
Teaching experience	153	2.71	1.01	1–4	–	0.16
Confidence to teach	153	2.65	0.65	1–3	–	ns
Group size	153	2.41	1.05	1–4	–	ns
Amount of homework	153	2.63	0.72	1–3	–	0.19
Quality of homework	153	1.90	0.30	1–2	–	−0.16
School's location	153	3.02	1.42	1–5	–	ns
Level of urbanization	153	2.19	0.88	1–3	–	ns
School size	153	2.22	0.67	1–3	–	ns

Note. [a] No Cronbach α for single variables; [b] indices for achievement score computed on the basis of student's five plausible values; [c] the mean of reliabilities for eight test booklets; [d] the proportion of female students/teachers; [e] presenting α factors for both components; [f] Pearson correlation with achievement score.

2002). There were also other questionnaire items available on home educational background, namely 'educational level of mother and father'. However, in Finland the percentage of missing values for these items was too high (more than 50 percent) to allow plausible computations to replace the missing values.

- *Educational aids in the home*. This refers to the availability of a computer, desk and dictionary at home (IEA, 2001). This construct reflects the educational level of the student's home. A high score means possession of all educational aids.

Student's mediating factors

- *Self-concept in mathematics*. This is an index based on five questions and reflects the student's self-confidence in studying mathematics (IEA, 2001). A high score of self-concept means that the student's confidence is strong.
- *Positive attitude towards mathematics*. This is an index based on five questions and reflects the student's attitudes towards mathematics (IEA, 2001). A high value indicates very positive attitude.
- *Perceived importance of mathematics*. This construct is based on four questions tapping on the student's perception about the importance of doing well in mathematics. A high value means that the student finds mathematics important.
- *Success attribution in mathematics*. This construct is based on four questions asking the student's opinion about what is needed in order to do well in mathematics. A high score on this construct means that success in mathematics is seen to require diligence rather than talent and good luck.
- *Maternal expectations*. This construct is based on three questions and reflects the student's perception of the extent to which his/her mother finds it important that the student does well at school in mathematics, science and mother tongue (Bos, 2002). A high score means that the student perceives a great pressure from his/her mother.
- *Peer expectations*. This construct has essentially the same contents as in the maternal expectations, but indicates the student's perception of the expectations of his/her friends.
- *Self-expectations*. This construct has essentially the same contents as the two indicators above, but concerns the student's perception of his/her own expectations.

Student's studying factors

- *Lesson climate*. This construct is based on three questions. Students were asked about their perception of the climate during mathematics lessons. A high score indicates a quiet and orderly atmosphere during lessons.
- *Teaching style*. This involves a dual approach with two initial factors constructed. The factor of teacher-centered teaching style as well as the one for student-oriented teaching style included seven questions each. The mean scores for these factors

were calculated and the 'student' mean was subtracted from the 'teacher' mean (Bos and Kuiper, 1999). Thus, a high overall score indicates that the class teaching was strongly teacher centered.

- *Studying mathematics at home.* This indicator is a recoded variable telling the number of hours the student spends on studying mathematics or doing related homework on schooldays. A high score means that a student uses a lot of time for studying.
- *Extra lessons in mathematics.* This indicator is a recoded variable telling the number of weekly hours the student spends on extra lessons in mathematics. A high score indicates that a student spends lot of time for extra lessons.
- *Out-of-school activities.* This indicator is a combination of three variables and expresses the number of hours per day the student uses for leisure activities. A high score means that a student uses a lot of time for watching television or videos, playing computer games and being with friends.

Teacher and instruction factors

- *Teacher's gender.* In this design, a positive link between teacher gender and student achievement score indicates that the students of male teachers perform better than the students of female teachers.
- *Teaching experience.* This recoded variable expresses the length of teacher's teaching experience on a four-step scale. A high value means that the teacher has a long teaching experience.
- *Confidence to teach mathematics.* This is an index based on 12 questions about different mathematics topics and reflects the teacher's confidence in his/her preparation to teach mathematics within these topics (IEA, 2001). A high score means that the teacher has strong confidence in teaching.
- *Group size.* This variable tells the number of students in the mathematics teaching group as reported by the teacher, and it has been recoded onto a four-step scale. A high score indicates that students in the bigger groups do better.
- *Amount of homework.* This is an index based on two questions and expresses how much and how often the teacher typically assigns mathematics homework to the students (IEA, 2001). A high score means that the teacher more often assigns homework to students.
- *Quality of homework.* This is an index based on three questions and it expresses how often the teacher assigns homework based on projects and investigations (IEA, 2001). A high score means that the teacher more often assigns this kind of homework to the students.

School factors

- *School's regional location.* The five region categories used in the sampling define the region where the school is located in Finland.

- *Level of urbanization*. The three categories used in the sampling indicate whether the school is located in a rural area, in a semi-urban area or in an urban area.
- *School size*. This variable indicates the total number of students enrolled in the school and it has been recoded to three categories.

A majority of the background variables included in multi-level modeling were student-level variables and many of them had been constructed by combining student responses on given Likert scale statements (for example, self-concept in mathematics, attitudes towards mathematics, perceived importance of mathematics). Table 9.1 shows descriptive statistics on the selected background factors.

Based on this information most of the constructs seem quite reliable. However, there were three constructs – perceived importance, success attribution and teaching style – with quite low reliability statistics, so that one should be particularly cautious when interpreting the results concerning these variables.

Method

Multi-level Models in the Analysis of School Data

It is widely accepted that attempts to reform educational policy and practice should be guided by the most efficient and powerful analytic techniques available (O'Dwyer, 2002). Statistical explanatory models are employed in order to investigate the complex network of relationships between student performance and various background factors at different levels of education. Substantial advances have been made in developing appropriate models and methods for analyzing hierarchically structured data.

During the last decade multilevel modeling techniques have become increasingly available to researchers (Bryk and Raudenbush, 1992; Goldstein, 1995; O'Dwyer, 2002). A starting point and focal interest in multi-level models derives particularly from the inherent structure of the population and from the need to account for that structure in the design of the statistical model (Goldstein, 1995). The strength of multi-level models in dealing with school data is based on the feature that they not only take account of the inherent structure of the data but also treat variables of different levels simultaneously within the same model. In recent years, hierarchical linear modeling (HLM, MLn) techniques, have been taken into wide use (e.g. Fullarton and Lamb, 2000; Bos, 2002; Howie, 2002).

In this study the HLM technique was employed to specify a two-level country model (Bryk and Raudenbush, 1992). The sampling procedure of TIMSS 1999 required that the participating countries sample at least one classroom per target grade in each school. Most countries, including Finland, sampled only one classroom at Grade 7. However, this design limits the number of levels that can be modeled and thus, in the Finnish context, a two-level model could be formulated in order to explain the variation in mathematics scores between students (within schools) and between schools (classrooms).

The Finnish data comprised test answers and background questionnaire information from a total of 2,920 students, 167 mathematics teachers and 159 school principals.

Data Analysis

In consideration of the data the purpose is to explain the variation of students' mathematics performance (test score) by the background variables. The results of the analysis show the degree to which the variables of different levels account for that variation and which particular factors stand out as (statistically) significant predictors.

The HLM analysis involved consecutive testing of the model, so that after each stage another group of variables was added to the model. The initial model (so-called null model) helped us estimate at the outset how the variation in students' scores was divided into between-students variance and between-schools variance, so that the model did not include any explanatory variables at this point. In developing the explanatory model, the first step was to add student's background variables to the model as explanatory variables. In the same way, the second step brought along student's mediating variables and the third step incorporated student's studying variables as well. The fourth model then encompassed some school-level variables in addition, namely the teacher and instruction variables. Finally, the fifth model completed the explanatory model by bringing along the three school variables. By examining the change occurring in the estimates of variance after adding each set of variables it was possible to analyze the effects of different – and different level – variables on students' mathematics performance.

The HLM program (Raudenbush *et al.*, 2000) provided two options for handling missing data at level 1 (student level): pair-wise and list-wise deletion of cases. In this study the list-wise deletion was applied. At level 2 (school level), HLM assumes complete data. Therefore the school data needed editing. In some schools the teaching group had had two mathematics teachers, but only one of them could be included in the data, that is, the one who had taught most of the lessons. The teacher and school variables included in the model still had some missing values. Some of them (regarding, for example, school size and group size) could be quite reliably substituted with information from other sources. In the end the analysis comprised 153 schools out of the 159 schools in the original sample and the six deleted schools were distributed evenly among strata.

Results

Tables 9.2 and 9.3 present the results of the analyzed HLM models. More specifically, Table 9.2 describes the variance in mathematics performance at student and school levels as well as the changes occurring in that variance when different background variables are controlled for. In the first phase, a fully unconditional (null) model was tested. The results of the null model (in boldface) reveal quite remarkably that the overall variation in mathematics achievement derives predominantly (90 percent) from between-students variance, while only about 10 percent of the variation comes from between-schools variance. This means that internationally the differences between schools in Finland were small indeed.

The next step of the analysis involved adding the student background variables to the model. These variables (student's gender, number of books, educational aids)

Table 9.2 Variance in mathematics achievement explained by two-level HLM models (TIMSS 1999/grade 7).

	Variance	Variance at each level (%)	Cumulative explained variance at each level (%)
Variance between students (level 1)	**5204.1**	**89.7**	
after controlling for:			
Student background variables	5028.4		3.4
Student's mediating variables	3929.5		24.5
Student's studying variables	3713.3		28.6
Teacher and instruction variables	3715.5		28.6
School variables	3715.5		28.6
Variance between schools (level 2)	**600.3**	**10.3**	
after controlling for:			
Student background variables	508.0		15.4
Student's mediating variables	438.6		26.9
Student's studying variables	409.8		31.7
Teacher and instruction variables	338.9		43.5
School variables	338.9		43.5
Total variance (level 1 + level 2)	**5804.4**	**100.0**	
after controlling for:			
Student background variables	5536.4		4.6
Student's mediating variables	4368.1		24.7
Student's studying variables	4123.1		30.0
Teacher and instruction variables	4054.4		30.1
School variables	4054.4		30.1

explained only a small share (3.4 percent) of the between-students variance, but a clearly greater amount (15.4 percent) of the variance between schools (which, as noted, was small). Student's gender had no statistically significant effect; in other words, boys and girls performed at an equal level.

Student's mediating variables (for example, self-concept, peer expectations) proved to be clearly the strongest predictors for the between-students variance and increased the percentage of explained variance by 21 percentage points. The mediating variables also helped to explain the variance between schools, but not as much as the variables concerning student's cultural and educational home background.

Student's studying variables (for example, teaching style, extra lessons, out-of-school activities) brought only a slight increase (about 4 percentage points) to the proportion explained both for the between-students and the between-schools variance.

Adding the teacher and instruction variables (for example, teaching experience, amount of homework) to the model increased to some extent (from 31.7 to 43.5 percent) the proportion explained for between-schools variance, but made no difference in view of between-students variance. Bringing the three school variables into the model did not enlarge the proportion of explained variance any further at either level.

In this analysis the 'final' model explained about 30 percent of the overall variation of the seventh-graders' mathematics achievement, which can be considered rather

Table 9.3 Two-level HLM estimates of mathematics achievement (TIMSS 1999/grade 7).

	Level 1 model–student background variables	Level 1 model–student's mediating variables	Level 1 model–student's studying variables	Level 2 model–teacher/instruction variables	Level 2 model–school variables
Intercept	525.9[a]	526.0[a]	526.0[a]	525.9[a]	525.8[a]
Student-level variables					
Background variables					
Student's gender	6.0	−7.0	−3.6	−3.4	−3.4
Number of books	8.8[a]	5.7[b]	5.0[c]	5.0[c]	4.9[c]
Educational aids	19.6[a]	11.7[a]	10.6[a]	10.2[b]	10.2[c]
Mediating variables					
Self-concept	–	47.3[a]	42.3[a]	42.2[a]	42.2[a]
Positive attitude	–	7.1	7.4	7.3	7.3
Perceived importance	–	5.6	5.0	4.9	4.9
Success attribution	–	12.0	7.4	7.3	7.2
Maternal expectations	–	2.3	4.4	4.3	4.3
Peer expectations	–	−13.5[b]	−11.5[c]	−11.5[c]	−11.5[c]
Self expectations	–	7.6	6.5	6.4	6.4
Studying variables					
Lesson climate	–	–	−3.6	−4.1	−4.1
Teaching style	–	–	8.6[c]	8.5[c]	8.4[c]
Studying at home	–	–	−3.2	−2.5	−2.5
Extra lessons	–	–	−15.8[b]	−16.1[b]	−16.1[b]
Out-of-school activities	–	–	−9.1[a]	−9.4[a]	−9.4[a]
School-level variables					
Teaching variables					
Teacher's gender	–	–	–	2.9	2.9
Teaching experience	–	–	–	4.7[c]	4.7[c]
Confidence to teach	–	–	–	−0.8	−0.8
Group size	–	–	–	2.8	2.8
Amount of homework	–	–	–	6.1[c]	6.0[c]
Quality of homework	–	–	–	−16.6	−16.4
School variables					
School's location	–	–	–	–	−0.4
Level of urbanization	–	–	–	–	−0.1
School size	–	–	–	–	−0.1

Note. [a] $p < 0.001$; [b] $p < 0.01$; [c] $p < 0.05$.

usual in this kind of study. For between-students variance the proportion explained was 29 percent, while for between-schools variance it was 44 percent.

Table 9.3 presents the individual variables – across the different variable clusters – that were included in the model during the consecutive phases of the HLM analysis as well as their effects on students' mathematics achievement. In the light of these results, quite a small number of statistically significant effects could be found. The results revealed that both the educational aids and cultural background of the home (number of books in the home) were clearly associated with students' achievement: the stronger the support to schoolwork from home, the better the student's results.

The strongest predictor for the seventh-graders' performance was self-concept in mathematics (that is, how confident the student is about his/her own learning of mathematics). The difference between confident and unconfident students' achievement scores was about 85 points, which is well above one standard deviation (72 in the Finnish data). Students' positive attitude toward mathematics had no significant relationship to student performance, which may be due to the notion that attitudes are strongly mediated through the mathematical self-concept as suggested by previous analyses (Kupari, 2001). Another interesting finding was that peer expectations appeared as an important predictor ($p < 0.05$) as well, so that the stronger these expectations were, the lower the student achieved.

As regards student's studying variables, teaching style was associated with student performance ($p < 0.05$). The results indicate that in terms of student performance the teacher-centered approach was more effective than the student-centered one. As expected, the connection of two other learning variables (extra lessons and out-of-school activities) was negative. The more the students needed additional instruction in mathematics outside school hours and the more time they spent for other activities after school – watching television and videos, playing computer games, spending time with friends – the lower was their mathematics achievement.

Among the teacher and instruction variables there were only two variables that statistically significantly explained student achievement, namely teacher's professional experience and amount of homework. Students whose teachers were more experienced and more often gave homework to their students performed better than the others. Finally, the results show that school-level variables had no effect on student performance in this population.

Conclusion

This analysis explored, on the one hand, how the variance occurring in mathematics achievement is distributed between the student level and the school/class level, and on the other hand, the statistical relationships of background variables at these levels on students' performance. The analysis is based on the Finnish TIMSS 1999 data and it employs two-level HLM models where the predominant sampling and analytic structure of the data consisted of students (level 1) nested within schools (level 2). Studies of this kind have been quite dominating during the last decade (Hill and Rowe, 1996).

This study brought out two interesting findings that have relevance to the development of Finnish education policy and mathematics instruction. First, the results showed that for the most part (90 percent) the variation of seventh-graders' mathematics performance is attributable to the variance between students, while only 10 percent of the overall variation comes from between-schools variance. When these figures were compared to corresponding results from national mathematics assessments carried out in the 1990s, it was found out that the variance between schools had remained about the same (Kuusela, 2002).

This finding is well in accordance with several earlier studies. For instance, the results of educational effectiveness studies (Reezigt *et al.*, 1999) show that most of the variance occurs at the student level, while relatively small percentages of the variance are attributable to the classroom and the school levels. However, classrooms and schools have influence on mathematics performance and their influence can be very important for individual students. According to Beaton and O'Dwyer (2002), the patterns of variance in mathematics achievement are quite different in different parts of the world. In some countries, the between-school variance ranges from 1 to 6 percent, indicating highly equal mathematics performance among the schools within these countries. Conversely, there are countries (Germany, the Netherlands) where the proportion of between-students variance ranges from 48 to 59 percent and between-schools variance is as high as 38 to 52 percent (Beaton and O'Dwyer, 2002; Bos, 2002). In the statistical model described above, the most parts of the unexplained variation consisted of between-students variance. Many studies (Lamb and Fullarton, 2000; Hill and Rowe, 1996) suggest that certain parts of the variation in students' mathematics achievement can be explained by the differences between classrooms. For this set of data, class level could not be included in the model, since there was only one classroom per school in the sample. When designing future studies, it would be good to ensure that also differences between classrooms could be examined. Generally, this means larger samples and sampling designs should allow for the joint estimation of effects at relevant levels.

Second, the analysis brought out a number of interesting background variables relevant to mathematics achievement. Students' self-concept in mathematics was by far the most significant predictor for their performance and seemed to be connected with their attitudes towards mathematics. Self-concept and attitudes are factors that can be influenced and hence set challenges to mathematics teachers and teacher education. Students' self-concept and attitudes can be influenced by reinforcing each student's confidence in their capability to learn mathematics and by providing them with successful experiences. In terms of learning it is also important to create a favorable atmosphere in which students feel unthreatened and are therefore willing to make a positive contribution to the lesson. In teacher preparation, the meaning of learner's self-concept and attitudes need to be duly addressed, as well as the topic of how teachers' own views and beliefs shape their teaching.

Another interesting finding was that peers' expectations appeared to have a negative effect on mathematics achievement, that is, the stronger the expectations perceived, the lower the students' performance. In contrast, neither the mother's nor the student's own expectations had any effect on the performance. This could be interpreted as pertinent to the age of the students (puberty). If there is pressure from friends to succeed in school in general, and in mathematics in particular, it may cause anxiety among many learners and thereby lead to poorer performance.

With regard to teacher's teaching style as perceived by students and its association with student performance, the finding that the teacher-centered approach produced better scores than the student-centered one was at first rather surprising. It was seemingly contradictory to current thinking according to which student-centered approaches lead to better learning and better results. Therefore, this issue requires some elaboration. First, we should take into account that the variables used here reflect students' experiences from the teachers' way of teaching and the lower reliability indices of the scales may indicate this. Hence, any interpretations should be made with caution. Moreover, many of the variables concerning mathematics instruction had no statistical relationship to student performance, which suggests that factors describing instructional characteristics are difficult to capture by means of survey-type methodology. Second, teacher-centered and student-centered practices both have good and effective qualities, and these approaches should not be seen as opposites or mutually exclusive. One explanation for the association described above might be that at this stage of schooling (seventh grade) a teacher-centered practice takes better care of the weaker learners, which then shows as a positive connection to student performance.

Finally, we could ask why the analysis brings out only a few variables as statistically significant predictors, although the TIMSS 1999 study involved extensive background questionnaires for students, teachers and headmasters, respectively, in order to find out explanatory factors for mathematics performance. There are many reasons and Reezigt *et al.* (1999) mention some of them. First, when teachers and headmasters fill out questionnaires 'social desirability' tends to come in play, which thus undermines the reliability of the responses. Second, some important background factors are sometimes measured by only a few items. To produce more reliable and valid factors and scales, a sufficient number of items should be used to measure the concepts. This is even more important in international studies to ensure the internal consistency and validity of the scales in all participating countries. Moreover, it is generally problematic to capture by means of questionnaires the characteristics of the instruction given by teachers, and this leads to failure when trying to build statistically and conceptually functional scales. It is also possible that some other variables are more important for student achievement than the factors included in the study.

In all, this study and its findings reveal that student achievement is affected by a highly complex combination of factors. Yet, although sorting out the relationships may be arduous and demanding, it is worth investing in. Through national and cross-national analyses nations can learn what benefits and weaknesses their educational systems have. On this basis they can then proceed to create the best possible conditions for their students' learning.

References

Beaton, A.E. and O'Dwyer, L.M. (2002) Separating school, classroom and student variances and their relationship to socio-economic status. In D.F. Robitaille and A.E. Beaton (eds.), *Secondary analysis of the TIMSS data* (pp. 211–31). Dordrecht: Kluwer.

Bos, K. (2002) *Benefits and limitations of large-scale international comparative achievement studies*. The case of IEA's TIMSS study. Enschede: University of Twente.

Bos, K. and Kuiper, W. (1999) Modeling TIMSS data in a European comparative perspective: Exploring influencing factors on achievement in mathematics in grade 8. *Educational Research and Evaluation*, 5(2), 157–79.

Bryk, S. and Raudenbush, S.W. (1992) *Hierarchical linear models: applications and data analysis methods*. Newbury Park, CA: SAGE.

Creemers, B.P.M. (1994) *The effective classroom*. London: Cassell.

Fullarton, S. and Lamb, S. (2000) Factors affecting mathematics achievement in primary and secondary schools: results from TIMSS. In J. Bana and S.A. Chapman (Eds.), *Mathematics education beyond 2000. Vol. 1.* (pp. 258–66). Perth, WA: Mathematics Education Research Group of Australasia Incorporated.

Goldstein, H. (1995) *Multilevel statistical models*. London: Edward Arnold.

Gonzales, E.J. and Miles, J.A. (eds.) (2001) *TIMSS 1999 user guide for the international database*. Chestnut Hill, MA: Boston College.

Hill, P.W. and Rowe, K.J. (1996) Multilevel modeling in school effectiveness research. *School Effectiveness and School Improvement*, 7(1), 1–34.

Howie, S. (2002) *English language proficiency and contextual factors influencing mathematics achievement of secondary school pupils in South Africa*. Enschede: University of Twente.

IEA (2001) *TIMSS 1999 user guide for the international database. Supplement three.* Chestnut Hill, MA: Boston College.

Köller, O., Baumert, J., Clausen, M. and Hosenfeld, I. (1999) Predicting mathematics achievement of eight grade students in Germany: An application of parts of the model of educational productivity. *Educational Research and Evaluation*, 5(2), 180–94.

Kupari, P. (2001) Exploring factors affecting mathematics achievement in Finnish education context. Paper presented in the European Conference on Educational Research, Lille, France.

Kupari, P., Reinikainen, P., Nevanpää, T. and Törnroos, J. (2001) *Miten matematiikkaa ja luonnontieteitä osataan suomalaisessa peruskoulussa. Kolmas kansainvälinen matematiikka-ja luonnontiedetutkimus TIMSS 1999 Suomessa* [*How do Finnish comprehensive school students perform in mathematics and science? The Third International Mathematics and Science Study TIMSS 1999 in Finland*]. Jyväskylä: University of Jyväskylä, Institute for Educational Research.

Kuusela, J. (2002) Links between school results and demographic factors. In R. Jakku-Sihvonen and J. Kuusela (eds.), *Evaluation of the equal opportunities in the Finnish comprehensive schools 1998–2001* (pp. 27–45). Helsinki: National Board of Education.

Lamb, S. and Fullarton, S. (2000) Classroom and teacher effects in mathematics achievements: results from TIMSS. In J. Bana and S.A. Chapman (eds.), *Mathematics education beyond 2000. Vol. 1.* (pp. 355–62). Perth, WA: Mathematics Education Research Group of Australasia Incorporated.

O'Dwyer, L.M. (2002) Extending the application of multilevel modeling to data from TIMSS. In D.F. Robitaille and A.E. Beaton (eds.), *Secondary analysis of the TIMSS data* (pp. 359–73). Dordrecht: Kluwer.

Plomp, T. (2001) The potential of challenges of international comparative studies of educational achievement. In K. Leimu, P. Linnakylä and J. Välijärvi (eds.), *Merging national and international interests in educational system evaluation* (pp. 23–39). Jyväskylä: University of Jyväskylä, Institute for Educational Research.

Raudenbush, S., Bryk, A., Cheong, Y.F. and Congdon, R. (2000) *HLM5. Hierarchical linear and nonlinear modeling*. Lincolnwood, IL: Scientific Software International.

Reezigt, G.J., Guldemond, H. and Creemers, B.P.M. (1999) Empirical validity for a comprehensive model on educational effectiveness. *School Effectiveness and School Improvement*, 10(2), 193–216.

Travers, K.J. and Westbury, I. (1989) *The IEA study of mathematics I: international analysis of mathematics curricula*. Oxford: Pergamon Press.

Välijärvi, J., Linnakylä, P., Kupari, P., Reinikainen, P. and Arffman, I. (2002) *The Finnish success in PISA – and some reasons behind it*. Jyväskylä: University of Jyväskylä, Institute for Educational Research.

Zabulionis, A. (1997) A first approach to identifying factors affecting achievement. In P. Vari (ed.), *Are we similar in mathematics and science? A study of grade 8 in nine central and eastern European countries* (pp. 147–68). Budapest: IEA.

10

The Role of Students' Characteristics and Family Background in Iranian Students' Mathematics Achievement

Ali Reza Kiamanesh

Introduction

Iran's educational system is highly centralized and students' learning goals are set at the national level. The Ministry of Education has the exclusive responsibility for developing the curriculum syllabi and textbooks from Grades 1 to 12. It is worth mentioning that Grade 12 in Iran is allocated to pre-university courses and students get their high school diploma at Grade 11. Students are required to take two school examinations in all subject matters during each academic year, which they must pass in order to be promoted to the next grade. Despite the full control that the Ministry of Education has over the curriculum and textbooks, teachers are responsible for assessing and evaluating students' achievement and individual schools use different test batteries in different academic years. Nevertheless, each year the Ministry of Education conducts national examinations for selected subjects in Grades 9 to 12 (these change each year). However, even under this arrangement, schools and teachers are responsible for scoring. Findings from these examinations are dependent on the difficulty level of the items and on scoring procedures. In other words, there is no central body in Iran responsible for:

- Assessing educational inputs and outputs at national level on a regular, periodic basis at different grades, particularly transitional grades (that is, fifth, eighth, last year of schooling, and pre-university).
- In-depth and multi-faceted study of the reasons behind the differences in achievement among genders, students in different geographical regions and with different cultural and economic backgrounds.

• Defining educational standards for different academic courses and grades.

When there are no reliable and valid data at the national or regional level, international studies such as the Third International Mathematics and Science Study (TIMSS 1995), TIMSS 1999 (the repeat of TIMSS 1995) and PIRLS (Progress in International Reading Literacy Study) are unique sources for evaluating Iran's educational system. In the above-mentioned studies students' achievement in mathematics, science and reading comprehension has been subjected to comprehensive analysis. The information obtained from these international studies could help policy-makers, curriculum specialists and researchers at the national level to better understand reasons behind the performance of their educational systems (Martin *et al.*, 2000; Mullis *et al.*, 2000). Actions taken by researchers, policy-makers and educators on the basis of these data could contribute to identifying the weaknesses and strengths of educational systems and developing intervention programs to improve educational effectiveness. Schmidt and Valverde (1995) argued that:

[B]y looking at the educational systems of the world we challenge our own conceptions, gain new and objective insights into education in our own country, and are thus empowered with a fresh vision with which to formulate effective educational policies and new tools to monitor the effects of these new policies. (cited in Kyriakides and Charalambous, 2004: 70)

This chapter aims to explore the factors that most contribute to Iranian students' mathematics achievement using the TIMSS 1999 data. Mathematics achievement involves a complex interaction of factors that have specific direct effects and/or indirect effects, through other factors, on school outcomes. Although the relationship between mathematics achievement and factors such as academic self-concept, home background, attitudes towards mathematics and attribution has been studied widely in other countries, it is important to investigate the issue in the Iranian national context. This would help fill the existing gap in the research carried out in Iran in this area. In addition, it could pave the way for more comprehensive research on the comparison of national and international research findings. In particular, the purpose of the present study is to:

• Identify a number of factors that represent the relationship among sets of inter-related variables.
• Examine the contribution of each factor to the explanation of the variance in the students' mathematics score and to determine the total variance that could be explained by these factors for the total sample and for each gender

Iranian Mathematics Achievement in International Perspective

The preliminary investigations of TIMSS 1995 and 1999 data in Iran indicated that the average performance of Iranian students in mathematics at the third, fourth, seventh and eighth grades was much lower than the international average. Even though the average performance of Iranian students at the eighth grade increased four scale-score points from 1995 to 1999 (from 418 to 422), the gap between the average perform-ance of Iranian students and the average performance of students from the 26 coun-tries that participated in both studies remained almost unchanged (103 scale-score points with standard error of 3.8 in TIMSS 1995 and 102 scale-score points with standard error of 3.3 in TIMSS 1999). In TIMSS 1995 and 1999 the performance of Iranian boys was significantly higher than that of girls and the existing gap remained the same in both studies (24 scale-score points).

Findings from the eighth grade students in TIMSS 1999 indicated that for the total sample and both genders there is a positive relationship between students' achieve-ment in mathematics and home background variables such as 'parents' level of edu-cation', 'number of books at home' and 'possessing dictionary, computer and study desk'. However, Iranian students 'who come from a family with the highest level of education of either parents' (8 percent of the students), 'possess all the three educa-tional aids' (5 percent of the students) and 'have more than 200 books at home' (9 per-cent of the students) score much lower than the international average score (Kiamanesh and Kheirieh, 2001).

Another finding is that the relation between the index of self-concept in mathemat-ics ability and mathematics achievement was positive and significant for the total sample and both genders. Nevertheless, the mathematics average achievement of Iranian students who have a high self-concept of their mathematics ability (14 percent of the students) was almost the same as the average achievement of 67 percent of the international students with medium self-concept in mathematics ability (482 com-pared to 486) and much higher than the average achievement of those who had low self-concept in mathematics ability (15 percent of the students) (Kiamanesh and Kheirieh, 2001).

Moreover, the TIMSS results indicated that Iranian students (the total sample and both genders) who have high positive perceptions or attitudes towards mathematics, compared to those with medium or low positive attitudes towards mathematics, showed better achievement in both mathematics and science (Kiamanesh and Kheirieh, 2001). Yet the mathematics average performance of 54 percent of Iranian students with high positive attitude towards mathematics was much lower than the international average performance of 11 percent of the students with low positive atti-tudes towards mathematics (439 compared to 473). In other words, Iranian students had a low performance in mathematics despite liking the subject. However, the inter-national data for TIMSS have shown that students who perform well in mathematics usually like the subject and have more positive attitudes towards mathematics (Mullis et al., 2000).

Literature on Background Variables Related to Mathematics Achievement

Following the findings of Coleman et al. (1966), suggesting that for the USA 'schools made no difference' and differences in school achievement reflected variations in family background, extensive research has been carried out both in developing and developed countries on in- and out-of-school variables affecting students' achievement. Some studies have indicated that classrooms and schools are important factors in students' achievement. For instance, Schmidt *et al*. (1999), comparing achievement across countries on the basis of the TIMSS data, reported that: 'classroom-level differences accounted for a substantial amount of variation in several countries including Australia' (cited in Fullarton, 2004: 17).

In addition, several studies have revealed that the educational level of students' parents (Beaton *et al*., 1996; Robitialle and Garden, 1989; Engheta, 2004), home educational resources (Mullis *et al*., 2000), socio-economic status (SES) of the family (Marjoribanks, 2002), home language versus language of test (Howie, 2002) and providing quality homework assistance by parents (Engheta, 2004) are among factors that can explain variance in academic achievement. Many of the variables that explain academic achievement are home- and family-related and thus are difficult to change and are outside the control of educators (Singh *et al*., 2002). Home is the backbone for children's personality development, and influences children directly and indirectly through the kind of relationship the family members have among themselves, as well as through helping them to get in contact with society (Weiss and Krappmann, 1993).

Considerable research has examined the relationship between students' characteristics, such as self-concept, attitudes towards mathematics as well as motivation, and students' subsequent academic performance. In general, across countries a consistent pattern of relationship between attitudes towards school subjects and achievement in the respective subjects has been confirmed through a large number of studies (McMillan, 1977; Aiken, 1976; Kulm, 1980; Keeves, 1992; Papanastasiou, 2002; Schereiber, 2000). Stodalsky *et al*. (1991) mentioned that students develop ideas, feelings and attitudes about school subjects over time and from a variety of sources. Various authors such as Papanastasiou (2002) and Schereiber (2000) found a positive relation between mathematics attitudes and mathematics achievement.

> Students' attitude towards a subject can influence achievement and vice versa: attitude is a determining factor in the amount of learning time and effort made in a particular subject. (Vermeer and Seegers, 1998, cited in Broeck *et al*., 2004: 89)

In contrast to these findings, Cain-Caston's (1993) study showed that for the third-grade students there was no significant relationship between students' attitude toward mathematics and students' achievement in this subject.

Press variables (or home-school interface), such as friends and maternal pressure for learning mathematics (Martin *et al*., 2000) are among the factors that construct students' attitudes towards and beliefs about mathematics (Kulm, 1980). Research evidence shows that if an important person encourages somebody to behave in a certain way, he or she will accept it. The influence of an important person is so strong that even the individual may change his or her attitude in agreement with that of the important person's (Berkowitz, 1986).

The relationship between mathematics self-concept and achievement is another area that has been investigated by researchers (Marsh, 1993; Hamachek, 1995). Low self-concept tends to appear together with students' underachievement. Franken (1994: 443) states that:

> [T]here is a great deal of research which shows that self-concept is perhaps the basis for all motivated behavior. It is the self-concept that gives rise to possible selves, and it is possible selves that create the motivation for behavior.

Most findings in this area showed that those who have higher self-concept, that is, perceiving themselves more confident in mathematics, have higher scores in mathematics (Wilhite, 1990).

Since motivation is a measure of the degree to which a person will expend effort to perform or learn, it may be an important factor in predicting achievement (Small and Gluck, 1994). Students are more likely to be motivated to engage in a task if they understand the goals of the task, the skills it will help them develop and the potential uses of those skills outside of school. Skaalvik (1994) and Skaalvik and Rankin (1995) found that motivation is correlated with achievement and academic performance. Furthermore, researchers have found that motivation leads to engagement in academic tasks, which is related to achievement (Banks *et al.*, 1978; DeCharms, 1984; Dweck, 1986).

According to the attribution theory, students attribute causes for successes and failures in achievement. Researchers (Frize *et al.*, 1983; Weiner, 1985) showed that attributions influence students' achievement. Weiner's (1992) attribution model categorizes perceived causes into three categories for students, that is, categories of locus, stability and controllability. The locus places responsibility for success or failure on external (environmental causes, like difficulty or luck) and on internal (personal causes, like ability or success) factors. The stability places causes for success or failure on stable or unstable factors over time, and finally the controllability places causes for success or failure on factors within the control of students or not within the control of students. Weiner has demonstrated that, in general, those who attribute success to ability and effort tend to fare better in school than those who implicate luck or other external factors. Although students may attribute their failure or success to the afore-mentioned variables, the efforts that they make in order to learn mathematics at school or do homework at home probably have an effect on their achievement.

Methodology

Sample

The data obtained from population 2 (eighth-grade Iranian students who were mostly 13 years old) in TIMSS 1999 were analyzed in this study. In total 5,301 Iranian students (2,096 girls and 3,205 boys) from 170 schools (one class per school) participated in TIMSS 1999. Using Pearson Product Moment Correlation Coefficient and research evidence, 31 items from the Student Questionnaire were selected for analysis. In the next step, through utilizing 'excluded cases pair-wise method' only those subjects who had answered all the above-mentioned 31 items were sampled. As a result, the number of subjects decreased to 3,101 students (1,242 girls and 1,859 boys). It is worth mentioning that mathematics achievement score (Rasch score with the mean of 150 and standard deviation of 10) was used as the criterion variable.

Data analysis

As stated earlier, 31 items from the Student Questionnaire were analyzed in the study. To determine whether there was an underlying structure, a factor analysis was performed. First, the correlation matrix was examined to determine its appropriateness for factor analysis. The On-Diagonal values in the anti-image correlation matrix or KMO values for each of the 31 items were more than 0.684. In addition, the value of the test statistic for sphericity based on a Chi-square transformation of the determinant of the correlation matrix was large (0.866) and the associated significant level was small (0.000). Given the obtained results, it was concluded that the data do not produce an identity matrix and are approximately multivariate normal. Furthermore, the correlation matrix contained sufficient covariation for factoring. For more information, see Table 10.1. The data for the total sample were then subjected to Principal Component Factor Analysis with Varimax Rotation.

Based on the Scree Test and Eigenvalues over one, eight factors were accepted as the most interpretable ones. These factors accounted for 55.264 percent of the variance. Table 10.2 shows the total variance.

The resulting factors were named on the basis of research carried out on the TIMSS data (Martin *et al.*, 2000; Papanastasiou, 2000, 2002; Koutsoulis and Campbell, 2001). The combinations of items with loadings greater than 0.50 were considered as separate factors and are defined as follows:

Table 10.1 KMO and Bartlett's test.

Kaiser-Meyer-Olkin measure of sampling adequacy		0.866
Bartlett's test of sphericity	Approx. chi-square	32085.141
	df	465
	Sig.	0.000

Table 10.2 Total variance explained.

Component	Initial eigenvalues			Rotation sums of squared loadings		
	Total	Variance (%)	Cumulative (%)	Total	Variance (%)	Cumulative (%)
1	5.286	17.052	17.052	3.083	9.946	9.946
2	3.154	10.173	27.225	2.517	8.120	18.065
3	2.475	7.985	35.209	2.492	8.038	26.103
4	1.524	4.916	40.126	2.129	6.869	32.971
5	1.347	4.346	44.471	1.971	6.358	39.330
6	1.211	3.907	48.378	1.763	5.688	45.018
7	1.123	3.622	52.001	1.667	5.379	50.397
8	1.012	3.263	55.264	1.509	4.867	55.264

Note. Extraction method: principal component analysis.

1. 'Students' attitudes towards mathematics' with five items (whether the student likes mathematics, thinks mathematics is an easy subject, likes finding a job that involves mathematics, enjoys learning mathematics and thinks they usually do well in mathematics), loadings between 0.597 to 0.764.
2. 'External motivation' with four items (whether student thinks they need to do well in mathematics in order to enter their desired school, to get their desired job, and to please themselves and their parents), loadings between 0.66 to 0.774.
3. 'Mathematics self-concept' with five items (whether the student thinks they will never really understand mathematics, they are just not talented in mathematics, mathematics is not one of their strengths, mathematics is more difficult for them than for many of their classmates, and they would like mathematics much more if it were not so difficult), loadings between 0.544 to 0.733.
4. 'Educational aids at home' with four items (the number of books at home and possessing dictionary, calculator and study desk at home), loadings between 0.563 to 0.72.
5. 'Press (home-school interface)' with three items (whether the student thinks that it is important for his/her mother, his/her friends, and themselves to do well in mathematics), loading between 0.715 to 0.768.
6. 'Attribution (belief)' with four items (whether the student thinks in order to do well in mathematics they need to memorize the textbook or notes, good luck, natural talent, or lots of hard work studying at home), loadings between 0.531 to 0.729.
7. 'Home background' with three items (the highest education level of the student's father and mother and number of people living at home), loading between 0.691 to 0.735.
8. 'Leisure time' with three items (watching video games, going to the movies, and watching sports programs), loadings between 0.653 to 0.710

The 31 items and their factor loadings are listed in Table 10.3.

Results

The correlation matrix between the studied factor scores and the mathematics score (Rasch score) showed that seven of the eight factor scores had significant correlations with the mathematics score. The factor scores of mathematics self-concept, attribution, students' attitudes towards mathematics, educational aids at home and home background had the highest correlations with the mathematics score for the total sample (0.29, 0.274, −0.227, −0.183, and 0.177, respectively).

In order to determine how much of the variance in the mathematics score could be explained by the above-mentioned factor scores, multiple regression analysis (stepwise solution) was utilized. The results of this analysis showed that seven of the eight factor scores under study totally accounted for approximately one quarter of the variance in the mathematics score (26.8 percent). The first factor that significantly entered into the regression equation was mathematics self-concept. Mathematics

Table 10.3 Factor loadings.

Factor 1: 'Attitude towards Mathematics'		Factor 2: 'External Motivation'		Factor 3: 'Mathematics Self-concept'		Factor 4: 'Educational Aids at Home'		Factor 5: 'Press (Home–School Interface)'		Factor 6: 'Attribution (Belief)'		Factor 7: 'Home Background'		Factor 8: 'Leisure Time'	
Items	Loadings	Items	Loadings	Items	Loadings	Items	Loadings	Items	Loadings	Items	Loadings	Items	Loadings	Items	Loadings
The student likes math	0.764	The student thinks he/she needs to do well in math to enter his desired school	0.774	The student thinks he/she is just not talented in math	0.733	Possessing dictionary at home	0.72	The student him/herself thinks it is important to do well in math	0.768	The student thinks he/she needs to memorize the textbook or notes to do well in math	0.729	The highest education level of the student's mother	0.735	The student watches video games	0.71
The student thinks math is an easy subject	0.749	The student thinks he/she needs to do well in math to please his/her parents	0.757	The student thinks he/she will never really understand math	0.693	Number of books at home	0.698	Students' mother thinks it is important for the student to do well in math	0.766	The student thinks he/she needs good luck to do well in math	0.624	The highest education level of the student's father	0.706	The student goes to the movies	0.695
The student likes a job that involves using math	0.734	The student thinks he/she needs to do well in math to please him/herself	0.754	The student thinks math is not one of his/her strengths	0.689	Possessing calculator at home	0.575	Student's friends think it is important to do well in math	0.715	The student thinks he/she needs natural talent to do well in math	0.607	Number of people living at home	0.691	The student watches sports programs	0.653
The student enjoys learning math	0.659	The student thinks he/she needs to do well in math to get his/her desired job	0.66	The student thinks math is more difficult for him/her than for many of his/her classmates	0.644	Possessing study desk at home	0.563	–	–	The student thinks he/she needs lots of hard work studying at home to do well in math	0.531	–	–	–	–
The student thinks he/she usually does well in math	0.597	–	–	The student thinks he/she would like math more if it were not so difficult	0.544	–	–	–	–	–	–	–	–	–	–

self-concept accounted for almost one third of the explained variance (8.4 percent). The second factor that entered into the regression equation after controlling mathematics self-concept was attribution and it explained 7.4 percent of the variance in mathematics score.

The next five factors that entered into the regression equation – students' attitudes towards mathematics, educational aids at home, home background, press, and external motivation – also explained significant proportions of the variance in the mathematics score (4.1, 3.3, 2.7, 0.8 and 0.2 percent, respectively). The last mentioned factor even though significant was negligible. Leisure time was the only factor that did not have a significant effect on mathematics achievement.

The overall multiple regression equation that assessed the joint significance of the complete set of predictor factors was significant [F (7, 3073) = 161.055, $p < 0.01$)]. Table 10.4 represents standardized regression coefficients and collinearity diagnostics for the seven independent factors and more specifically the beta weight as well as the estimate of tolerance and the variance inflation factor (VIF). The lowest and highest tolerance and VIF values for the seven factors were 0.993 and 0.999 as well as 1.001 and 1.007, respectively. All the tolerance and VIF values are acceptable and it can be concluded that each factor is completely uncorrelated with the other independent factors.

The inspection of the respective scatter plots for the standardized predicted values against standardized residuals shows that the relationship between the dependent variable and the seven factors is linear and the variance of the residuals at every set of values for the dependent variable is equal.

It is worth mentioning that the standardized regression coefficients (beta) for students' attitudes towards mathematics (–0.207), educational aids at home (–0.177), and press (-0.89) factor scores were negative. The negative weight for betas is the result of the coding procedure. For instance, in the questions for educational aids, the answer 'yes' or 'possessing the aids' has been coded 1 and the answer 'no' or 'not possessing'

Table 10.4 Standardized regression coefficients and collinearity diagnostics for seven independent factors.

Factors in the equation	Beta	t	Sig.	Correlations			Collinearity statistics	
				Zero-order	Partial	Part	Tolerance	VIF
Mathematics self-concept	0.271	17.501	0.000	0.290	0.301	0.270	0.995	1.005
Attribution	0.265	17.172	0.000	0.274	0.296	0.265	0.999	1.001
Attitudes towards mathematics	−0.207	−13.374	0.000	−0.227	−0.235	−0.206	0.993	1.007
Educational aids at home	−0.177	−11.463	0.000	−0.183	−0.202	−0.177	0.998	1.002
Home background	0.165	10.677	0.000	0.177	0.189	0.165	0.998	1.002
Press	−0.089	−5.738	0.000	−0.081	−0.103	−0.089	0.998	1.002
External motivation	0.041	2.687	0.007	0.032	0.048	0.041	0.999	1.001

has been coded 2. In addition, in attitude questions those who generally agreed with the statements had high positive attitudes (code 1) and those who generally disagreed with the statements had low positive attitudes (code 4) towards mathematics. The same case holds true for press factor questions.

The results of multiple regression analysis (stepwise solution) for girls showed that, from among the eight factor scores under study, seven had significant effects on the girls' mathematics score and totally explained 32.4 percent of the variance in the girls' mathematics score. The most important factors affecting the girls' mathematics achievement were attribution and mathematics self-concept, which accounted for 9.6 and 9.2 percent of the variance in the girls' mathematics score, respectively. Students' attitudes towards mathematics, educational aids at home, home background, press, and external motivation, also explained 4.3, 4, 2.9, 1.5, and 0.9 percent of the remaining variance in the girls' mathematics score, respectively.

Regarding the results for boys, it should be stated that 25.7 percent of the variance in the boys' mathematics score was explained by seven out of the eight factor scores. The most important factors that explained the variance in the boys' mathematics score were mathematics self-concept and attribution. These factors explained 8.1 and 6.4 percent of the variance in the boys' mathematics score, respectively. The other five factors which significantly contributed to the explanation of the variance in the boys' mathematics score after controlling the first two factors were students' attitudes towards mathematics, educational aids at home, home background, press, and leisure time. These five factors explained 3.9, 3.3, 2.5, 0.7, and 0.8 percent of the variance in the boys' mathematics score, respectively (see Table 10.5).

It is worth mentioning that the mathematics self-concept standardized regression coefficients (betas) for the total sample, girls and boys were 0.271, 0.266 and 0.275, respectively. In addition, students' attitudes towards mathematics standardized regression coefficients for the afore-mentioned groups were -0.207, -0.215 and -0.195, respectively. And finally, standardized regression coefficients for attribution were 0.265, 0.27, and 0.26, and for educational aids at home were -0.177, -0.195 and -0.184, respectively.

In general, the TIMSS 1999 data revealed that leisure time and external motivation factors either had low or no effect on predicting Iranian students' mathematics achievement. External motivation had a low significant effect on the mathematics performance of the total sample as well as girls' mathematics achievement (0.2 and 0.9 percent of the total variance, respectively). However, it did not affect boys' mathematics achievement. In contrast, leisure time had a low significant effect on the performance of boys and did not affect the performance of the total sample as well as girls' mathematics achievement.

Findings and Discussion

Education is a complex process and many factors directly or indirectly affect school outcomes. As a result, it is difficult to define properly the major factors influencing students' achievement. This study, similar to abundant research carried out in educa-

Table 10.5 Amount of variance explained by each of the factors under study.

Factors in the equation	Total sample		Girls		Boys	
	Entering order	Percent	Entering order	Percent	Entering order	Percent
Mathematics self-concept	1	8.4	2	9.2	1	8.1
Attribution	2	7.4	1	9.6	2	6.4
Attitudes towards mathematics	3	4.1	3	4.3	3	3.9
Educational aids at home	4	3.3	4	4	4	3.3
Home background	5	2.7	5	2.9	5	2.5
Press	6	0.8	6	1.5	6	0.7
External motivation	7	0.2	7	0.9	–	–
Leisure time	–	–	–	–	7	0.8
Total percent explained	–	26.8	–	32.4	–	25.7

tion, revealed a significant effect of mathematics self-concept (Wilhite, 1990; Marsh, 1993; Franken, 1994; Hamachek, 1995; Kiamanesh and Kheirieh, 2001), home background (Weiss and Krappmann, 1993; Coleman *et al.*, 1966; Beaton *et al.*, 1996; Robitialle & Garden, 1989; Kiamanesh and Kheirieh, 2001; Howie, 2002; Marjoribanks, 2002; Singh *et al.*, 2002; Engheta, 2004), students' attitudes towards mathematics (Aiken, 1976; McMillan, 1977; Kulm, 1980; Keeves, 1992; Schereiber, 2000; Papanastasiou, 2002), attribution (Frize *et al.*, 1983; Weiner, 1985; Weiner, 1992), press variables (Kulm, 1980; Berkowitz, 1986; Martin *et al.*, 2000) and external motivation (Banks *et al.*, 1978; DeCharms, 1984; Dweck, 1986; Skaalvik, 1994; Skaalvik and Rankin, 1995; Small and Gluck, 1994) on mathematics achievement.

In the present study it was shown that mathematics self-concept, attribution and students' attitudes towards mathematics are the three important factors that account for the most variance in the mathematics achievement scores for the total sample as well as both genders in Iran. However, comparing the proportion of the total variance explained in the total sample (28.6 percent) as well as both genders (girls 32.4 percent and boys 25.7 percent) and the proportion of variance accounted for by each factor in each group, it should be stated that attribution and press factors have slightly more effect on girls' mathematics achievement than that of boys. In addition, mathematic self-concept and attitudes towards math have slightly more effect on boys' mathematics achievement than that of girls. Conversely, the total variance explained by the two background factors – home background and educational aids at home – for the total sample as well as both genders is 6, 6.9 and 5.8 percent, respectively. This also shows that family background factors have slightly more effect on boys' performance than that of girls.

The proportion of the variance accounted for by three students' characteristics factors (mathematics self-concept, attribution and students' attitudes towards mathematics) was much more than the variance explained by the combination of two background factors (home background and educational aids at home) for the total sample (19.9 compared to 6 percent) as well as for girls (23.1 compared to 6.9 percent) and boys (18.4 compared to 5.8 percent). The aforementioned dominant students' characteristics can directly and indirectly be influenced by teachers and schools. Findings from the related literature such as 'self-concept can be formed and developed in the classroom' (Marsh, 1993; Hamachek, 1995); 'self concept is a complex system of learned beliefs, attitudes, and opinions that each person holds' (Purkey, 1988); 'self is the basis for all motivated behavior' (Franken, 1994); 'motivation is correlated with achievement and academic performance' (Skaalvik, 1994; Skaalvik and Rankin, 1995); 'students will be motivated if they know the goals of tasks' (Small and Gluck, 1994); 'attribution influences students' achievement' (Weiner, 1985); 'an individual may change his attitude in agreement with that of an important person' (Berkowitz, 1986), 'teachers with the most experience, scientific training and interest influence the students' attitude more' (Martin, 1996); and 'it is possible to change the self-concept' (Franken, 1994) are all indicative of the fact that schools and teachers 'can make a difference' and play a strong positive role in creating a learning environment for students.

This conclusion is in line with Fuller's (1987) conclusion that in developing countries schools have a great influence on students' achievement after accounting for the effect of home background. This argument makes sense when we compare the cultural and educational stimulus enrichment in the 'have' and 'have not' families. The students who come from deprived families do not have access to educational reinforcement at home. In these families parents are not well educated and hence are not able to help students with their schoolwork or even provide them with an appropriate learning environment. In this situation schools are the main source of providing learning opportunities.

The findings of the present study are important for the Iranian education system due to the fact that 'it is possible to change the self-concept' and 'attitude may change in agreement with that of an important person (especially the teacher's)'. These findings are also important since improving the teaching procedures in the classroom is much easier to achieve than changing background factors affecting students' performance. The implications of these findings could serve as a guideline for teacher educators so that they provide teachers with appropriate training before they enter the profession as well as in-service training programs to help them understand the importance of teaching and how to teach effectively. These results could also serve as a basis for transition from teacher-centered methodology, which is the prevailing teaching methodology in Iranian schools, towards a more participatory and active learning methodology. This transition seems an indispensable part of the reforms to be undertaken since participatory and active learning strategies result in teachers who achieve higher educational standards.

In addition, the findings could be a guideline for educational practitioners and curriculum developers so that they can ensure that the utilized educational policies, methodologies and activities would help students improve their academic self-concept, beliefs and positive attitudes toward school subjects including mathematics.

Regarding the findings of the international studies and the outcomes of other research studies of Iranian education, there is an urgent need in Iran to improve the quality of education by eliminating existing barriers and restrictions. The first step might be to acquaint the country's political and cultural authorities and educational policy-makers with the current state of educational outputs and the factors influencing the teaching–learning process. Besides the publicity activities, the following steps might prove helpful:

- Providing suitable conditions for recruitment, training, maintenance and continuing education of teachers.
- Changing teaching methods and moving towards widespread use of active and process-oriented methods by creating positive attitudes and awareness among educators and teachers, and preparing appropriate environment and facilities suited to the application of these methods in schools.
- Providing principles and teachers with real authority and responsibility at school level.
- Reform methods of examination and assessment and move from 'answer-based' and 'score-based' examinations towards 'comprehension-based' and 'learning-oriented' ones.

References

Aiken, L.R. (1976) Update on attitudes and other affective variables in learning mathematics. *Review of Educational Research*, 46, 293–311.

Banks, C. W., McQuater, G. V. and Hubbard, J. L. (1978) Toward a reconceptualization in the socio-cognitive bases of achievement orientations in Blacks. *Review of Educational Research*, 28(3), 381–97.

Beaton, A.E., Mullis, I.V.S., Martin, M.O., Gonzalez, E.J., Kelly, D.L. and Smith, T.A. (1996) *Mathematics achievement in the middle school years: IEA's third international mathematics and science study* (TIMSS). Chestnut Hill, MA: Boston College.

Berkowitz, L. (1986) *A Survey of Social Psychology*. New York: CBS College Publishing.

Broeck, A.V.D., Damme, J.V. and Opdenakker, M.C. (2004) The effects of student, class and school characteristics on mathematics achievement: Explaining the variance in Flemish TIMSS-R data. In C. Papanastasiou (ed.), *Proceedings of the IRC-2004 TIMSS* (Vol. I, pp. 87–98). Nicosia: Cyprus University.

Cain-Caston, M. (1993) Parent and student attitudes towards mathematics as they relate to third grade mathematics. *Journal of Instructional Psychology*, 20(2), 96–102.

Coleman, J., Campbell, E., Hobson, C., McPartland, J., Mood, A., Weinfeld, F. and York, R. (1966) *Equality of educational opportunity*. Washington, DC: Department of Health, Education, and Welfare.

DeCharms, R. (1984) Motivation enhancement in educational settings. In R.E. Ames and C. Ames (eds.), *Research on motivation in education: Vol. I, Student motivation* (pp. 275–310). Orlando, FL: Academic Press.

Dweck, C.S. (1986). Motivational processes affecting learning. *American Psychologist*, A1(10), 1040–8.

Engheta, C.M. (2004) Education goals: Results by the TIMSS-99 for participating G8 countries. In C. Papanastasiou (ed.), *Proceedings of the IRC-2004 TIMSS* (Vol. II, pp. 172–86). Nicosia: Cyprus University.

Franken, R. (1994) *Human motivation* (3rd ed.). Pacific Grove CA: Brooks/Cole Publishing Co.

Frize, I.H., Francis, W.D. and Hanusa, B.H. (1983). Defining success in classroom settings. In J.M. Levin and M.C. Wang (eds.), *Teachers and students perceptions: Implications for learning* (pp. 3–28). Hillsdale, NJ: Erlbaum.

Fullarton, S. (2004) Closing the gaps between schooling: Accounting for variation in mathematics achievement in Australian schools using TIMSS 1995 and TIMSS 1999. In C. Papanastasiou (ed.), *Proceedings of the IRC-2004 TIMSS* (Vol. I, pp. 16–31). Nicosia: Cyprus University.

Fuller, B. (1987) What school factors raise achievement in the Third World? *Review of Educational Research*, 47(3), 255–292.

Hamachek, D. (1995) Self-concept and school achievement: Interaction dynamics and a tool for assessing the self-concept component. *Journal of Counseling & Development*, 73(4), 419–25.

Howie, S.J. (2002) English language proficiency and contextual factors influencing mathematics achievement of secondary school pupils in South Africa. Enschede: Doctoral dissertation, University of Twente.

Keeves, J.P. (1992) *Learning science in a changing world: Cross-national studies in science achievement: 1970 to 1984*. The Hague: IEA.

Kiamanesh, A.R. and Kheirieh, M. (2001) *Trends in mathematics educational inputs and outputs in Iran: Findings from the third international mathematics and science study and its repeat*. Tehran: Institute for Educational Research Publication.

Koutsoulis, K.M. and Campbell, J.R. (2001) Family processes affect students' motivation, and science and math achievement in Cypriot high schools. *Structural Equation Modeling*, 8(1), 108–27.

Kulm, G. (1980) Research on mathematics attitude. In R.J. Shum Way (ed.), *Research on Mathematics Education* (pp. 336–87). Reston, VA: National Council of Teachers of Mathematics.

Kyriakides, L. and Charalambous, C. (2004) Extending the scope of analyzing data of IEA studies: Applying multilevel modeling techniques to analyze TIMSS data. In C. Papanastasiou (ed.), *Proceedings of the IRC-2004 TIMSS* (Vol. I, pp. 69–86). Nicosia: Cyprus University.

Marjoribanks, K. (2002) *Family and school capital: Towards a context theory of students' school outcomes*. Dordecht: Kluwer Academic Publishers.

Marsh, H.W. (1993) Academic Self-concept: Theory measurement and research. In J. Suls (ed.), *Psychological Perspective on the Self* (Vol. 4, pp. 59–89). Hillsdale, NJ: Erlbaum.

Martin, V.M. (1996) Science literature review. In R.A. Garden (ed.), *Science performance of New Zealand Form 2 and Form 3 students* (pp. 19–37). Wellington: Research and International Section, Ministry of Education.

Martin, M.O., Mullis, I.V.S., Gregory, K.D., Craig, H. and Shen, C. (2000). *Effective schools in science and mathematics. IEA's third international mathematics and science study.* Boston, MA: TIMSS International Study Center, Boston College.

McMillan, J.H. (1977) The effect of effort and feedback on the formation of student attitudes. *American Educational Research Journal,* 14(3), 317–30.

Mullis, I.V.S., Martin, M.O., Beaton, A.E., Gonzalez, E.J., Gregory, K.D., Garden, R.A., O'Connor, K.M., Chrostowski, S.J. and Smith, T.A. (2000). *TIMSS 1999: International mathematics report, finding from IEA's report of the third international mathematics and science study at the eight grade.* Boston, MA: TIMSS International Study Center, Boston College.

Papanastasiou, C. (2000) School, effects of attitudes and beliefs on mathematics achievement. *Studies in Educational Evaluation,* 26, 27–42.

Papanastasiou, C. (2002) School, teaching and family influence on student attitudes toward science: Based on TIMSS data Cyprus. *Studies in Educational Evaluation,* 28, 71–86.

Purkey, W. (1988) *An overview of self-concept theory for counselors.* Eric Clearing House on Counseling and Personnel Services, Ann Arbor, Mich. (An Eric/CAPS Digest: ED 304630Services, Ann Arbor, Mich. (An Eric/CAPS Digest: ED 304630).

Robitaille, D.F. and Garden, R.A. (eds.) (1989) *The IEA study of mathematics: Contexts and outcomes of school mathematics.* Oxford: Pergamon.

Schereiber, B.J. (2000) Advanced mathematics achievement: A hierarchical linear model. PhD dissertation, Indiana University.

Schmidt, W.H. and Valverde, G.A. (1995) Cross-national and regional variations in mathematics and science curricula. Paper presented at the Annual Meeting of the Merican Educational Research Association, San Francisco.

Schmidt, W.H. *et al.* (1999) *Facing the consequences: using TIMSS for a closer look at US mathematics and science education.* Dordrecht: Kluwer.

Singh, K., Granville, M. and Dika, S. (2002) Mathematics and science achievement: Effects of motivation, interest, and academic engagement. *Studies in Educational Evaluation,* 28, 71–86.

Skaallvik, E.M. (1994) Attribution of perceived achievement in school in general and in math and verbal areas: Relations with academic self-concept and self-esteem. *British Journal of Educational Psychology,* 64(1), 133–43.

Skaalvik, E.M. and Rankin, R.J. (1995) A test of the internal/external frame of reference model at different levels of math and verbal self-perception. *American Educational Research Journal,* 32(1), 161–84.

Small, R.V. and Gluck, M. (1994) The relationship of motivational conditions to effective instructional attributes: A magnitude scaling approach. *Educational Technology,* 34 (8), 33–40.

Stodalsky, S.S., Salk, S. and Glaessner, B. (1991) Student views about learning math and social studies. *American Educational Research Journal,* 28 (1), 89–116.

Weiner, B. (1985) An attributional theory of achievement, motivation, and emotion. *Psychological Review,* 92, 543–73.

Weiner, B. (1992). *Human motivation: Metaphors, theories, and research.* Newbury Park, CA: SAGE Publications.

Weiss, K. and Krappmann, L. (1993, March) *Parental support and children's social integration*. Paper presented at the biennial meeting of the Society for Research in Child Development, New Orleans, LA (ERIC Document Reproduction Service No. ED 361 077).

Wilhite, S.C. (1990) Self-efficacy, locus of control, self-assessment of memory ability, and study activities as predictors of college course achievement. *Journal of Educational Psychology*, 82, 696–700.

11

Multi-level Factors Affecting the Performance of South African Pupils in Mathematics

Sarah J. Howie

Introduction

This research is a secondary analysis of the performance of the South African pupils in the Third International Mathematics and Science Study 1999 in which pupils wrote tests (in either English or Afrikaans) in mathematics and science. South African pupils also had to write an English test, which was included as a national option. TIMSS 1999 was conducted in 1998/1999 in South Africa and internationally under the auspices of the International Association for the Evaluation of Educational Achievement (IEA). The South African pupils' performance was significantly below that of all 37 other participating countries, including other developing countries such as Morocco, Tunisia, Chile, Indonesia and the Philippines. This was also the case in 1995 where South Africa performed below all 40 other countries in the study. This was partly the inspiration to explore the factors that had an effect on the South African pupils' performance in mathematics. This chapter presents some of the final results of a three-year research project (Howie, 2002) and includes the final partial least square (PLS) analysis and multi-level analysis of the effect of contextual factors on student and school levels.

The research question addressed by this chapter is: 'Which factors at school level, class level and student level have an effect on South African pupils' performance in mathematics?'

The factors relating to the pupils' performance in mathematics were explored in relation to the background information that was collected from the pupils, teachers and principals of the schools included in the study.

A number of factors have been reported pertaining to the poor performance of pupils in South Africa, such as: inadequate subject knowledge of teachers; inadequate communication ability of pupils and teachers in the language of instruction; lack of instructional materials; difficulties experienced by teachers to manage activities in classrooms; the lack of professional leadership; pressure to complete examination driven syllabi; heavy teaching loads; overcrowded classrooms; poor communication between policy-makers and practitioners and lack of support due to a shortage of professional staff in the ministries of education (see for instance, Adler, 1998; Arnott and Kubeka, 1997; Kahn, 1993; Monyana, 1996; Setati and Adler, 2000; Taylor and Vinjevold, 1999).

Internationally, research addressing factors (such as textbooks, teacher quality, time, leadership, organization and management, decision-making, within-school hierarchy and communication and class size, home environment, socio-economic status [SES], education of parents, books in home, self concept and attitudes towards mathematics) related to achievement in mathematics were found using data from, for example, Belgium (Van den Broeck and Van Damme, 2001), and Eastern Europe (Vari, 1997), but most were found in the USA (Sojourner and Kushner, 1997; Teddlie and Reynolds, 2000, among others). No studies were found either nationally or internationally that attempt to link English language proficiency to mathematics achievement at secondary level using such a comprehensive dataset with data on pupil, class and school levels.

Theoretical Framework

The conceptual framework for this study (see Figure 11.1) is based upon the framework by Shavelson et al. (1987) and the IEA thinking on curriculum. However, a number of adaptations have been made to the original frameworks to better suit the research questions posed by this study.

This model allowed for the exploration of contextual factors within different levels influencing pupils' achievement in mathematics within the context of South Africa. A very brief summary of the model description is given here while a complete description can be found in Howie (2002).

The model shown in Figure 11.1 presents the education system in terms of inputs, processes and outputs. The curricula for academic subjects play a central role in an education system. The IEA believes that the curriculum is the key in the evaluation of educational achievement. They differentiate between the intended, the implemented and the attained curriculum. In the model (Figure 11.1), the central positioning of the intended, implemented and attained curricula and their linkage between elements within the model illustrate the key role of the curricula. The model serves as an important theoretical and conceptual basis for the analysis of the TIMSS 1999 data. As the data were collected on a number of education levels, namely, school, class and pupil levels, the model serves as a guide to explore the causal links for the pupils' achievement.

Research Design

The research comprises both exploratory and analytical phases of the study focusing on the secondary analysis of the TIMSS 1999 data related to mathematics achieve-

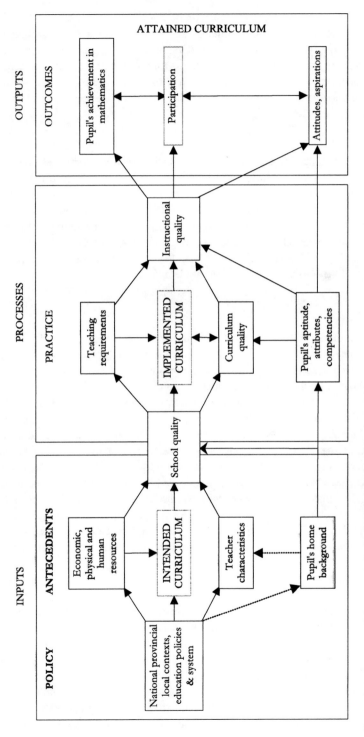

Figure 11.1 Conceptual framework factors related to mathematics achievement.
Source. Howie (2002, adapted from Shavelson *et al.*, 1987).

ment. The data were explored to investigate the reasons for the pupils' performance and to explore the interrelationships of achievement and the background variables revealed by pupils, teachers and the school principal. In particular, the exploratory part of the study was to determine the factors that influence mathematics achievement and performance of South African pupils and to ascertain the effect of South African pupils' language and communication skills on their achievement in mathematics.

Sample
A two-stage stratified cluster sample of 225 schools was randomly selected and strati-fied according to province, type of education (government or private) and medium of instruction (English and Afrikaans). One class was randomly selected within each school. Tests and questionnaires were administered (by the Human Sciences Research Council based in Pretoria, South Africa) to more than 9,000 pupils. Questionnaires were also administered to 200 school principals and 400 teachers of mathematics and science at Grade 8 level. After the data cleaning a representative sample of 194 schools and 8,146 pupils was used in the data analysis.

Instruments
In addition to the TIMSS 1999 instruments, namely test booklets, pupil questionnaire, teacher questionnaires and school questionnaire, an English language proficiency test was included specifically for South African pupils. This instrument had previously been validated by the Human Sciences Research Council and standardized for Grade 8 English Second Language pupils in South African schools (HSRC, 1990). At the time of the TIMSS 1999 study, this test was the only standardized South African second language test at the Grade 8 level that could be found. Questions were also included in the TIMSS 1999 pupils' and teachers' questionnaires, to ascertain the extent and level to which the pupils are exposed to English. They included pupils' home language, ethnic group, the language spoken predominantly by the pupils in the mathematics class, the language used by the mathematics teacher in class, media languages pupils are exposed to and the language of their reading materials. In this research, data from the test booklets, pupil questionnaires, mathematics teacher questionnaires, school questionnaire and the national option were analyzed.

Data Analysis
After examining frequencies, building constructs and reviewing correlation matrices, PLS analysis and multi-level modeling were applied.

Given that there are a number of variables reported in the literature to influence pupils' achievement, as well as the vast number of variables in the database, and that some of these were intricately interrelated, PLS analysis (Sellin, 1989) was used as a first step to analyze those student-, class- and school-level factors on individual levels that influenced pupils' achievement in mathematics. This type of analysis allows one to estimate or predict both the direct and indirect effects of a set of independent variables on a dependent variable (with each path taking into account the effects of all the other variables).

Due to the fact that data were collected on three levels – student level, class level and school level, multi-level modeling (Institute of Education, 2000) was applied. In this study, multi-level modeling was used to distinguish between the variance in mathematics achievement uniquely explained by student-level factors, as opposed to the variance uniquely explained by the class- and school-level factors and to investigate the individual effects of variables inserted in the model once the multi-level structure of the data is taken into account. As only one class per school was sampled, only two levels could be analyzed. This was due to the original TIMSS 1999 design where class and school are considered one level.

Results

Results of the Mathematics Tests

Overall, South African pupils achieved 275 points out of 800 (standard error, 6.8) in the mathematics test, while the international average was 487. This result is significantly below the mean scores of all other participating countries, including the two other African countries of Morocco and Tunisia, as well as that of other developing or newly developed countries such as Malaysia, the Philippines, Indonesia and Chile.

From a comparative analysis of South African pupils' results with other countries in TIMSS 1999 (see Howie, 2001), some interesting observations were made. More than 70 percent of pupils from South Africa, Indonesia, Morocco, the Philippines and Singapore did not always speak the language of the test at home (see Figure 11.2). Nonetheless, the mean achievement scores vary considerably across this group of countries and there are also some interesting trends in the data. In South Africa, pupils who speak the language of the test (namely, English or Afrikaans) more frequently also attain higher scores on the mathematics test. When comparing those pupils who almost always or always speak the language of the test to those that never speak the language of the test, the former achieve scores that are more than 140 points higher than the latter. However, Indonesia for instance is described as a highly diverse country with more than 600 languages and 200 million people (Baker and Prys-Jones, 1998: 375) and yet apparently their pupils do not appear to have been as disadvantaged by writing the test in a second language. A similar pattern was also observed for Morocco and the Philippines in mathematics. In Singapore there does appear to be a difference, yet those who never speak the language of the test at home still outperform pupils from 33 other countries. In the African context, children in Morocco who never speak the language of the test at home performed better in mathematics than those in South Africa.

This issue needs to be explored further, as it appears from the data that pupils from other developing countries do not seem to be disadvantaged to the same extent by writing tests in their second or third language in mathematics or science; however it is not clear why this is so. Important lessons for South Africa may lie in the answers (see Figure 11.2).

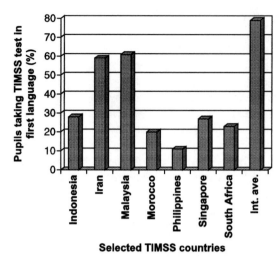

Selected TIMSS countries

Figure 11.2 Percentages of pupils from selected TIMSS countries taking the test in their first language.

Results of the English Language Test

In addition to the mathematics and science tests conducted in TIMSS 1999, an English language proficiency test was included that aimed to assess pupils' writing-related skills and language usage in English. The test comprised 40 multiple-choice items. Thirty of the 40 items had four options, while the remaining 10 items had two answer options. The overall mean score for the language test was 17 out of 40 (42.5 percent; $n = 8,349$). The minimum score attained was 0 and the maximum score 40. In general, the scores for boys and girls were comparable.

As the test was designed for second language English speakers, it is not surprising that native English speakers performed the best of all language groups (25 points out of 40), although one might have expected the scores to have been higher given this fact. The Afrikaans speaking children attained the next highest score with 21 points out of 40. The scores were more or less consistent across the pupils whose main language was one of the nine African languages.

PLS Exploration of the Contextual Factors on Student, Classroom and School

Three hypothesized models on student, class and school levels were analyzed using PLS analysis to explore the direct and indirect effects of individual variables on all three levels. The results of these analyses were scrutinized and thereafter the class- and school-level models were combined into one model and reanalyzed. The main results are summarized here and the detailed explanation and discussion of these PLS results can be found in Howie (2002). As an illustration of a PLS model, the student-level model is depicted in Figure 11.3 (see Appendix for explanation of the variables). The numbers on the arrows are the so-called beta-coefficients, indicating on a scale [−1,1]

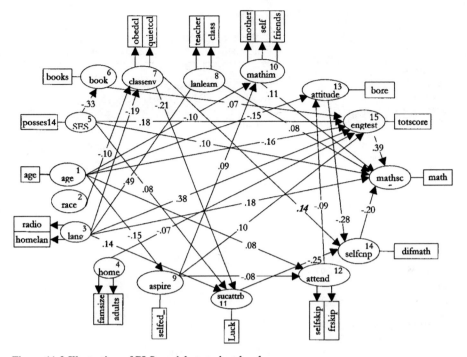

Figure 11.3 Illustration of PLS model at student level.

the strengths and the direction of the relationship. Sellin (1990) suggests that for large samples (such as this sample of pupils) a coefficient with an absolute value > 0.07 represents an effect of the factor on mathematics achievement, whilst he recommends (like the school-level model) a critical value of 0.10 for models with smaller samples.

Student-level Factors
Data pertaining to the pupils' home background, their personal characteristics, their aptitude and competencies were explored. A high percentage of variance (50 percent) in the pupils' mathematics score was explained (see Table 11.1). Six factors were found to have a direct effect on South African pupils' performance in mathematics, namely the pupils' proficiency in English (*engtest*), their own self concept in terms of mathematics (*selfcnpt*), the language pupils spoke at home (*lang*), their socio-economic status (SES) at home, and whether or not they, their friends and their mothers, thought that maths was important (*mathimpt*) and language of learning in the classroom (*lanlearn*).

School-level Only Factors
Some important aspects of school quality related to school leadership, parent involvement, school profile, physical resources, human resources, autonomy, learning envi-

Table 11.1 Inner model results of PLS at student level.

Factor	Beta	Correlation with math	Tolerance	R^2
MATHIMPT				0.02
Age	–0.10	–0.11	0.0227	
Aspire	0.09	–0.10	0.0227	
LANLEARN				0.24
Lang	0.49	0.49	0.0000	
BOOK				0.11
SES	–0.33	–0.33	0.0000	
CLASSENV				0.07
Lang	–0.19	–0.24	0.2938	
Race	–0.10	–0.20	0.2938	
ATTEND				0.02
Age	0.08	0.09	0.0227	
Aspire	–0.08	–0.10	0.0227	
ATTITUDE				0.03
Age	–0.15	–0.16	0.0089	
Attend	–0.09	–0.10	0.0089	
ASPIRE				0.02
Age	–0.15	–0.15	0.0000	
SELFCNPT				0.18
Classenv	0.14	0.20	0.0688	
Succattrb	-0.25	–0.31	0.0736	
Attitude	–0.28	–0.29	0.0057	
SUCATTRB				0.10
Lang	0.14	0.23	0.2176	
SES	0.08	0.17	0.1867	
Classenv	–0.21	–0.26	0.0625	
ENGTEST				0.38
Age	–0.16	–0.30	0.0697	
Lang	0.38	0.53	0.2212	
Home	–0.07	–0.22	0.0691	
SES	0.18	0.42	0.2598	
Book	0.07	0.26	0.1216	
Aspire	0.10	0.19	0.0318	
MATHSCR				0.50
Lang	0.18	0.52	0.4089	
SES	0.10	0.41	0.2482	
Mathimpt	0.11	0.17	0.0479	
Lanlearn	0.08	0.40	0.2808	
Selfcnpt	–0.20	–0.35	0.1008	
Engtest	0.39	0.63	0.3787	
Mean R^2				0.16

ronment and school administration were explored in the data from the school principal's questionnaire. Two important antecedents related to the location of the school and the home language of the pupil were included in the model. Sixty-two percent of the variance in the pupils' scores in mathematics was explained by three factors at the school level (see Table 11.2), namely, the school location (*communit*) (with urban

Table 11.2 Inner model results of PLS at school level.

Factor	Beta	Correlation	Tolerance	R^2
REPEATER				0.69
Teachrat	0.83	0.83	0.0000	
SCHOOL				0.21
Communit	0.45	0.45	0.0000	
CLASS				0.12
Communit	–0.19	–0.21	0.2628	
Resource	0.16	0.20	0.0464	
Firslang	–0.18	–0.22	0.1025	
School	0.16	0.01	0.2193	
MATH				0.62
Communi	0.18	0.41	0.1281	
Union	–0.17	–0.25	0.0275	
Firslang	0.66	0.74	0.1126	
Class	–0.08	–0.26	0.0708	
Mean R^2				0.41

schools achieving higher results), the negative influence that the teachers' union has on the curriculum (*union*), and an aggregated pupil variable, the extent to which the pupils in the class spoke the language of instruction as their first language (*firslang*). (For more detail see Howie, 2003.)

Classroom-level Only Factors
From the mathematics teacher questionnaire, a number of classroom-level factors were also explored and these resulted in including the following factors in the model (see Table 11.3): teachers' gender, teaching experience, teachers' level of education, time spent on activities, lesson preparation, teaching load, time on task, teachers' attitudes, success attribution, teachers' beliefs, teaching style, resources, limitations and class size. In total, this model explained 46 percent of the variance in the pupils' mathematics scores by seven factors – the teachers' attitudes (*attitude*), their beliefs about mathematics (*beliefs*), the extent of their teaching and other workload (*tload*), the size of the class they are teaching (*classize*), their gender (*gender*), resources (*resource*) and their dedication towards lesson preparation (*dedic*). These results are elaborated further in Howie (in press).

Combined School–Class-level Factors
As only one class per school was included in the study, class effects could not be studied independently from school effects. Therefore the school-level model and the class-level model were combined and the predictors of mathematics achievement were selected from both models and combined with four aggregated student-level antecedent factors into one model. Therefore factors related to teachers' characteristics, pupils' home background, their aptitude, their attitudes, school quality, teaching

Table 11.3 Inner model results of PLS at class level.

Factor	Beta	Correlation	Tolerance	R^2
ATTITUDE				0.35
Tload	−0.27	−0.35	0.06	
Tbackgr	−0.09	−0.20	0.08	
Teduc	−0.14	−0.28	0.10	
Classize	0.14	0.22	0.05	
Beliefs	0.32	0.41	0.11	
Limit	−0.21	−0.17	0.01	
DEDIC				0.06
Attitude	−0.24	−0.24	0.00	
STSUCC				
Exper	0.16	0.11	0.03	
Beliefs	0.27	0.25	0.03	
ACTIV				0.08
Tload	−0.14	−0.18	0.03	
Teduc	−0.10	−0.19	0.07	
Beliefs	0.28	0.32	0.07	
TSTYLE				0.10
Exper	0.11	0.06	0.03	
Beliefs	0.32	0.34	0.03	
BELIEFS				0.12
Tload	−0.09	−0.13		
Gender	0.09	0.08	0.01	
Exper	−0.21	−0.18	0.03	
Teduc	−0.26	−0.24	0.04	
Limit	0.08	0.06	0.02	
MATHSCR				0.46
Tload	0.19	0.37	0.16	
Tbackgr	0.08	0.24	0.08	
Gender	−0.11	−0.18	0.03	
Resource	0.11	0.16	0.05	
Classize	−0.13	−0.29	0.09	
Beliefs	−0.23	−0.44	0.20	
Attitude	−0.29	−0.53	0.31	
Dedic	0.10	0.28	0.12	
Mean R^2				0.22

Note. Factors in italics signify that the path coefficients fall below the required $p = 0.10$ criterion.

requirements, curriculum quality and instructional quality were all explored in one model. Finally, six factors were found (see Table 11.4) that had direct effects on pupils' achievement in mathematics and that explained 47 percent of the variance in the mathematics score. These were the location of the school (*communit*), class size (*classize*), the attitude of the teacher (*attitude*), teachers' beliefs about mathematics (*beliefs*), the teachers' workload (including teaching) (*tload*) and their dedication toward lesson preparation (*dedic*). See Howie (2005) for further elaboration.

Table 11.4 Inner model results of PLS at school and class level.

Factor	Beta	Correlation	Tolerance	R^2
AGGSELFC				0.49
Aggeng	−0.70	−0.70	0.0000	
CLASSIZE				0.11
Communit	−015	−0.28	0.4006	
Agglang	−0.13	−0.26	0.4034	
Aggses	−0.12	−0.30	0.5807	
BELIEFS				0.21
Agglang	−0.46	−0.46	0.0000	
ATTITUDE				0.41
Agglang	−0.56	−0.63	0.1760	
Tload	−0.13	−0.33	0.1232	
Classize	*0.07*	*0.23*	*0.0682*	
TLOAD				0.14
Gender	*0.04*	*−0.02*	*0.0249*	
Aggses	0.38	0.38	0.0249	
LIMITS				0.05
Agglang	*0.04*	*−0.11*	*0.4028*	
Aggses	−0.24	−0.21	0.4028	
AGGENG				0.73
Communit	0.19	0.48	0.1749	
Agglang	0.66	0.75	0.1473	
Limits	−0.16	−0.27	0.0458	
DEDIC				0.09
Agglang	−0.10	0.15	0.5015	
Tload	0.15	0.21	0.1461	
Attitude	−0.15	−0.21	0.4106	
Aggselfc	−0.19	−0.21	0.4106	
RESOURCE				0.02
Agglang	*0.05*	*0.11*	*0.4028*	
Aggses	*0.09*	*0.12*	*0.4028*	
MATHSCOR				0.47
Gender	*−0.06*	*−0.16*	*0.0352*	
Communit	0.18	0.42	0.1852	
Tload	0.18	0.37	0.1502	
Limits	*−0.09*	*−0.24*	*0.0925*	
Classize	−0.10	−0.30	0.1203	
Beliefs	−0.24	−0.43	0.1931	
Attitude	−0.25	−0.53	0.3187	
Dedic	0.10	0.26	0.0767	
Mean R^2				0.27

Note. Factors in italics signify that the path coefficients fall below the required $p = 0.10$ criterion.

Results from the Multi-level Analysis

From the PLS analysis, the factors that had a direct effect on math achievement were identified and included into the multi-level analysis (see Table 11.5): language pupils spoke at home, SES, pupils' English test score, pupils' own self concept in terms of

Table 11.5 Multi-level analysis of the South African TIMSS data with the mathematics test score as dependent variable (weighted data).

Fixed effect	Null model (Model 0)		Student model (Model 6)	Student–school model (Model 15)
Student-level				
Intercept	288		278	299.50
Home language			5.35 **	3.27
Socio-economic			1.20 **	0.88*
English test			4.07 **	4.00**
Self-concept			–6.32 **	–6.29**
Importance of maths			6.39 **	6.35**
Radio language			4.75 **	3.95**
School-level				
Status				–17.27**
Beliefs about maths				–4.46**
Location				8.00**
Class language				2.59**
Enrolment				0.00
Work time				0.52**
Class size				–0.27
Lesson planning				8.02*
Teaching time				0.10
Random effects				
School-level variance	6,520	(55%)	2,451	1,336
Student-level variance	5,342	(45%)	4,570	4,560
Difference in deviance	–		1,340#	132.80#
Explained proportion of variance in math achievement (when compared with null model):				
School-level variance	0.61	0.79		
Student-level variance	0.44	0.50		

Notes. Intercept represents the grand mean of the mathematics scores (differs slightly from the South African mean score of 275 on the international test due to the fact that a number of schools could not be included in the analysis).

$N = 7,651$ pupils in 183 schools (one class per school); * t-value > 1.96, this resembles a confidence interval of 95%; ** t-value > 2.58, this resembles a confidence interval of 99%; # the difference between the deviances is significant ($p < 0.001$).

mathematics, whether or not they, their friends and their mothers thought that mathematics was important, language on the radio they most often listen to, attitude of the teacher towards the profession, teachers' beliefs about mathematics, location of the school, language spoken most often in the class by pupils and teachers, teachers' total workload, size of the class, teachers' dedication toward lesson preparation, and teachers' teaching load. A final variable was included because it was believed to be important from a political perspective, namely the number of pupils enrolled in a school. Ultimately, 183 schools and 7,651 South African pupils were included in the multilevel analysis.

In the first step of the analysis the only independent variable included was the school of the learner (this is the so-called null model). The null model (see Table 11.5) shows that more than half of the variance in the mathematics achievement scores is situated on the school level (55 percent) while 45 percent of the variance can be situated on student level.

Student Model – Model 6

In the next step of the multi-level analysis the six student variables were entered successively in the model. The results are summarized in Table 11.5, column 'Student model – Model 6', in which the numbers represent the regression slopes associated with the various variables (for example, 5.35 is the regression coefficient associated with 'home language'). Table 11.5 shows that all regression coefficients are significant, meaning that mathematics achievement tends to be better when the scores are higher on the variables home language and SES.

The variables English tests, self-concept, importance of mathematics and radio language are described in the Appendix. However, one explanatory remark should be made: as self-concept has a negative scale, the negative regression coefficient should be interpreted as 'the more difficulties a learner has with mathematics, the lower the achievement score'.

A final remark on the student model pertains to 'difference in deviance' (see Table 11.5). The deviance is a measure for the appropriateness as to whether a model is a good representation of the reality represented by the data. The fact that the difference in deviance between the student model and the null model is highly significant means that the student model is a highly significant improvement compared to the null model.

Student–School Model – Model 15

The next phase in the multi-level analysis was to enter the nine school-level variables, resulting in the 'student–school model', also called model 15 as the model is based on 15 independent variables. The results are summarized in the final column of Table 11.5. In total, 11 of the 15 factors were found to be significant predictors of South African pupils' achievement in mathematics. Enrolment, class size, teaching time and home language appeared not significant in the student–school model. Once the school-level variables were entered, the effect of home language, which was significant in the student model, lost its significance as a consequence of the school-level variables, such as location of the school. In contrast, the English test score again appears to be one of the most significant (see Howie, 2002). Another result that can be derived from the full model is that SES is no longer significant (see Howie, 2002). The strength and significance of the school-level variables compensated for the pupil variables, resulting in home language and SES losing their significance in the mult-level analysis.

The quality of the student–school model as the best one in representing the relationship between the independent variables and achievement can be shown from two other parts of Table 11.5. At first, the difference in deviance between the student–school model (Model 15) and the student model (Model 6) is significant, indicating that the student–school model is superior. Furthermore, as can be concluded from the lowest part of Table 11.5, the strength of the student–school model is also illustrated by the variance in mathematics achievement explained by this model, being 79 percent of the variance at school level, and 50 percent of the variance at student level. Once all the predictors are added to the model, most of the school-level variance in pupils' achievement scores could be explained in the student–school model. This is not the case for the student-level variance, as a large percentage of the variance on student level (50 of the 45 percent in the null model) could not be explained by the predictors (including a number of language related variables) used in this model. This result may be due to the fact that other variables that are not included in this study are important as well. For example, cognitive ability was not measured in this study, but was included in the Belgian-Flemish study as a national option. Van den Broek and Van Damme (2001) show that in Belgium this variable explains a great deal of variance on student level and explains in fact more than any other single variable in their multi-level model. Clearly more research is needed here for South Africa.

However, the predictors did explain, for the South African data, a high percentage of the variance between schools. This finding means that a large part of the differences between schools in pupils' mathematics achievement can be attributed to these variables. The student–school model indicates that significant predictors for how pupils in different schools perform in mathematics are the pupils' performance in the English test, the SES (to a lesser extent), the pupils' self concept, the pupils' perception of the importance of mathematics, their exposure to English, how pupils' mathematics teachers perceive their professional status, pupils' mathematics teachers beliefs about mathematics, the location of the school, the extent to which English is used in the classroom, the amount of time teachers spend working and the amount of time teachers spend in lesson planning. They are also significant predictors of how well pupils perform in the same school (within-school variance), but to a lesser extent. Noteworthy is that two of these variables have a negative effect, teachers' perception of their status and their beliefs about mathematics. The stronger the teachers' ideas about mathematics and the perception about the status of the profession are, the poorer their pupils perform in mathematics. This observation should not be looked at in isolation, but in conjunction with the other variables that have a significant effect on mathematics achievement, but further discussion is beyond the scope of this chapter.

Extension of the Student–School Model

Due to the amount of the explained proportion variance on school level by the English test score (*engtest*), the student–school model was extended with random slopes (that is, with a random slope for each school) of the average English test score (*engtest*). This is a significant improvement of the full model (Model 15), as the deviation from the full model is highly significant ($p < 0.001$). The extension of the full model with random slopes results in SES being no longer significant. This means that the other school variables explain the variance in SES (as was concluded for home language [*hlang*] when discussing the full model). The inclusion of the random slopes (that is, per school its average value on *engtest* instead of taking the national average) results in lower estimates on all school variables, which shows that language proficiency is related to all school variables in another way. Figure 11.4 shows the final model graphically extended with a random slope for each class on the English test (*engtest*), with the other axis representing the mathematics achievement score.

If one looks at the pattern of the slopes in Figure 11.4, it would appear as if the impact of the English test (*engtest*) on mathematics achievement is less in classes with a low average score on the English test. In other words, schools where pupils did poorly in the English test, their proficiency hardly made any difference to their mathematics score. Conversely, the better classes of pupils performed on the English test, the stronger the relationship of this outcome was with mathematics. In other words, the correlation between the English test and the mathematics score is higher for classes with an on average high score on the English test. There appears to be a curvilinear relationship between English and Mathematics, which means that language proficiency matters more when the English proficiency of classes is higher.

Another observation should be made here. Figure 11.4 shows that there are schools with a high average score on the English test and yet a low average performance on mathematics, combined with a low correlation between the two variables. This is an indication that there are, in addition to English proficiency, other variables (either within the model, for example, location may be a candidate to investigate, or outside

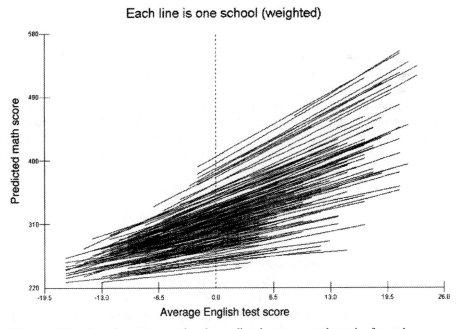

Figure 11.4 Random slopes representing the predicted score on mathematics for each school/class (based on the English test scores).

the present study) that are related to mathematics achievement. One possibility is that some informal tracking of pupils may be implemented at schools. So, there may be classes where the majority of pupils speak English at home, but may be grouped into a low ability mathematics class. Possible evidence of this practice was highlighted during the data collection for TIMSS 1999, where in at least six schools principals prevented field workers from testing the sampled class on the grounds that it was a low achieving group and insisted that the 'A' class be tested. Pupils from these schools were subsequently withdrawn from the TIMSS 1999 sample.

Conclusion and Reflection

This research shows that, in South Africa, pupils tended to achieve higher scores in mathematics when their language proficiency in English was higher and were more likely to attain low scores in mathematics when their scores on the English test were low. Those who spoke English or Afrikaans at home tended to achieve higher scores in mathematics. Alternatively, children from homes where African languages were used were more likely to achieve lower scores. Pupils in classes where the pupils and the teachers mostly interacted in the official media of instruction (English or Afrikaans) were more likely to achieve better results in mathematics.

The student–school model shows the effect of the location of the school in rural or urban areas on mathematics achievement, which is not surprising given the under-development in rural areas in South Africa. An interesting outcome was the strength of teachers' attitudes, beliefs and dedication as predictors of pupils' achievement. While it is difficult to determine whether this is a cause or consequence, this finding may mean that the teachers with stronger beliefs have less content knowledge and less understanding about the philosophy of mathematics, perhaps due to their training or lack thereof. Further investigation of teachers' beliefs in relation to pupils' achievement is needed.

Teachers with feelings of being appreciated by society and their pupils were more likely to produce pupils with lower results. This finding could be due to the fact that these are teachers in more rural areas who are highly regarded by their communities, as they are well educated compared to others in those communities. However, schools in these same communities are also poor and lack resources and pupils come from poor homes and have much less exposure to the languages of learning than their urban counterparts. So, although teachers feel affirmed, they are also challenged by the conditions in their schools, which results in pupils attaining lower scores. Teachers' dedication and commitment appear to play a key role in pupils' performance and the amount of time that teachers spent working and preparing lessons resulted in pupils being more likely to achieve higher results.

The lack of effect of class size as a predictor of achievement confirms previous findings in studies of developing countries. In the case of South Africa, those classes with large numbers of pupils (on average 50 pupils) are also those at schools with poor conditions described earlier.

These findings are significant against the background of the situation in many South African schools, particularly in those where there are African pupils taught by African teachers. In these schools the conditions are the worst: limited resources and facilities, large percentages of under-qualified teachers, pupils from poor socio-economic backgrounds and instruction occurring in a secondary language.

The findings give rise to a *number of reflections* on the relationship between language and achievement in relation to the language policy implemented by the South African government. The difficulty of not being able to communicate fluently in a common language is leading to increased frustration for the teacher, disorientation on the part of the child, a slow rate of learning, disciplinary problems and teacher-centered instruction. Although teachers are aware of the national language policy, they may have very different interpretations of it (Setati, 1999). The majority of parents, pupils and teachers perceive English as the gateway to global opportunities and therefore many want the pupils to participate in their education through the medium of English. If, as the majority of parents seem to desire, English is increasingly used, then the necessary support mechanisms need to be put into place, including intensive language training of second language teachers who will be teaching through the medium of English. If, on the other hand, the decision is made to teach in mother tongue beyond Grade 5, the ramifications are enormous, considering that in urban environments almost all the official languages are found as well as increasingly more foreign languages (from elsewhere in Africa and beyond). Segregating children and indeed teachers in terms of language (as clearly offering 11 languages of learning will be impossible within a single school) will be an enormous task and may reverse the cultural integration beginning to take place. Of course, in rural areas where one language is almost always clearly dominant, it may be more feasible. The danger of considering only one approach in the language policy, albeit an important one (for example only considering cultural identity or the political perspective), is that the multiple functions of schooling will be ignored. The curriculum and pupils' education have to fulfill the desires of society, namely, that of educating the young towards being responsive citizens in a democratic society, of attaining certain basic knowledge and skills, being prepared for the workplace and/or further education and acquiring adequate social and interpersonal skills. The government has to take a firm lead in finding the appropriate balance between these perspectives, since if this decision is left to the schools, the status quo will remain.

In conclusion, the strength of the language component represented in a number of variables that have strong effects on mathematics achievement is clear. In the past few years, some progress has been made to address shortcomings in South African schools. Although significant progress has been made with regard to administrative restructuring, policy development and infrastructural improvements, nonetheless the quality of education that the majority of pupils are receiving is far from satisfactory. The challenges abound within the education system of this country and, in addition to the issues of access and equity, the most important challenge awaiting now is that of quality.

Appendix

Appendix 1. Student-level variables.

Number in Figure 3	Latent variables	Manifest variables	Description	Scoring
1	AGE	AGE_1	Age of student	Number of years
2	RACE	RACE_1	Race of student: African, Coloured, Indian, White, Asian	1. African 2 Asian 3 Coloured 4 Indian 5 White
3	LANG	RALANG_1	Language on favorite radio station	1 All other langs 2 Afrikaans 3 English
		HOMELANG	Language spoken most often at home	0 Other languages 1 African languages 2 Afrikaans 3 English
4	HOME	FAMSIZE	Number of people living in the home	–
		PARENT (COMPOSITE)	Whether pupils have two parents	0 = no 1 = yes
5	SES	POSSES10 (COMPOSITE)	Computer, electricity, tap water, tv, CD player, radio, own bedroom, flush toilets, car (9 items)	0 = no 1 = yes
6	BOOK	BOOKS	Number of books in the home	1 = 0–10 2 = 11–25 3 = 26–100 4 = 101–200 5 = more than 200
7	CLASSENV	QUIETCL OBEDCL	Class climate items Extent to which the pupils report that pupils in their classes are quiet in class, are obedient in class	Scale of (–)1–4 (+) Strongly agree to strongly disagree scale of (–)1–4 (+) Strongly agree to strongly disagree
8	LANLEARN	LANLEARN (COMPOSITE)	Extent to which both speak language of speak language of instruction in maths class	1 Language spoken at home if not English/Afrikaans 2 Sometimes English/Afrikaans 3 Most of the time English/Afrikaans 4 Always English/ Afrikaans
9	ASPIRE	SELFED_1	Aspirations to education	1 Some secondary 2 Finished secondary 3 Finished technikon 4 Some university 5 Finished university

10	MATHIMPT	MATHIMPT (COMPOSITE)	Extent to which pupils' mother, friends think that maths is important (three items)	Scale of (+) 1–4 (–) Strongly agree to strongly disagre
11	SUCATTRB	LUCK	If the student attributes success to luck	Scale of (–)1–4 (+) Strongly agree to strongly disagree
12	ATTEND	ATTEND (COMPOSITE :SELFSKIP, FRSKIP)	Extent to which student or pupils' friends skip school (two items)	Scale of (+) 0–4 (–) never, once or twice three or four times five or more
13	ATTITUDE	BORE	If student finds maths boring	Scale of (–)1–4 (+) Strongly agree to strongly disagree
14	SELFCNPT	DIFMATH	The extent to which student reports having difficulty with maths (five items)	Scale of (–) 1–4 (+) Strongly agree to strongly disagree
15	ENGTEST	TOTSCORE	Student mean score on English language proficiency test	Score out of 40 points
16	MATHSCR	MATH	Student mean score on TIMSS-R 1999 mathematics test	Score out of 800 points

References

Adler, J. (1998) A language for teaching dilemmas: unlocking the complex multi-lingual secondary mathematics classroom. *For the Learning of Mathematics*, 18, 24–33.

Arnott, A. and Kubeka, Z. (1997) *Mathematics and science teachers: Demand, utilisation, supply, and training in South Africa*. Johannesburg: Edusource.

Baker, C. and Prys-Jones, S. (1998). *Encyclopedia of bilingualism and bilingual education*. Clevedon: Multilingual Matters.

Howie, S.J. (2001) *Mathematics and Science Performance in Grade 8 in South Africa 1998/1999: TIMSS 1999 South Africa*. Pretoria: HSRC report.

Howie, S.J. (2002) English language proficiency and contextual factors influencing mathematics achievement of secondary school pupils in South Africa. Enschede: Doctoral dissertation, University of Twente.

Howie, S.J. (2003) Conditions in schools in South Africa and the effects on mathematics achievement. *Studies in Educational Evaluation*, 29(3), 227–41.

Howie, S.J. (2005) Contextual factors on school and classroom level related to pupils' performance in mathematics in South Africa. In S.J. Howie and T. Plomp (eds.), TIMSS findings from an international perspective: trends from 1995 to 1999. Special issue of *Educational Research and Evaluation*, 11(2), 123–40.

Human Sciences Research Council (HSRC) (1990) *Achievement Test English Second Language Standard 6*. Ref. 2564/2. Pretoria: HSRC, Institute for Psychological and Edumetric Research.

Institute of Education (2000) *MLwiN: a visual interface for multilevel modeling. Version 1.10*. Multilevel Models project, University of London.

Kahn, M. (1993) *Building the base: report on a sector study of science and mathematics education*. Pretoria: Commission of the European Communities.

Monyana, H.J. (1996) Factors related to mathematics achievement of secondary school pupils. M.Ed. Thesis. University of South Africa. Unpublished.

Sellin, N. (1989) PLSPATH, *Version 3.01. Application Manual*. Hamburg: University of Hamburg.

Sellin, N. (1990) On aggregation bias. In K.C. Cheung, J.P. Keeves, N. Selling and S.C. Tsoi (eds.), The analysis of multivariate data in educational research: studies of problems and their solutions. *International Journal of Educational Research*, 14, 257–68.

Setati, M. (1999) Innovative language practices in the classroom. In N. Taylor and P. Vinjevold (eds.), *Getting learning right*. Johannesburg: Joint Education Trust.

Setati, M. and Adler, J. (2000) Between languages and discourses: Language practices in primary multilingual classrooms in South Africa. *Educational Studies in Mathematics*, 43, 243–69.

Shavelson, R.J., McDonnell, L.M. and Oakes, J. (1987) *Indicators for monitoring mathematics and science education: a sourcebook*. Santa Monica, CA: The RAND Corporation.

Sojourner, J. and Kushner, S.N. (1997) Variables that impact the education of African-American students: parental involvement, religious socialisation, socio-economic status, self-concept and gender. Paper presented at the Annual Meeting of the American Education Research Association. Chicago, Illinois, 24–28 March.

Taylor, N. and Vinjevold, P. (eds.) (1999) *Getting learning right*. Johannesburg: Joint Education Trust.

Teddlie, C. and Reynolds, D. (2000) *The international handbook of school effectiveness research*. London: Falmer Press.

Van den Broeck, A. and Van Damme, J. (2001) The effects of school, class and student characteristics on mathematics education in Flemish TIMSS-R data. Paper presented at the European Conference on Educational Research, Lille, 5–8 September.

Vari, P. (1997) *Monitor'95. National assessment of student achievement*. Budapest: National Institute of Public Education.

12

Factors Affecting Korean Students' Achievement in TIMSS 1999

Chung Park and Doyoung Park

Introduction

This chapter investigates the factors affecting Korean students' mathematics and science achievement, which is unexpectedly high as compared to the other countries participating in TIMSS 1999 (Martin, Mullis, Gonzalez, Gregory, Smith *et al.*, 2000; Mullis *et al.*, 2000). In addition to the TIMSS 1999 results, the PISA study, undertaken by the OECD (2000) to assess high school students' interdisciplinary intellectual ability, also showed a similar result for Korean students' academic achievement. These findings indicate that the overall level of Korean students' academic achievement is relatively high.

Despite these exceptional performances in international assessments, there has been a growing concern within Korea in recent years about decreasing academic achievement resulting from its weakened and unreliable public schooling (Park *et al.*, 2002). Furthermore, Korea obtained very low achievements in affective assessment, and exhibits a huge discrepancy between boys' and girls' academic achievements. These phenomena seem to be inconsistent with Korean students' high academic achievement and lead us to ask what are the main factors to have affected Korean students' high academic achievement.

In order to investigate these factors, we analyzed the effect of the variables used in the TIMSS school, teacher and student questionnaires on Korean students' achievement. As this study was an exploratory investigation, we used all of the school-, teacher- and student-level variables in TIMSS conceptual frameworks (Martin and Kelly, 1996). The factors in this study were classified into two groups: school-level and student-level exploratory variables. Because Korea showed less than 10 percent variance between schools (Martin, Mullis, Gonzalez, Gregory, Hoyle *et al.*, 2000), this study attempts to confirm the international results of school effects in mathematics and science achievement for Korea.

The research questions addressed in this study are:

1. How much do school factors, teacher characteristics and student variables explain the variance of science and mathematics achievement?
2. What factors in each group are most associated with high science and mathematics achievement?

In order to discern the exploratory factors in the Korean data, the first step was to collate the frequencies of all the possible school, class and student variables. The data were explored to make constructs and then a correlation matrix was made. The correlation matrix was used to identify possible variables related to achievement, to build constructs and to prepare a basic model for further analysis.

Data Analysis

Methods of Analysis

Answering the above questions involved building 10 hierarchical linear models (HLM), which have a two-level data structure consisting of students (level 1) nested within schools or classes (level 2). As only one class per school was sampled, only two levels could be analyzed.

Student-level explanatory variables from the TIMSS 1999 student questionnaire were classified into three input variables (background variables, after-school activities and learning motivation) and two process variables (classroom climate and classroom activities). School-level variables from the TIMSS 1999 school and teacher questionnaires were classified into two input variables (school factors and teacher characteristics) and one process variable (classroom instruction and practice).

The basic model was estimated first, that is, a fully unconditional one-way ANOVA model with no predictor variables was specified. This model broke down the total variance of achievement scores into within-school (level 1) and between-school (level 2) variance to determine the source of variations in academic achievement. After the continuous predictors were centered on their grand means, 10 random-intercept models, in which only the intercepts of level 1 variables vary randomly on level 2, were estimated, with successive inclusion of level 1 and level 2 variable sets to examine the pure effects of process as well as that of input factors.

Data Collected from TIMSS 1999

The present study involved multi-level data, that is, the responses of 150 principals, 193 mathematics teachers, 180 science teachers and 6,130 students (3,080 boys and 3,050 girls) from 150 schools sampled by the two-stage stratified-cluster method for TIMSS 1999 data in February 1999 in Korea. The Korea Institute of Curriculum and

Evaluation report by Kim *et al.* (1999) has more detailed descriptions of the data collection procedures.

Variables
National scaled scores for mathematics and science, with a mean of 150 and a standard deviation of 10, were used as dependent variables, and 51 variables were used as predictor variables. Twenty-eight variables consisted of a combination of several items and their reliability indices ranged from $\alpha = 0.31$ to 0.95. Composite variables made out of dichotomous items were found to have lower reliability. For a more detailed description of the variables, refer to Appendices A–D.

Results

An ANOVA model with no predictor variables was estimated as a preliminary step. As can be seen in Table 12.1, level 1 variances (σ^2) indicating achievement variation within schools were 89.60 (95.57 percent) for mathematics and 90.75 (94.59 percent) for science achievements. Hence, level 2 variances (σ^2) representing achievement variation between schools were only 4.16 (4.43 percent) for mathematics and 5.19 (5.41 percent) for science achievement. These results imply that the effects of school-level predictors would be hard to detect in this sample because there was little level 2 variation.

$$\text{School-level proportion of variance explained \%} = \frac{\tau_{00}(\text{model}_{k-i}) - \tau_{00}(\text{model}_k)}{\tau_{00}(\text{model}_0)} \times 100$$

$$\text{Student-level proportion of variance explained \%} = \frac{\sigma^2(\text{model}_{k-i}) - \sigma^2(\text{model}_k)}{\sigma^2(\text{model}_0)} \times 100$$

Results from Models 1–5 with Level 1 (Student-level) Variables
To calculate the explained variances of levels 1 and 2 and to find out significant variables, each set of variables was cumulatively introduced to the models. As shown in Table 12.1, personal background explained 5.35 percent of level 1 and 48.96 percent of level 2 mathematics variance, and 5.10 percent of level 1 and 57.63 percent of level 2 science variance. Because personal background approximately accounted for half the variance of mathematics and science achievement at school-level, the achievement variation between schools seemed to be related mainly to the fact that some schools consisted of students with high family socio-economic status (SES) and their mother's academic pressure.

According to the result of the fixed effects in the final model 10 (Table 12.2), all of the predictors, that is, students' gender, family SES and mother's academic pressure, showed positive effects on science achievement. In particular, Korean boys

Table 12.1 Summary of each model's variance components.

Model number and number of predictors[a]	Mathematics variance				Science variance			
	Level 1 (σ^2)		Level 2 (σ^2)		Level 1 (σ^2)		Level 2 (σ^2)	
	Value	explained (%)	Value	explained (%)	Value	explained (%)	Value	explained (%)
0. No predictor	89.60	0.00	4.16	0.00	90.75	0.00	5.19	0.00
(ANOVA)	84.81	5.35	2.12	48.96	86.12	5.10	2.20	57.63
1.3 Student background (level 1)	78.26	7.31	1.41	17.01	78.18	8.75	1.60	11.47
2.5 After-school activities (level 1)	59.67	20.75	1.07	8.28	63.89	15.75	1.47	2.57
3.6 Learning motivation (level 1)	58.74	1.03	1.10	–0.76	63.01	0.79	1.49	–0.47
4.5 Classroom discipline (level 1)	57.57	1.31	1.27	–4.00	62.85	0.18	1.33	3.12
5.8 In-class activities (level 1)	57.56	–0.02	0.67	3.68	62.83	0.00	1.00	1.44
6.7 School environment (level 2)	57.57	–0.01	0.57	2.53	62.84	–0.01	0.84	3.14
7.8 School curricula (level 2)	57.54	0.03	0.19	9.07	62.85	–0.01	0.70	2.65
8.9 School climate (level 2)	57.52	0.03	0.06	3.14	62.83	0.02	0.43	5.18
9.12 Teacher characteristics (level 2)	–	–	–	–	–	–	–	–
10.15 Instructional characteristics (level 2)	–	–	–	–	–	–	–	–
Total variance explained (%)	35.81		98.59		30.77		91.72	

Note. [a] Values on left side of this column indicate the model number and its associated predictor variables. For example, 1.3 represents Model 1 with three predictor variables.

scored 1.34 points higher in science than did Korean girls. Mathematics achievement was positively associated only with family SES and mother's academic pressure. That is, there was no gender difference in mathematics.

Compared to the variance explained by personal background, after-school activities explained slightly more level 1 mathematics and science variance (7.31 and 8.75 percent, respectively) but much less level 2 variance (17.01 percent for mathematics and 11.47 percent for science). Table 12.2 shows that high achieving students are more likely to spend their extra time reading books and getting tutoring/instituting, and less likely to play with friends.

Meanwhile, learning motivation was found to be the most powerful predictor of level 1 variance, explaining 20.75 and 15.75 percent of mathematics and science, respectively. Thus, differences in students' achievement within schools may be strongly associated with the variation in their learning motivation regardless of their

Table 12.2 Summary of fixed effects in model 10.

Fixed effects for achievement	Mathematics Coefficient (SD)		Science Coefficient (SD)	
Overall mean achievement (intercept)	151.99	(1.70)[c]	147.02	(1.70)[c]
Level 1 variables				
Student background				
Gender (boy)	0.21	(0.26)	1.34	(0.29)[c]
Family SES	0.18	(0.04)[c]	0.20	(0.04)[c]
Mother's academic pressure	0.36	(0.08)[c]	0.40	(0.09)[c]
After school activities				
Tutoring/institute time	0.19	(0.09)[a]	0.72	(0.11)[c]
Studying time	0.11	(0.16)	0.15	(0.18)
Reading time	0.76	(0.14)[c]	1.26	(0.14)[c]
TV/video watching time	0.12	(0.12)	−0.09	(0.12)
Playing time	−0.83	(0.11)[c]	−0.86	(0.12)[c]
Learning motivation				
Educational aspiration	1.97	(0.15)[c]	1.78	(0.16)[c]
Perception of the importance of study	0.08	(0.08)	0.07	(0.09)
Preference of the subject	0.78	(0.22)[c]	0.98	(0.23)[c]
Using computer to learn math/science	−0.39	(0.08)[c]	−0.29	(0.08)[c]
Confidence about math/science	1.24	(0.05)[c]	1.41	(0.07)[c]
Attitude toward math/science	0.00	(0.08)	0.04	(0.08)
Classroom discipline				
Friends' perception of studying	−0.19	(0.06)[b]	−0.33	(0.07)[c]
Friends' misbehavior	0.29	(0.06)[c]	0.17	(0.07)[a]
Skipping class	−1.04	(0.02)[c]	−1.11	(0.22)[c]
Stolen something	−0.81	(0.15)[c]	−0.63	(0.16)[c]
Alienation from friends	−0.27	(0.27)	0.03	(0.29)
In-class activities				
Teacher's demonstration	0.39	(0.14)[b]	0.20	(0.14)
Note-taking	−0.83	(0.12)[c]	0.18	(0.14)
Teacher's review of homework	−0.11	(0.08)	0.07	(0.09)
Students' discussion of homework	−0.12	(0.09)	−0.20	(0.09)[a]
Quiz/test	0.37	(0.15)[a]	−0.27	(0.16)
Student's work on project or self-study	0.36	(0.09)[c]	0.10	(0.07)
Using OHP	−0.46	(0.11)[c]	−0.15	(0.09)
Using computer	−0.19	(0.12)	0.21	(0.10)[a]
Level 2 variables				
School environment				
Location	0.97	(0.22)[c]	0.46	(0.26)
Shortage of general facilities	−0.12	(0.05)[a]	−0.12	(0.06)[a]
School curricula				
Small grouping in class	0.82	(0.27)[b]	0.21	(0.33)
Teacher characteristics				
Final degree	−1.60	(0.43)[c]	0.36	(0.44)
Instructional characteristics				
Time ratio of homework review	0.02	(0.04)	−0.08	(0.04)[a]

Notes. [a] $p < 0.05$, [b] $p < 0.01$, [c] $p < 0.001$. For the presentation, insignificant variables in level 2 were omitted.

personal background and participation in after-school activities. A relatively large amount of level 2 mathematics variance (8.28 percent) was also accounted for by students' learning motivation.

The fixed effects in the final model (Table 12.2) revealed that educational aspiration and confidence in mathematics and science had the largest effect on students' achievement (1.97 for mathematics and 1.78 for science). Also students' achievement was positively affected by their preference for the two subjects. Unexpectedly, studying with computers had a negative effect on mathematics and science achievement. Internet use may distract students.

The variance explained by classroom discipline and in-class activities was minimal and negligible. However, attention should be given to some significant variables providing unexpected results. Although the positive effect of friends' learning value and the negative effect of friends' misbehavior had been expected, their directions were reversed. It is difficult to explain this result, but one possible explanation may be that low achievement students tend to overestimate their friend's perception of learning, while high achievement students tend to overestimate their friends' misbehavior.

Among in-class activities, note-taking and overhead projector (OHP) usage showed a significant but negative effect on mathematics achievement. In other words, the more frequently students took notes and mathematics teachers used OHP, the worse the students performed on mathematics achievement tests. It can be inferred from these results that a simple note-taking without understanding would not contribute to the mathematics proficiency. The positive effect of teachers' demonstration on achievement needs to be considered in conjunction with teachers' OHP usage. Because most Korean students are accustomed to teachers' demonstration of problem solving on the blackboard, teachers' alternative OHP usage in classes might distract students' understanding of problems or tasks.

In contrast, science achievement showed positive association with teachers' computer usage in classes. Given the fact that science teaching in Korea utilizes multimedia devices and computer use may be one of the effective ways in teaching and learning science. It was also an unexpected result that students' reporting and frequent discussions of homework had a negative effect on science achievement. It seems that students might not really join the discussion but just chat with friends, as if it were free time.

Results from Models 6–10 with Level 2 (School-level) Variables

Regarding these models, most of the 51 variables showed no statistical significance, with the exception of school location, facility shortage, teacher's final degree and time ratio for homework check. Such a result is understandable, given that there was not much variance at the school-level as the ANOVA model suggested. Hence, it was expected that level 2 variables would not show significant association with academic achievement.

Nonetheless, it should be noted that school location (remote = 1, downtown = 4) exercised a relatively large effect on mathematics and science achievements (0.97 for mathematics and 0.46 for science). Korean students are automatically allocated to the middle school nearest to their home and as such this makes the school location represent an approximate average of students' SES. Therefore, it seems that students' family and economic background is related to the achievement differences between schools to some degree.

Small grouping in class was also positively associated with differences in mathematics achievement between schools. That is, schools that apply small group activities tend to have higher school average than schools that do not. This may also be closely related to the positive effects of two student-level variables (in class quiz/test and problem solving in class). More specifically, high achievement in mathematics could be attributed to schools that apply small groupings in classes and to mathematics teachers who use quizzes or tests and their demonstrations in class.

Surprisingly, mathematics teachers' final degree had a strong negative effect on the school mean mathematics achievement (-1.60). In other words, the higher the mathematics teachers' degree, the lower the school achievement in mathematics. Taking into consideration the tendency of teachers to enter graduate schools to obtain some credits required for their promotion, we can infer that mathematics teachers who pursue higher degrees just for their promotion are likely to have less enthusiasm for teaching, and this may have negative effect on their teaching.

Summary and Discussion

This study was conducted to show the extent to which individual factors were related to Korean students' achievement in mathematics and science by using TIMSS 1999 data. To achieve this purpose, 39 school factors, 12 teacher characteristics and 27 student variables were used as predictors to explain the variances of students' mathematics and science achievement.

The results presented in this chapter show that the extent to which the factors affected mathematics and science achievement was similar in both subjects except for gender and teaching characteristics. In what follows, two major findings will be described.

First, there was almost no variation in achievement among Korean schools. Within the school variance 4.43 percent in mathematics and 5.41 percent of science achievement are attributed to school variables; as is consistent with that of Martin *et al.* (2000a). Korea was excluded in their study because of the small variance among schools (below 10 percent). While there were few school effects in Korea due to small variance of achievement between schools, the location of school seemed to be the most important factor that explained the variance of achievement between schools. The effect of the location of the school reflects that schooling cannot moderate students' home background and then maintains the inequality of students' SES. Even if school effectiveness and school-related variables seem to be unimportant, further study using specific variables to find effective school practices among schools is still needed to provide more detailed information.

One of the reasons that the variance of achievement among schools is so small in Korea may be found in its centralized system of schooling. The Equalization Policy under this centralized school system in Korea was designed to reduce school gaps as well as regional gaps in the government's educational service. Additionally, nine years of compulsory education under a national curriculum is compulsory. The government attempted to give more autonomy to the local schools as far as their curriculum management was concerned following implementation of the 7th National Curriculum Reform in 1997. Yet, most of the junior high schools in Korea still continue to follow the instructions given by both the Department of Education and their local boards of education, so that it was very hard to identify characteristics or differences among the schools across Korea.

In addition, students in Korea were arbitrarily assigned to schools, whether public or private, in accordance with their residential districts. Teachers in public schools are supposed to move to a different school every four years. The government's support of private schools is not much different from that of public schools in terms of budget management and curriculum organization. All these facts seem to have contributed to the small variances in achievement between schools in Korea.

Second, the 95 percent variance in mathematics achievement scores at the student level can be divided into 5.35 percent (student's home background), 7.31 percent (after-school activities), 20.75 percent (learning motivation), 1.31 percent (in class activities) and 1.03 percent (classroom discipline). The 5 percent variance from school factors can be divided into 69.49 percent (student level variables), 16.86 percent (school factors), 9.07 percent (teacher characteristics) and 3.14 percent (in class activities) (see Figure 12.1).

In the case of science achievement, the variance at the student level can be divided into 5.1 percent (student's home background), 8.75 percent (after-school activities), 15.75 percent (learning motivation), 0.18 percent (in class activities) and 0.97 percent (classroom norm). The variance discernible from school factors can be divided into 74.2 percent (student-level variables), 9.57 percent (school factors), 2.65 percent (teacher characteristics) and 5.18 percent (in class activities) (see Figure 12.2).

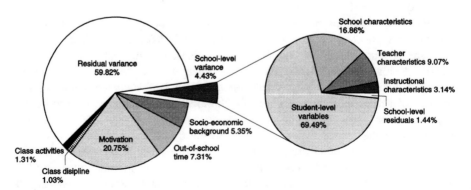

Figure 12.1 Proportions of mathematics achievement variance.

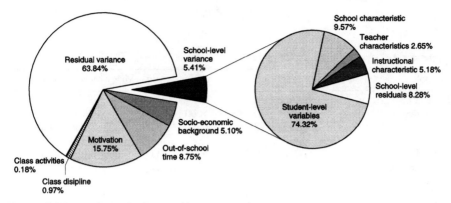

Figure 12.2 Proportions of science achievement variance.

The most active factor in Korean students' achievement was student's learning motivation: namely, educational aspiration for the future, confidence in subject and preference for subject. The fact that 'educational aspiration for the future' is the most active factor reflects the academic sectarianism in Korean society and explains the very low affective achievement result. Korean students outperformed in cognitive achievement but under-performed in affective achievement. Confidence in and preference for the subject may help students' future learning. Given these results, it can be argued that it is desirable to study methods to increase Korean students' intrinsic motivation for learning mathematics and science. While learning motivation is the most important factor in academic achievement, it also moderates the effect of home background on achievement and learning contextual factors, such as grouping in class, teaching method for each subject and reading time.

Classroom activities and teacher characteristics, compared to school factors, also accounted for a substantial amount of variation, although to different degrees for science and mathematics. While teachers' demonstration and problem solving in the class positively affect mathematics achievement, both note-taking and computer use in class had a negative affect. The effect of computer use in class however has a positive effect on science achievement.

This implies that classroom practices may influence achievement, and that indicates that properly tailored classroom practices can be employed to accommodate students' ability. To improve student learning in classroom, we need a further detailed study of those. Since the teachers' role for students' learning is so important, a substantial effort to enhance the qualifications of teachers is needed.

Finally, there was a substantial difference in science achievement between boys and girls, compared to the difference in mathematics. Differences between boys and girls in science achievement remain, while the difference in mathematics seems to have disappeared between TIMSS 1995 and 1999. A further study is needed to identify the reasons for boys faring better than girls in science achievement and the reasons for the disappearance of gender difference in mathematics.

Appendices

Appendix A. Level 1 mathematics variables.

Variable	N	M	SD	Min	Max	Description
Student background						
Gender	6,111	0.5020	0.5000	0.00	1.00	0 = girl, 1 = boy
Family SES	6,019	8.3814	3.0380	2.00	14.00	2 item composite, $\alpha = 0.79$
Mother's academic pressure	6,037	9.7134	1.5521	3.00	12.00	3 item composite, $\alpha = 0.85$
After school activities						
Tutoring/institute time	5,967	1.9591	1.3040	1.00	5.00	
Math study time	6,044	1.8994	0.7785	1.00	5.00	
Reading time	6,018	1.8769	0.8289	1.00	5.00	
TV/video time	6,062	3.3971	1.0059	1.00	5.00	
Playing time	6,051	2.4449	0.9887	1.00	5.00	
Learning motivation						
Educational aspiration	6,093	4.6921	0.7868	1.00	5.00	
Learning value	6,060	9.5477	1.7371	3.00	12.00	3 item composite, $\alpha = 0.88$
Math preference	6,084	2.5809	0.8041	1.00	4.00	
Math studying with computer	6,079	1.2168	1.3164	0.00	4.00	
Confidence about math	6,059	13.2220	2.7986	5.00	20.00	5 item composite, $\alpha = 0.79$
Attitude toward mathematics	6,040	8.7233	2.1150	4.00	16.00	4 item composite, $\alpha = 0.76$
Classroom discipline						
Friends' perception of studying	6,060	8.6017	1.9518	3.00	12.00	3 item composite, $\alpha = 0.94$
Friends' misbehavior	6,011	5.1865	1.9798	3.00	12.00	3 item composite, $\alpha = 0.61$
Skipping class	6,067	1.2060	0.5760	1.00	4.00	
Stolen something	6,054	1.7332	0.8436	1.00	4.00	
Alienation from friends	6,054	1.1176	0.4085	1.00	4.00	
In-class activities						
Teacher's demonstration	6,087	3.3841	0.8154	1.00	4.00	
Note-taking	6,094	2.7630	0.9883	1.00	4.00	
Teacher's review of homework	6,051	5.2712	1.5926	2.00	8.00	2 item composite, $\alpha = 0.65$
Discussion of homework	6,049	3.1585	1.3283	2.00	8.00	2 item composite, $\alpha = 0.48$
Quiz and test	6,084	2.1506	0.7845	1.00	4.00	
Work on project or self-study	6,058	4.6834	1.4381	2.00	8.00	2 item composite, $\alpha = 0.48$
Using OHP	5,994	2.7100	1.1326	2.00	8.00	2 item composite, $\alpha = 0.68$
Using computer	6,049	2.5502	0.9731	2.00	8.00	$\alpha = 0.47$
Mathematics achievement score	6,111	150.0871	9.9853	116.51	175.09	Scaled by IRT

Appendix B. Level 1 science variables.

Variable	N	M	SD	Min	Max	Description
Student background						
Gender	6,111	0.5020	0.5000	0.00	1.00	0 = girl, 1 = boy
Family SES	6,019	8.3814	3.0380	2.00	14.00	2 item composite, $\alpha = 0.79$
Mother's academic press	6,037	9.7134	1.5521	3.00	12.00	3 item composite, $\alpha = 0.85$
After-school activities						
Tutoring/institute time	5,798	1.6054	1.0977	1.00	5.00	
Math study time	6,017	1.6837	0.7198	1.00	5.00	
Reading time	6,018	1.8769	0.8289	1.00	5.00	
TV/video time	6,062	3.3971	1.0059	1.00	5.00	
Playing time	6,051	2.4449	0.9887	1.00	5.00	
Learning motivation						
Educational aspiration	6,093	4.6921	0.7868	1.00	5.00	
Learning value	6,060	9.5477	1.7371	3.00	12.00	3 item composite, $\alpha = 0.88$
Math preference	6,091	2.5472	0.7500	1.00	4.00	
Math studying with computer	6,070	1.3725	1.4087	0.00	4.00	
Confidence about science	6,060	10.6295	2.1564	4.00	16.00	4 item composite, $\alpha = 0.79$
Attitude toward science	6,051	8.9316	2.1359	4.00	16.00	4 item composite, $\alpha = 0.75$
Classroom discipline						
Friends' perception of studying	6,060	8.6017	1.9518	3.00	12.00	3 item composite, $\alpha = 0.94$
Friends' misbehavior	6,011	5.1865	1.9798	3.00	12.00	3 item composite, $\alpha = 0.61$
Skipping class	6,067	1.2060	0.5760	1.00	4.00	
Stolen something	6,054	1.7332	0.8436	1.00	4.00	
Alienation from friends	6,054	1.1176	0.4085	1.00	4.00	
In-class activities						
Teacher's demonstration	6,076	3.0583	0.8769	1.00	4.00	
Note-taking	6,076	3.2581	0.9462	1.00	4.00	
Teacher's review of homework	6,038	4.9232	1.5590	2.00	8.00	2 item composite, $\alpha = 0.66$
Discussion of homework	6,020	3.1940	1.3917	2.00	8.00	2 item composite, $\alpha = 0.59$
Quiz and test	6,065	2.1716	0.8390	1.00	4.00	
Work on project or self-study	5,964	7.4323	2.0390	3.00	12.00	3 item composite, $\alpha = 0.54$
Using OHP	6,006	3.2051	1.5202	2.00	8.00	2 item composite, $\alpha = 0.62$
Using computer	6,046	2.7686	1.3151	2.00	8.00	2 item composite, $\alpha = 0.64$
Science achievement score	6,111	150.0769	9.9936	103.52	194.15	Scaled by IRT

Appendix C. Level 2 mathematics variables.

Variable	N	M	SD	Min	Max	Description
School characteristics						
Environment						
School location	150	3.4067	0.6864	2.00	4.00	1 = remote, 4 = downtown
Student number per teacher	150	24.4024	4.9566	10.56	51.83	
Teaching continuity	150	1.4653	0.6010	1.00	4.00	
Number of computers	150	39.7973	23.3945	0.00	150.00	
Internet-accessible computers	150	1.7667	1.5987	0.00	5.00	
Shortage of facilities	150	9.2603	2.8404	4.00	16.00	4 item composite, $\alpha = 0.70$
Poor quality of teachers	150	2.2517	0.8811	1.00	4.00	
Curricula						
Teaching days per year	150	223.8811	4.2828	193.00	232.00	
Lessons per week	150	27.9028	5.1296	15.00	45.00	
School's math curriculum	150	0.6600	0.4753	0.00	1.00	0 = no, 1 = yes
School's math text	150	0.5000	0.5017	0.00	1.00	0 = no, 1 = yes
Ability grouping class	150	0.6400	0.4753	0.00	1.00	0 = no, 1 = yes
Small group class	150	0.3867	0.4886	0.00	1.00	0 = no, 1 = yes
Enriched class	150	0.2600	0.4401	0.00	1.00	0 = no, 1 = yes
Supplementary class	150	0.2467	0.4325	0.00	1.00	0 = no, 1 = yes
Organizational climate						
Principal's responsibility	150	4.5385	2.2726	1.00	13.00	
Deputy-head's responsibility	150	12.1275	1.3275	7.00	14.00	
Teachers' responsibility	150	8.4698	2.3984	2.00	14.00	
Commission's responsibility	150	9.1879	2.2024	0.00	14.00	
Staff's cooperation	150	2.4133	0.7253	0.00	3.00	3 item composite, $\alpha = 0.33$
Principal's work hours	150	132.4880	87.2877	6.00	500.00	
Parental participation	150	4.5918	2.1261	0.00	10.00	10 item composite, $\alpha = 0.67$
Students' absence	150	0.5556	0.6832	0.00	4.00	
Students' misbehavior	150	36.4931	15.0716	18.00	86.00	18 item composite, $\alpha = 0.95$
Teacher characteristics						
Final degree	150	3.1389	0.3209	3.00	4.00	1 = middle, 4 = master/doctor
Trained year	150	3.9384	2.3203	0.00	19.00	
Teaching career	150	12.6839	8.2290	1.00	41.00	
Classes per week	150	19.3706	2.3820	12.00	26.00	
Office work per week	150	5.0015	3.8302	0.00	20.00	
Teaching work per week	150	24.5249	8.0361	1.00	60.00	
Familiarity with national doc	150	7.2647	1.7458	3.00	12.00	3 item composite, $\alpha = 0.83$
Familiarity with regional doc	150	9.6951	1.9995	4.00	15.00	4 item composite, $\alpha = 0.79$

Readiness for teaching	150	42.5453	4.7573	27.00	48.00	12 item composite, $\alpha = 0.91$
Discussion with math teachers	150	4.4311	1.5881	1.00	7.00	
Teachers' effects	150	11.0267	2.5928	4.00	16.00	4 item composite, $\alpha = 0.71$
Job satisfaction	150	2.0711	0.8835	0.00	3.00	3 item composite, $\alpha = 0.45$
Instructional characteristics						
Teaching time per week	150	180.7192	41.8790	35.00	586.00	
OTL (units covered)	150	32.8233	9.5971	13.00	58.00	
Text use	150	3.2434	0.7445	1.00	4.00	
Computer use	150	2.5529	0.7590	2.00	7.00	2 item composite, $\alpha = 0.31$
Whole class activity	150	5.4020	1.2316	2.00	8.00	2 item composite, $\alpha = 0.45$
Individual activity	150	4.9911	1.1828	3.00	8.00	2 item composite, $\alpha = 0.59$
Small group activity	150	3.9249	1.0927	2.00	8.00	2 item composite, $\alpha = 0.51$
Lecture	150	33.1220	16.7310	0.00	80.00	
Exercise	150	35.7433	12.1952	11.00	80.00	
Homework check	150	6.4779	3.7255	0.00	20.00	
Quiz and test	150	6.6167	3.9378	0.00	20.00	
Students' individual differences	150	4.5223	1.0477	2.00	8.00	2 item composite, $\alpha = 0.52$
Students lacking motivation	150	8.6232	1.7468	3.00	12.00	3 item composite, $\alpha = 0.74$
Shortage of teaching facilities	150	7.0491	1.9369	–	12.00	3 item composite, $\alpha = 0.75$
Teachers' low morale	150	2.0000	–	–	4.00	

Appendix D. Level 2 science variables.

Variable	N	M	SD	Min	Max	Description
School characteristics						
Environment						
School location	150	3.4067	0.6864	2.00	4.00	1 = remote, 4 = downtown
Student number per teacher	150	24.4024	4.9566	10.56	51.83	
Teaching continuity	150	1.5035	0.6332	1.00	4.00	
Number of computers	150	39.7973	23.3945	0.00	150.00	
Internet-accessible computers	150	1.7667	1.5987	0.00	5.00	
Shortage of facilities	150	9.2603	2.8404	4.00	16.00	4 item composite, $\alpha = 0.70$
Poor quality of teachers	150	2.2432	0.8946	1.00	4.00	
Curricula						
Teaching days per year	150	223.8811	4.2828	193.00	232.00	
Lessons per week	150	27.9028	5.1296	15.00	45.00	

Appendix D. Level 2 science variables (continued).

Variable	N	M	SD	Min	Max	Description
School's science curriculum	150	0.6600	0.4753	0.00	1.00	0 = no, 1 = yes
School's science text	150	0.4267	0.4962	0.00	1.00	0 = no, 1 = yes
Ability grouping class	150	0.2333	0.4244	0.00	1.00	0 = no, 1 = yes
Small group class	150	0.3800	0.4870	0.00	1.00	0 = no, 1 = yes
Enriched class	150	0.2133	0.4110	0.00	1.00	0 = no, 1 = yes
Supplementary class	150	0.1667	0.3739	0.00	1.00	0 = no, 1 = yes
Organizational climate						
Principal's responsibility	150	4.5385	2.2726	1.00	13.00	
Deputy-head's responsibility	150	12.1275	1.3275	7.00	14.00	
Teachers' responsibility	150	8.4698	2.3984	2.00	14.00	
Commission's responsibility	150	9.1879	2.2024	2.00	14.00	
Staff's cooperation	150	2.4133	0.7253	0.00	3.00	3 item composite, $\alpha = 0.33$
Principal's work hours	150	132.4880	87.2877	6.00	500.00	
Parental participation	150	4.5918	2.1261	0.00	10.00	10 item composite, $\alpha = 0.67$
Students' absence	150	0.5556	0.6832	0.00	4.00	
Students' misbehavior	150	36.4931	15.0716	18.00	86.00	18 item composite, $\alpha = 0.95$
Teacher characteristics						
Final degree	150	3.2011	0.3746	3.00	4.00	1 = middle, 4 = master/doctor
Trained year	150	4.0861	2.0515	0.00	15.00	
Teaching career	150	12.7244	7.6092	1.00	35.00	
Classes per week	150	19.4378	2.6593	12.00	27.00	
Office work per week	150	5.3889	3.6352	0.00	24.00	
Teaching work per week	150	27.1091	9.7399	4.00	80.00	
Familiarity with national doc	150	7.3100	1.5640	3.00	12.00	3 item composite, $\alpha = 0.82$
Familiarity with regional doc	150	9.8865	1.8621	4.00	14.00	4 item composite, $\alpha = 0.77$
Readiness for teaching	150	30.4671	3.9073	17.00	39.00	10 item composite, $\alpha = 0.76$
Discussion with other teachers	150	4.6444	1.6309	1.00	7.00	
Teachers' effects	150	11.4185	2.4117	5.00	16.00	4 item composite, $\alpha = 0.68$
Job satisfaction	150	2.3345	0.8582	0.00	3.00	3 item composite, $\alpha = 0.57$
Instructional characteristics						
Teaching time per week	150	160.6729	43.4324	45.00	360.00	
OTL (units covered)	150	23.8994	8.8949	5.50	54.50	
Text use	150	2.8670	0.9987	1.00	4.00	
Computer use	150	2,7700	0.8612	2.00	6.00	2 item composite, $\alpha = 0.60$
Whole class activity	150	5.3174	0.9918	3.00	8.00	2 item composite, $\alpha = 0.43$

Individual activity	150	4.0849	0.8232	2.00	7.00	2 item composite, $\alpha = 0.43$
Small group activity	150	3.8878	0.8917	2.00	8.00	2 item composite, $\alpha = 0.43$
Lecture	150	34.6477	17.5179	0.00	80.00	
Exercise	150	14.9656	8.9307	0.00	45.00	
Experimentation	150	24.7253	14.3422	0.00	80.00	
Homework check	150	5.5830	4.6927	0.00	39.67	
Quiz and test	150	5.2564	4.0844	0.00	30.00	
Students' individual differences	150	4.8673	0.9554	2.00	7.50	2 item composite, $\alpha = 0.58$
Students lacking motivation	150	8.3393	1.5976	4.00	12.00	3 item composite, $\alpha = 0.72$
Shortage of teaching facilities	150	6.7789	1.9376	3.00	12.00	3 item composite, $\alpha = 0.85$
Teachers' low morale	150	2.1925	0.8407	1.00	4.00	

References

Kim, S., Lee, C., Lim, C., Yoo, J. and Seo, D. (1999) *Korean national technical report for TIMSS 1999. Research report of KICE*. Seoul: KICE (in Korean).

Martin, M.O. and Kelly, D.L. (eds.) (1996) *Third International Mathematics and Science Study Technical Report: Design and Development*. Boston, MA: Boston College, Center for the Study of Testing, Evaluation, and Educational Policy.

Martin, M.O., Mullis, I.V.S., Gonzalez, E.J., Gregory, K.D., Hoyle, C. and Shen, C. (2000) *Effective schools in science and mathematics: IEA's Third International Mathematics and Science Study*. Boston, MA: Boston College, Center for the Study of Testing, Evaluation, and Educational Policy.

Martin, M.O., Mullis, I.V.S., Gonzalez, E.J., Gregory, K.D., Smith, T.A., Kelly, D.L., Chrostowski, S.J., Garden, R.A. and O'Connor, K.M. (2000) *TIMSS International Science Report: Findings from IEA's Repeat of the Third International Mathematics and Science Study at the Eighth Grade*. Boston, MA: Boston College, Center for the Study of Testing, Evaluation, and Educational Policy.

Mullis, I.V.S., Martin, M.O., Gonzalez, E.J., Gregory, K.D., Garden, R.A., O'Connor, K.M., Chrostowski, S.J. and Smith, T.A. (2000) *TIMSS International Mathematics Report: Findings from IEA's Repeat of the Third International Mathematics and Science Study at the Eighth Grade*. Boston, MA: Boston College, Center for the Study of Testing, Evaluation, and Educational Policy.

OECD (2000) *Knowledge and Skills for Life: First results from OECD programme for International student Assessment* (PISA) 2000. Paris: OCED.

Park, C., Chae, S., Kim, M. and Choi, S. (2002) *How does Korean students' achievement in the international comparison study*. Seoul: KICE (in Korean).

Part 3

Background Variables and Achievement: Multiple Country Studies

13

Exploring the Factors that Influence Grade 8 Mathematics Achievement in The Netherlands, Belgium Flanders and Germany: Results of secondary analysis on TIMSS 1995 Data

Klaas T. Bos and Martina Meelissen

Introduction

Since 1964, the International Association for the Evaluation of Educational Achievement (IEA) has organized several international comparative achievement studies in different core subjects, such as mathematics, science and reading. IEA recognizes two main goals of its achievement studies (Plomp, 1998):

1. to provide policy-makers and educational practitioners with information about the quality of their education system in relation to relevant reference systems;
2. to assist in understanding the reasons for observed differences between education systems.

Each goal requires its own kind of comparison. The first goal asks primarily for international comparisons of test scores at a descriptive level on international achievement tests. It also includes a comparison between countries of contextual indicators referring to educational processes at different levels (student, class/teacher, school and country levels). Examples of these indicators are students' attitude towards the tested subject, the pedagogical approach of the teacher, the educational infrastructure of the school and characteristics of the national educational policy.

The second goal refers to seeking explanations for the described differences in achievement within and – more importantly – across nations. This goal can be approached by analyzing the possible relationships of the context indicators with achievement in an international comparative context. This objective is far from simple. Studies of the effectiveness of education showed that the identification of factors on different levels (student, teacher/class and school level), influencing students' achievement within a country is complicated enough as it is (Scheerens and Bosker, 1997). Finding explanations for differences in achievement between nations means that an extra level is added to an already very complex model.

This chapter focuses on the second goal of the IEA's studies: understanding the reasons for the differences in students' achievement between relevant reference countries. In order to do this the Grade 8 TIMSS 1995 data from three neighbor educational systems (Germany, Belgium Flanders and the Netherlands) were explored. Table 13.1 shows that Belgium Flanders performed significantly better (based on a Bonferroni procedure for multiple comparisons; $p < 0.05$; Gonzalez, 1996) in the mathematics test than the Netherlands (the scores were standardized for all participating countries with a mean of 500 and a standard deviation of 100, Adams and Gonzalez, 1996). Germany performed significantly worse compared to Belgium Flanders and the Netherlands.

The goal of this study was to find out whether it is possible to explain these cross-national differences in students' achievement in mathematics at Grade 8, with the available TIMSS data.

Design and Research Question

Based on the differences in achievement between the Netherlands, Belgium Flanders and Germany in the Grade 8 TIMSS 1995 mathematics test, the following research question is addressed in this article: To what extent can variances in the overall

Table 13.1 Composition of research group and mean (SD) of TIMSS mathematics weighted test scores in grade 8 Belgium Flanders, Germany and the Netherlands.

Education system	Number of students	Number of teachers (schools)	Mathematics achievement TIMSS test (Spring 1995)	
			M	s.e.
Belgium Flanders (Bfl)	2,748	147 (147)	586	5.7
Germany (Ger)	2,020	94 (94)	515	4.5
Netherlands (Nld)	1,814	88 (88)	551	6.7
Total (pooled dataset)	6,582	329 (329)	555	5.6

Notes. The column 'students' contains the number of students that could be linked to the teachers who are presented in the next column.

TIMSS mathematics test scores for the upper grade in TIMSS population 2 (Grade 8) in the Netherlands, Belgium Flanders and Germany be explained by variances in the scores on variables at student and class/teacher level?

Related questions to the main research question are:

- Which variables at student and classroom levels are associated with mathematics achievement in the Netherlands, Belgium Flanders and Germany at Grade 8?
- What can be learned from the differences and similarities across the three education systems, in the outcomes of student and class variables that turned out to be predictors of achievement in mathematics?

The dependent variable in this study is students' results on the TIMSS 1995 mathematics achievement test for Grade 8 secondary education. Information on the development, international reliability and validity of the TIMSS achievement test can be found in Beaton *et al.* (1996) and Kuiper *et al.* (1999). The analyses of TIMSS data in this study are limited to the exploration of the student and class/teacher characteristics. The school questionnaire data were not included in the explorations, because the school data set from the Netherlands contain too many missing values (fewer than 75 percent of the selected principals returned the questionnaire).

To address the research questions, it is important to identify which variables measured in TIMSS are potentially influencing mathematics achievement in the three education systems. Particularly, the investigation is directed to the variables that can be manipulated by policy-makers, school managers and teachers in order to improve student achievement. The first step in the exploration of the TIMSS data was to develop a conceptual framework that could be used as a guide for the explorative data analysis.

Organizing Conceptual Framework

The conceptual framework that is used in TIMSS was based on the one of its predecessors – the Second International Mathematics Study (SIMS, Travers and Westbury, 1989). In that framework three curriculum levels are distinguished: the intended (what students should know), the implemented (what students are taught) and the attained (what students have learned). Each level is influenced by curricular antecedents and contextual factors. Country or system characteristics are supposed to be related to the intended level; school, teacher and class characteristics are related to the implemented level and students' characteristics are related to the attained curriculum.

The three curriculum level framework of TIMSS has been reviewed by Bos (2002). He concluded that the framework was too general. Although the framework assumed that on each curriculum level factors are directly or indirectly influencing achievement, no concrete definitions of the factors were available. He also criticized the background questionnaires applied in TIMSS because of their unknown correspondence with factors mentioned in the framework. It was not very clear which questions in the various questionnaires were operationalizations of the mentioned factors.

In order to conduct the explorative data analysis in this study, the TIMSS framework needed to be developed further. A more specific framework would also provide insight into what the TIMSS data could offer (in terms of availability of relevant context factors) for finding explanations for the differences in achievement between the three countries. Therefore, the three curriculum level conceptual framework of TIMSS was extended by potentially effectiveness enhancing factors derived from a literature search on instructional and school effectiveness (Creemers, 1994; Scheerens and Bosker, 1997).

In Figure 13.1 this organizing conceptual framework is presented. The framework is identified as 'organizing', because it was primarily used to categorize theoretically important factors in clusters that can guide the search for operationalizations of factors belonging to the different clusters. The clusters were organized (categorized) in such a way that assumptions about relationships between the clusters can be derived from it. The framework is not meant to formulate hypotheses about the relationships of *individual* factors within and across the clusters.

The framework is based on three curricular dimensions and the four levels of education. The curricular antecedents (mostly unchangeable by policy-makers, school or teacher) and context (changeable) factors are conditional for the curricular content factors.

Potentially Effective Factors Indicated in TIMSS Instruments

The student and teacher questionnaires from TIMSS were examined to identify items or sets of items which – based on their content – could be operationalizations of factors in the organizing conceptual framework. In Table 13.2 the results are presented in the form of a list of potentially effective educational factors and their possible indicators available in TIMSS instruments (see also Bos, 2002). The format of the list is in accordance with the curricular dimensions of the organizing framework. For some of the factors in the framework no items could be found in the TIMSS instrumentation. These factors were excluded from Table 13.2. Particularly, at class level no indicators were available in TIMSS instruments for the factors about characteristics of the curricular materials such as 'explicitness and the ordering of goals and content', 'structure and clarity of content' and 'use of advance organizers'.

For a few other factors, only so-called 'proxy items' could be found in the TIMSS data. These items do not directly refer to a factor but are possible indicators of a factor (for example, 'number of books at home' is regarded as an indicator for the educational level of the parents; more appropriate measures could not be found in the TIMSS data).

Data Analysis

The data analysis conducted to address the research question consisted of three steps:

1. Exploring the factors (composites) of Table 13.2, resulting in scale scores and their associated statistical reliability. The lower bound for the Cronbach α to keep

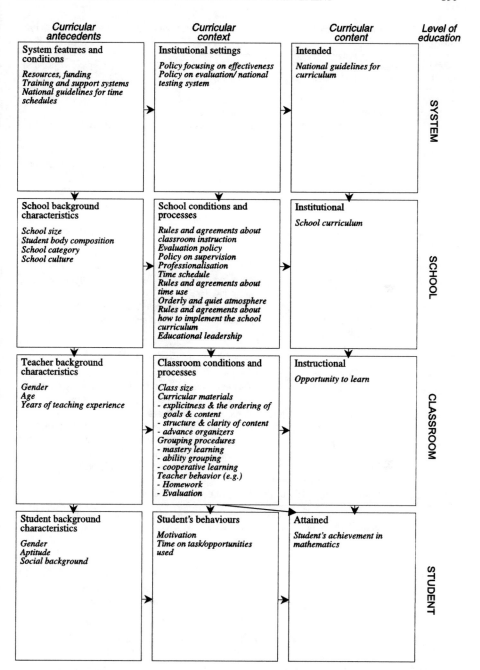

Figure 13.1 Organizing conceptual framework filled in with potential effectiveness enhancing factors derived from literature search on instructional and school effectiveness.

Table 13.2 Potentially effective educational factors and their possible indicators available in TIMSS instruments.

Level of education	Potentially effective factors in framework	Indicators in TIMSS instruments: (sets of) items and description
Student		
Antecedent	Gender	Student's sex
	Social background	Out-of-school activities; number of books at home
Context	Motivation	Attitude towards mathematics; success attribution; maternal academic expectation; friends' academic expectation
	Time on task	Number of minutes math/week; amount of homework per day; number of after school math classes per week
Content	Attained curriculum	Achievement in mathematics (TIMSS test)
Class		
Antecedent	Teacher background characteristics	Teacher's gender; teaching experience; % mathematics lessons a teacher is assigned
	Evaluation materials for student outcomes, feedback & corrective instruction	Textbooks used for mathematics; assessment features (standardized test vs. more subjective assessment)
	Class size	Number of students in tested class
Context	Cooperative learning	Frequency of 'working in pairs or small groups'
	Teaching style	Teacher's teaching style as perceived by students (student oriented versus teacher centered)
	Management and orderly and quiet atmosphere	Perceived class climate; perceived school climate (safety); limitations to teach the tested class related to student/resources/parental features
	Homework	Frequency of homework; follow up
	Evaluation, feedback and corrective instruction	Use of assessment results for different goals
Content	Implemented curriculum	Content coverage mathematics

Sources. IEA Three curriculum level conceptual framework (Travers and Westbury, 1989; Creemers, 1994; Scheerens and Bosker, 1997).

a scale in the data analysis was 0.60 (in datasets of all education systems). This rather low bound was allowed because the data analysis in this article is explorative and because of the international character of the TIMSS instruments (Pelgrum and Anderson, 1999). The difference of the mean score of the composite variables across each pair of the three education systems was tested by means of the Bonferroni post hoc pairwise multiple comparison test ($p < 0.01$). Also in this first step, correlation matrices of the bivariate correlations of the explored composites and singletons and mathematics achievement and matrices of the intercorrelations of explored composites and singletons were analyzed.

2. Exploring direct and indirect relationships between background factors and mathematics achievement by means of explorative path analysis for each level.

For the analyses on teacher level, students' achievement scores were aggregated.

3. Estimating an explorative hierarchical linear model by means of multi-level analysis in which the variance in mathematics achievement explained by factors (variables) at student level is separated from the variance explained by factors at class level.

Steps 1 and 2 can be seen as preparatory steps for step 3 and their results are reported in Bos (2002). However, one remark has to be made about the exploration of composites (step 1) in the datasets of the three countries. It turned out that for some composites the Cronbach α differed substantially across the three countries. For example, the statistical reliability of 'perceived school climate' (safety) is in Germany quite strong (0.75) and in the Netherlands rather weak (0.59). Other composites are only statistically reliable in one of two of the countries. The differences in reliability of composites between countries is an indication of one of the major difficulties that occur when survey data across countries is compared; the international content validity of the data. In TIMSS a lot of attention is given to the translation of the English instruments in the languages of the tested countries. But, even if the translation of the instruments is carried out with great care, social and cultural differences across countries could lead to different interpretations of the questions by the respondents. In the case of 'perceived school climate', students will judge the seriousness of violence in school in different ways, dependent on what is regarded as (un)usual in a country (Bos, 2002). In TIMSS, attention for international content validity is mostly directed to the TIMSS-test, and far less to the questionnaires.

Step 3 is required because the TIMSS data were collected at three different levels of education (student, class/school and country levels). The most appropriate technique to analyze the data is hierarchical linear modeling (HLM) in which this nested design of the datasets is taken into account (Snijders and Bosker, 1999). The major advantage of HLM techniques (for example, multi-level analysis) over unidimensional techniques, such as partial least squares techniques (PLS), is the estimation of the effects of variables on the dependent variable at one level (for example, student level) taking into account at the same time the effect of variables on the dependent variable at another level of the hierarchical data structure (for example, class level). Multi-level analysis results in better estimation of the amount of variance in the output variable every variable in the model can explain. The direct effects of each variable at each level can be estimated, but also direct effects of class/school variables on student variables can be estimated. Next to direct interaction effects, indirect interaction effects of class/school variables on the effect of a student variable on mathematics achievement can be specified. An example of the latter is the effect of the amount of mathematics topics a teacher has covered in his lessons (class level) on the relationship between student's attitude towards mathematics and mathematics achievement.

In this study the MLn program (Woodhouse, 1995) was used to specify two-level country models: students at level 1 and schools (classes) at level 2. A hierarchical random intercept (effects) model was specified for each country. In TIMSS studies schools are grouped with classrooms because within each selected school only one intact classroom was selected.

Requirements of the Datasets

The basic requirements for the datasets that were explored are that the data are standardized and weighted and that the datasets do not contain any missing value. Therefore, the cases with a missing value on the dependent variable – the mathematics score on the international TIMSS test – were removed from the dataset. Furthermore, cases with a valid answer on the dependent variable, but missing values on a majority of the other variables, were removed as well. For some cases, imputation was applied: the mean score of all the cases on the variable (or the class mean of the variable) was used to replace the missing value(s), provided that the percentage missing values did not exceed 20 percent. Variables with more than 20 percent missing values were excluded from further analyses.

The estimation of a HLM was conducted on the standardized and weighted datasets. In TIMSS studies two weighting variables were made available: the total student weight and the senate weight (Gonzalez and Smith, 1997). The senate weight was applied in pooled datasets (combined data set of all three education systems). If a weighted estimate of the mean score on a certain variable is required for the total population of three systems, each system should contribute equally to the international estimate. The senate weight is proportional to the total student weight by the ratio of 1,000 divided by the size of the population. The sum of the senate weights within each system is 1,000.

Test of Significance

The *fixed effects* were tested on their difference from zero (significant effect) by means of a t-ratio for the γ-coefficients, which can be interpreted as standardized path coefficients (Snijders and Bosker, 1999). The t-ratio is defined as the proportion of the estimated γ-coefficient and its standard error. In this explorative study two-tailed tests were applied, which means that the t-value should be greater than 1.96 with a p-value in excess of 0.05 to keep the corresponding effect in the analysis. The *deviance test* is used for tests concerning *the random part* of HLM (Snijders and Bosker, 1999).

Results

Table 13.3 shows the results of the estimation of hierarchical linear models on the pooled dataset. First, a model was estimated without identification of students by their country (model 1). Thereafter, models 2 and 3 were estimated in which students were identified by their country by means of a dummy variable (for example, the students from Germany were assigned code '1' on their dummy variable and all of the others students were assigned code '0' on the same dummy variable). The results of the final models 1 and 3 were compared to find factors that possibly could explain differences in mathematics achievement across the three education systems.

Table 13.3 Final estimation of fixed effects in two-level models on mathematics achievement in pooled data-set, weighted, standardized data and γ-coefficient.

Level	Model 0	Model 1	Model 2	Model 3	Model 1a
Intercept	−0.05	−0.01	−0.12	−0.10	0.02
Belgium Flanders	–	–	0.40	0.45	–
Germany	–	–	−0.36	−0.27	–
1. Student					
Student's gender	–	0.07	–	0.07	0.07
Home educational background	–	0.05	–	0.05	0.05
Out-of-school activities related to leisure time	–	−0.01 (n.s.)	–	−0.01 (n.s.)	−0.01 (n.s.)
Maternal academic expectation	–	n.s.	–	n.s.	–
Friends' academic expectation	–	−0.07	–	−0.07	−0.07
Perceived safety in school	–	0.02	–	0.02	0.02
Class climate	–	n.s.	–	n.s.	–
Attitude towards mathematics	–	0.16	–	0.16	0.16
Working hard doing homework	–	0.06	–	0.06	0.06
Teaching style perceived by students	–	−0.04	–	−0.04	−0.04
2. Classroom					
Class size	–	–	–	–	0.03
		n.s.	–	0.13	(n.s.)
Percentage of girls in classroom[a]	–	−0.05 (n.s.)	–	−0.04 (n.s.)	–
Home educational background[a]	–	0.16	–	0.19	0.14
Out-of-school activities related to leisure time[a]	–	−0.28	–	−0.19	−0.37
Perceived safety in school[a]	–	0.12	–	0.15	0.12
Class climate[a]	–	0.09	–	n.s.	–
Attitude towards mathematics[a]	–	−0.08	–	n.s.	–
Teaching style perceived by the students[a]	–	−0.08	–	n.s.	–
Teacher's gender	–	n.s.	–	n.s.	
Experienced limitations in teaching due to characteristics of (problem) students	–	0.06	–	0.07	0.07
Time on task	–	0.11	–	n.s.	–
Percentage mathematics lessons of all lessons of the teacher's assignment	–	0.15	–	0.10	0.16
Content coverage mathematics	–	0.08	–	0.09	0.06
Homework frequency	–	n.s.	–	n.s.	–
Amount of homework	–	0.07	–	n.s.	–
Kind of tests	–	n.s.	–	n.s.	–
Use of assessment results	–	n.s.	–	n.s.	–

Notes. [a] aggregated student variables at classroom level; γ-coefficients significant ($p < 0.05$; two-tailed tested); n.s. = fixed effect not significant.

Model 0 contains none of the student and classroom variables and is called the fully unconditional model. In model 1, the final set of student and class variables are included that have a significant effect just after they were included within the step-up method on which the model was built. Some of the variables in the final model turned out to be non-significant ($p < 0.05$) after adding one or more of the other variables. In Table 13.3, the coefficients of these non-effective variables are indicated by 'n.s.' between brackets. Variables that have no significant effect *just after* their inclusion to the model are indicated by 'n.s.' without brackets.

The student variable with the greatest positive effect in pooled model 1 (without dummy variables for countries) is 'attitude towards mathematics' (α coefficient = 0.16). Other student variables with a relative important effect on mathematics achievement are student's gender, (boys outperformed girls) and friends' academic expectation (the higher the expectations of friends to do well in mathematics, the lower the test score).

The influence of students' social background characteristics within a class (aggregated scores) on mathematics achievement seems more important than the scores of individual students on these variables. At the classroom level, the variables with relatively strong positive effects (α coefficient > 0.10) are aggregated student variables: home educational background and students' perceived safety in school. Furthermore, there is a relatively strong negative relation between the factor 'out of school activities related to leisure time' and achievement; the more time students spend outside school hours on leisure activities (playing with friends, watching television, sporting, and so on) the lower the score on the TIMSS test.

'Genuine' class variables with a relatively positive strong effect are 'time on task' (number of minutes of mathematics per week), 'the percentage of mathematics lessons of all lessons of the teacher's assignment' and 'content coverage mathematics'.

Model 2 is the empty model in which two dummy variables were included for the education systems Belgium Flanders and Germany. The final model with these dummy variables is model 3. Model 3 was estimated with the same list of variables used for the estimation of model 1 in which dummy variables for the countries were not included.

In the final model 3, the list of student and classroom variables with a significant effect ($p < 0.05$) is essentially the same as the one of final model 1. In both models the same *student* variables turned out to have an effect on mathematics achievement with identical coefficients. Hence, in every country, each of the effective student variables contributes to the explanation of variance in student achievement scores.

However, the effect of some of the classroom variables is different across model 1 and model 3. In model 3, 'class size' is a relatively strong factor (0.13), and in model 1 this factor has no significant effect. This difference indicates that the effect of class size on achievement exists within each of the countries. The mean class size in Belgium Flanders is significantly lower than in Germany and the Netherlands (not included in table). The two latter have equal mean class sizes in Grade 8. It also turned out that the bivariate Pearson correlation coefficient between class size and mathematics achievement is rather high in all countries. Considering the effect of class size and the frequency and correlation results of the three countries and the higher mean test score of Belgium Flanders compared to the other two countries, it could be argued that class size influences mathematics achievement and that smaller classes (Belgium Flanders) might be enhancing achievement more than larger classes.

The estimated models show more variables that are important to consider within and across countries. Six other classroom level variables than class size show an effect in model 1, but they show no effect in model 3: 'class climate' (aggregated student variable), 'attitude towards mathematics' (aggregated student variable), 'teaching style perceived by the student' (aggregated student variable), 'time on task' and 'amount of homework'. With respect to the influence of these variables on mathematics achievement, the country in which the students live seems unimportant. For example, the results of model 3 show that, within countries, the average score per class on 'class climate perceived by students' has no effect on mathematics achievement. The classroom variables 'time on task' and 'amount of homework' have an effect in the pooled dataset without the identification of the countries. The effect disappears in model 3. However, the mean 'minutes of instructional mathematics time' differs greatly across the three systems. The mean score is the lowest in the Netherlands (149 minutes), and the highest in Belgium Flanders (224 minutes). Also, in Belgium Flanders and the Netherlands, the bivariate correlation between 'time on task' and 'mathematics achievement' is significantly different from 0 and positive, and Belgium Flanders outperformed the Netherlands on the TIMSS mathematics test. Consequently, an increase in the number of minutes of mathematics per week could affect the achievement of Dutch students in mathematics positively.

For some other classroom variables an effect is shown in both model 1 and model 3. This indicates that the variables seem effective within separate countries. Five of these variables show an 'increasing' effect from model 1 to model 3, varying from 0.01 to 0.09, indicating that within the three countries these variables might be even more effective than across the countries. For example, 'experienced limits in teaching due to characteristics of problem students' shows an effect of 0.06 in model 1 and of 0.07 in model 3. Both within and across countries, these limitations show a negative effect on mathematics achievement.

Also, 'the percentage of mathematics lessons of all lessons of the teacher's assignment' shows a decrease in coefficient from model 1 (0.15) to model 3 (0.10). The decrease indicates that, in each country, the more lessons a teacher is teaching mathematics weekly, the better students' achievement in mathematics. In other words, mathematics lessons taught by teachers who are subject specialists seem to be more effective than mathematics lessons taught by more broadly oriented teachers. This factor has different mean scores (not in table) for the Netherlands (86 percent of the assignment concerns mathematics lessons), Belgium Flanders (73 percent) and Germany (52 percent). As the mean score of this factor, as well as the mean TIMSS test scores in both the Netherlands and Belgium Flanders are higher than in Germany, this factor seems to be especially important for Germany.

The comparison between final model 1 and final model 3, in relation to model 0, can be made in a more accurate way if model 1 is estimated with only the variables that turned out to have a significant effect in model 3. The resulting model is labeled 'model 1a' (see final column of Table 13.3).

With respect to the student variables, the results of model 1a and model 1 are the same. The differences in effect size of the seven variables inserted at class level between model 1a and model 3 differ slightly from the differences described between model 1 and model 3, but only in strength, not in direction.

Proportion of Variance Explained by the Two-level Model 1, Model 1a and Model 3

The results of the MLn analysis provide estimates of the proportion of variance associated with each level: the final estimation of variance components. The comparison of these estimates belonging to a particular model with figures of the fully-unconditional model can provide an indication of the amount of variance explained by the predicting variables at each level (Snijders and Bosker, 1999).

Table 13.4 shows for model 1, model 1a and model 3, respectively, the proportion of variance that was explained at the student and classroom levels. The percentage of variance that could be explained at the two levels is presented in the first two columns. Thereafter, the percentage of variance explained after the dummy variables were included is presented (only applicable for model 3). Next, the percentage of variance explained after all of the selected *student* variables were added is shown (the resulting model is called the unconditional level 2 model). The next two columns show the proportion of variance at the student and class levels that was explained after adding the level 2 variables to the unconditional level 2 model. Finally, the proportion of variance at the two levels that was explained by the final level 2 model is presented.

The variance at the class level is not only related to variances in class variables. Students within the same class are more similar to one another than they are to students from different classes. Thus, student background variables such as 'home educational background' and 'perceived safety in school' are also to be seen as class characteristics. The effects of variables located at the student level on mathematics achievement are combined student and class effects. Therefore, the label 'total' is inserted for the variance explained by level 1 (student) variables.

All differences in deviance turned out to be significant ($p < 0.001$). Table 13.4 shows also an increase in proportion of explained variance from final model 1a to final model 3. This increase indicates that country specific factors might play a role in relation to cross-national differences in student achievement. The identification of the country in which the students live ties variance in achievement scores between classes.

These results confirm that students from different countries performed differently on the TIMSS mathematics achievement test (see Table 13.1) and classes differ more across countries than students do. The description of the results of Table 13.3 show some examples of student and class variables that contribute to the explanation of country differences in mathematics achievement.

Conclusion and Discussion

In this chapter, results of explorative TIMSS data analysis of three education systems (Germany, Belgium Flanders and the Netherlands) were compared. Both similarities

Table 13.4 Proportion of variance in students' mathematics achievement scores explained at student and class levels in fully unconditional two-level model and final level 2 model of model 1 and model 1a (without dummy variables for countries) and model 3 (with dummy variables for countries); pooled dataset.

Pooled dataset	Proportion of variance								
	To be explained from fully unconditional model 0		Explained by adding dummy variables for countries[a]		Explained by adding level 1 variables		Explained by adding level 2 variables		Explained by final level 2 model
	between students	between classes	total	between classes	total	between classes	total	between classes	total	between classes
Model 1 (without dummy variables)	49%	51%	n.a.	n.a.	8%	9%	33%	60%	41%	69%
Difference in deviance	–	n.a.	–	493.8 (df = 8)	–	364.6 (df = 12)	–	858.4 (df = 20)[b]	–	–
Model 1a (without dummy variables)	49%	51%	n.a.	n.a.	8%	9%	30%	56%	38%	65%
Difference in deviance	–	n.a.	–	493.8 (df = 8)	–	234.8 (df = 7)	–	728.6 (df = 15)[b]	–	–
Model 3 (with dummy variables)	49%	51%	10%	19%	0%	11%	33%	44%	43%	74%
Difference in deviance	–	69.9 (df = 2)	–	508.6 (df = 8)	–	327.9 (df = 8)	–	905.4 (df = 18)[b]	–	–

Notes. All differences in deviance significant ($p < 0.001$); df = degrees of freedom.
[a] n.a. = not applicable; [b] difference in deviance between fully unconditional model 0 and final level 2 model.

and differences were found with regard to predictors at student and class/school level of mathematics achievement at Grade 8. The results can be reflected upon to find answers to the question whether the TIMSS data can serve the 'understanding of differences between countries' goal of IEA studies. The question was why the differences exist and how they can be explained. In order to enable country comparisons, the datasets of the three education systems were pooled and two hierarchical linear models were estimated.

For the three countries, it turned out that besides attitude towards mathematics, several – non-changeable – student background variables (student's gender, friends' academic expectations and home educational background) are related to mathematics achievement both at student and class levels (aggregated scores of student variables).

A limited number of 'changeable' genuine class/school variables seem to be related to achievement. The strength of influence of these factors on achievement differs across the three countries. Students' perceived safety in the school is an important 'changeable' influencing factor on class level in all three countries. Two other examples of class factors which account for some part of the variance in student achievement within and across countries are the number of minutes of mathematics scheduled per week for Grade 8 (indicator for 'time on task') and the percentage of mathematics lessons of all lessons of the teacher's assignment.

A number of changeable factors included in the organizing framework – based on several studies on the effectiveness of education – did not show effects on mathematics achievement in any country. Examples of such factors measured in TIMSS are factors regarding evaluation, feedback and corrective instruction such as 'kind of tests used' and the 'use of assessment results'.

The question is whether these factors, in contrast to many other studies, are not really related to mathematics achievement. There might be other reasons for this outcome. One possible reason is that the operationalization of these factors in TIMSS is not appropriate. Usually, the operationalization of factors resulting in concrete questionnaire items is based on a well developed conceptual framework. Because the framework of TIMSS was too general and little information was available about the correspondence between the framework and the instruments, this process was turned around in this study. Based on their content, the items in the questionnaires were first explored in order to find possible factors that belonged to the (improved) framework.

Furthermore, the exploration of composites also showed that the statistical reliability of some of the factors differed substantially between the countries. This result indicated that questionnaire items could have different meanings for respondents from different countries. In TIMSS, the international validity of the achievement test was analyzed thoroughly, but little attention was paid to the international validity of the questionnaire items (Beaton et al., 1996). If the international validity and reliability of survey questionnaires are largely ignored, the results of data analysis will be less reliable and valid within countries and less comparable across countries.

Finally, the comparison between the organizing framework and the factors for which TIMSS data were available, also showed that several important 'changeable' class factors are missing or are very poorly operationalized in TIMSS. Particularly, factors located at the level of the curricular context were not covered in TIMSS. Examples are characteristics of curricular materials (explicitness and ordering of goals and content, structure and clarity of content, and advance organizers) and some features regarding teacher behavior (for example, high expectations about student achievement, clear goal setting, and immediate exercise after presentation of new content). By adding relevant factors to and by improving the operationalization of relevant factors in the TIMSS questionnaires – based on a well developed conceptual framework – the benefits of TIMSS for policy-makers, school management, teachers and other participants in the educational field would be increased.

The possibilities that the TIMSS data offer to understand cross-national differences in relations between potentially effective factors on mathematics achievement are limited by the reasons mentioned above. If the conceptual foundation of the TIMSS instruments could be improved, TIMSS would provide better opportunities to explain differences between countries in achievement. However, the search for reasons why one country is outperformed by another country still remains very complicated because of the many factors involved and the social and cultural differences between countries.

Acknowledgements

The authors are grateful to Ch. Matthijssen for his contribution to the data analysis. The research for this chapter was conducted by the authors at the University of Twente, the Netherlands.

References

Adams, R.J. and Gonzalez, E.J. (1996) The TIMSS test design. In M.O. Martin and D.L. Kelly (eds.), *Third International Mathematics and Science Study. Technical Report. Volume I: Design and Development*. Boston, MA: Center for the Study of Testing, Evaluation and Educational Policy.

Beaton, A.E., Mullis, I.V.S., Martin, M.O., Gonzalez, E.J., Kelly, D.L. and Smith, T.A. (1996) *Mathematics Achievement in the Middle School Years. IEA's Third International Mathematics and Science Study*. Boston, MA: Center for the Study of Testing, Evaluation and Educational Policy.

Bos, K.Tj. (2002) Benefits and limitations of large-scale international comparative achievement studies. The case of IEA's TIMSS Study. Dissertation. Enschede: OCTO, University of Twente.

Creemers, B.P.M. (1994) *The effective classroom*. London: Cassell.

Gonzalez, E.J. (1996) Reporting student achievement in mathematics and science. In M.O. Martin and D.L. Kelly (eds.), *Third International Mathematics and Science Study. Technical Report. Volume II: Implementation and Analysis*. Boston, MA: Center for the Study of Testing, Evaluation and Educational Policy.

Gonzalez, E.J. and Smith, T.A. (eds.) (1997) *User guide for the TIMSS international database*. Boston, MA: Boston College.

Kuiper, W.A.J.M., Bos, K.Tj. and Plomp, Tj. (1999) Mathematics achievement in the Netherlands and appropriateness of the TIMSS mathematics test. *Educational Research and Evaluation*, 5(2), 85–104.

Pelgrum, W.J. and Anderson, R.E. (eds.) (1999) *ICT and the emerging paradigm for life long learning: a worldwide educational assessment of infrastructure, goals and practices*. Amsterdam: IEA.

Plomp, Tj. (1998) The potential of international comparative studies to monitor the quality of education. *Prospects*, 28(1), 45–59.

Scheerens, J. and Bosker, R.J. (eds.) (1997) *Foundations of educational effectiveness*. London: Routledge.

Snijders, T.A.B. and Bosker, R.J. (1999) *Multilevel analysis. An introduction to basic and advanced multilevel modeling.* London: SAGE Publications.

Travers, K.J. and Westbury, I. (1989) *The IEA study of mathematics I: international analysis of mathematics curricula.* Oxford: Pergamon Press.

Woodhouse, G. (ed.) (1995). *A guide to MLn for new users.* London: Institute of Education, University of London.

14

Participation and Achievement in Science: An Examination of Gender Differences in Canada, the USA and Norway

Kadriye Ercikan, Tanya McCreith and Vanessa Lapointe

Introduction

Efforts to encourage and improve participation and achievement in science have a well-established history. In an effort to increase the involvement and success of various groups in scientific endeavors, including different gender and ethnic groups, much research has been done related to the involvement of factors that are not directly related to schooling. Such research has included the investigation of student attitudes, self-beliefs and expectations, parental education levels, home support for learning, and the expectations of peers, teachers and parents – all of which have been shown to affect student learning in science and student perceptions of competence in science (Simpson *et al.*, 1994). Furthermore, research findings in this area highlight the idea that factors not directly related to curriculum and instruction can account for signifi-cant variability in participation and achievement in science (Greenfield, 1997; Simpson and Oliver, 1985). Within this literature, the investigation of attitude has been at the forefront. In this research, attitudes to science include students' sense of confidence, beliefs about science and whether they like science or not. Researchers have established that differences in attitudes to science exist between gender groups (Jones *et al.*, 2000; Trankina, 1993), various ethnic groups (Fleming and Malone, 1983; Greenfield, 1996; Simpson and Oliver, 1985) and different countries (Hendley and Parkinson, 1995). Given that attitude is a differentiating variable for many groups, the predictive power of attitudes to science has been studied in terms of future science course selection, achievement and future science career paths. Most authors report some relationship between attitudes to science and achievement or future science involvement (Joyce, 2000; Joyce and Farenga, 1999; Rennie and Dunne, 1994; Weinburgh, 1995).

In addition to attitude, expectations, such as those of self, peer, parent or teacher, have also been explored as potential factors associated with group differences in achievement and participation in science. Among the various types of expectations, parental expectations are particularly influential, with the child's exposure to such spanning all stages of development and socialization. As such, parents may foster in their child(ren) the internalization of social ideals such as the notion that science is traditionally a field in which males, and not females, excel and explore (Butler-Kahle and Lakes, 1983; Joyce, 2000). By the time children reach school age, exposure to parental expectations has potentially played a significant role in the definitions that children form of science, and hence the extent to which children deem participation in science acceptable (Osbourne and Whittrock, 1983).

Once a child enters the school system, they may be exposed to other sources of expectations, such as those of peers and teachers. If teachers are compelled to believe, based on sociological ideals or otherwise, that certain groups excel in specific subject areas, the resultant teacher behavior has the power to alter student performance and choice with regards to achievement and participation (Kagan, 1992; Ware and Lee, 1988). Student exposure to peer expectations also increases as a child enters school. Researchers have found that students are affected by the collective ambition and direction of his/her peer group when applying themselves towards academic success, and when making choices about participation in various subject areas (Wentzel, 1999).

It logically follows that self-expectations, which may be fed by parent, peer and teacher expectations, are also forefront in discussions about predicting achievement and participation. Researchers have found that different groups, such as women or minorities, have lower levels of science-based self-efficacy that prevent them from enrolling in science-based post-secondary programs, have lower expectations for participation in advanced science courses, and have lower expectations for success in science classes and careers (Lorenz and Lupart, 2001; Sayers, 1988). As a result, there is an under-representation of these groups in advanced science-based post-secondary study as well as science-based careers (Lee, 1998).

Expectations are related to another non-student variable – genderized perceptions of science. Gender-based sociological ideals in most western cultures have tradition-ally dictated science as a masculine domain. Through expectations, exposure, and other related aspects of a child's development and socialization, these ideals become internalized. The end result is the masculinization of science as a subject. In this way, women are discouraged from entering into the advanced study of science, and are often less motivated to achieve in science, as they perceive it to be too difficult for them and more suitable for boys (Jones et al., 2000).

An additional factor that has been investigated in terms of participation in and suc-cess with science-related endeavors is socio-economic status (SES). Previous research documents a positive relationship between SES and achievement (Ercikan et al., 2005; Schibeci and Riley, 1986). Since achievement often impacts decisions, or is a pre-requisite for future involvement in more advanced science courses, SES as it relates to achievement is inherently intertwined with participation. While SES is measured in a variety of ways, one method involves a secondary probe, such as parental education level, or the amount of support in the home for learning (dictionaries, calculators, computers and books). Such research documents a direct link between SES and achievement and participation (Schibeci and Riley, 1986).

Perhaps one of the most widely researched topics is the role of gender in science achievement and participation. Gender-based differences in attitudes to science, self-expectations for performance in science, confidence in science, and perceptions of science, among other variables, have all been well documented (Jones et al., 2000; Simpson and Oliver, 1985; Simpson et al., 1994). These differences, in turn, have been shown to play a role in the qualitatively different tangents that define males versus females in terms of success with science in the classroom, and pursuit of the advanced study of science or science-based careers (Simpson and Oliver, 1985; Ware and Lee, 1988).

This chapter focuses on the investigation of key variables that are not directly related to schooling as predictors of achievement in science and participation in advanced science courses. The variables included in this study mirror those discussed above – namely, attitudes towards science (including the extent of liking expressed for specific sub-content areas of science), confidence in science, the expectations of self, friends, parents and teachers, home support for learning, level of parental education, and the perceived gender-linkage of science as a domain. The study focuses on gender and country differences in Canada, the United States of America (USA) and Norway, based on the Third International Mathematics and Science Study (TIMSS) (Martin and Kelly, 1996). In all three of these countries, the science literacy scores based on TIMSS for males from Population 3 were significantly higher than those for females. Second, the degree of gender differences varied amongst these countries, with the USA sample having the smallest gender differences, Canada, with mid-level differences, and Norway demonstrating the highest degree of gender differences in Population 3 (Mullis et al., 2000).

An important aspect of the present study is that it addresses the issue of science as a discipline composed of many sub-content areas. While gender differences in achievement and participation have been documented in science as a unified subject area, it has become increasingly apparent that these gender differences vary in scope and magnitude across the various sub-content areas that compose the more general area of science. These sub-content areas can include, but are not limited to, biology, physics, chemistry and earth sciences. Of these, it appears that the physical sciences are viewed as more suitable for males, while the biological or life sciences are viewed as more suitable for females (Jones et al., 2000; Vockell and Lobonc, 1981). Thus, females may deem it acceptable to excel in a life science discipline, but not in a more masculine physical sciences discipline. As such, the current authors expect to find differences based on the sub-content area of science in question, and thus will analyze the relationship between attitude and science achievement across three of the aforementioned sub-content areas (biology, physics and chemistry) for each of the included countries.

Method

Data

This study focuses on data from the science literacy test. In addition to achievement tests, students and principals of participating schools were administered background questionnaires. Information collected in these questionnaires include: (1) students' participation in advanced mathematics and science courses; (2) students' beliefs and attitudes towards mathematics and science; (3) self and others' expectations of the students regarding attending university; and (4) home environment.

Analyses

Science achievement and participation in advanced science courses were analyzed as dependent variables in relation to several independent variables. The independent variables were: students' attitudes to science, student's degree of liking of the relevant science sub-content area (biology, chemistry and physics), highest level of parent's education, expectations of self, peers, parents and teachers, students' confidence in science, home support for learning, and whether students intended to pursue studies in mathematics or science. An analysis of the relationship between achievement and the various independent variables was done using multiple regression analysis, while an analysis of the relationship between participation in advanced science courses was done using discriminant function analysis. Attitudes and home support for learning were measured by a set of questions with ordinal responses. Non-linear principal components analysis (Gifi, 1990) was used to combine responses to questions measuring attitudes and home support for learning to create composites of ordinal variables.

Multiple regression analyses for science achievement

There were three stages of multiple regression analyses. The first stage included the dummy variable Country (described below) as an independent variable. If the regression results indicated that the variable Country was a significant predictor of science achievement in the model, then the remaining regression analyses would be completed separately by countries. That is, if knowing what country a student is from can be used to predict achievement scores in science, then it is more appropriate to examine models at a country level, rather than across countries. Using the same sort of reasoning, a second set of multiple regression analyses were completed where Gender was entered as a dummy variable as an independent variable in the model. If the results indicated that gender was a significant predictor of science achievement in the model, then the remaining regression analyses would be completed separately by gender. Interactions between any pairs of the independent variables were not considered in this study. The following independent and dependent variables were used in the regression analyses.

Dependent variable:

• *Science achievement*: Science Literacy Scores on TIMSS.

Independent variables:

- *Students' overall attitudes towards science learning*: Four point ordinal responses to the following items were combined using a non-linear principal components analysis: (1) I like earth science; (2) I like chemistry; (3) I like physics; (4) I like biology. This scale of attitudes towards science had a coefficient alpha of 0.800.
- *Highest level of parent's education*: Highest education level attained by the mother or the father as reported by the student.
- *Whether others and students themselves had high expectations of the students*: Students' separate responses to the questions about whether the student perceives that their mother, father, teacher and friends as well as the students themselves, expect them to go to a four-year university on a full-time basis immediately after completing secondary school. These were identified as mother's expectations, father's expectations, teacher's expectations, friends' expectations, and self expectations.
- *Home support for learning*: Ordinal responses to the following items were combined using a non-linear principal components analysis, and the combined scale was interpreted as home support for learning: (1) number of books at home; (2) whether the student owns a calculator at home; (3) whether the student has a computer at home. This scale of home support for learning had a coefficient alpha of 0.485.
- *Confidence in science*: Four point ordinal responses to 'I usually do well in science' was recoded as 0 if the response was Strongly Disagree or Disagree, and 1 if the response was Strongly agree or Agree.
- *Intention to study mathematics or science*: The variable was recoded into two categories: one indicating that the student intends to study in a mathematics or science related field, and the other indicating that the student intends to study in an area outside of mathematics or science.

Discriminant function analyses

Discriminant function analyses were conducted to examine and describe differences between groups of students who had taken advanced level science courses, as well as the association of different factors with participation in advanced science courses. The dependent variable in this study was students' responses to whether they have taken advanced science courses in a given content area, namely biology, chemistry and physics.

The independent variables were the same as those used in the multiple regression analyses, with the sole difference being that liking of a particular content area within the domain of science was used rather than the overall attitude to science measure. Separate discriminant function analyses were run for each content area using the corresponding liking variable, in combination with the rest of the independent variables. For example, a separate discriminant function analysis was conducted with participation in advanced biology courses as the dependent variable and students' responses to whether they liked biology or not was used as one of the independent variables, along with all the independent variables used in regression analyses, except for the attitude towards science variable.

Results

Mean Differences between Gender Groups

The descriptive statistics in Table 14.1 display the mean and standard deviation of science literacy scores by country and gender. The means and standard deviations were obtained using sampling weights in order for each country sample to be representative. There were significant differences between males and females in all three countries. The largest difference between the gender groups was obtained for Norway, approximately two-thirds of one standard deviation of the male sample, and the smallest difference was obtained between the USA gender groups, less than one-fourth of a standard deviation of the USA male sample mean.

Gender Differences in Participation Rates

The participation rates in advanced science courses, calculated using sampling weights, by country and gender are presented in Table 14.1. Large differences are observed in the participation rates of males and females in biology and physics in Canada and Norway. In biology, females had larger participation rates, whereas in physics males had larger rates of participation in both of these countries. In the USA, differences were small and males had only slightly higher participation rates in these science areas. In chemistry, the differences in participation rates between the gender groups were small and males had higher participation rates in all three countries.

Multiple Regression Analysis for Science Achievement

When country differences in the regression model were tested using two dummy variables for country membership, significant differences were identified. This result indicates that a single regression model is not meaningful for the three different countries. Further, the regression analysis which tested the significance of gender differences in the regression model identified that there were significant differences between the two groups in each country. These findings indicate that a single regression model for the two gender groups is not appropriate. The results of the separate multiple regression analyses for each country and gender group are presented in Table 14.2.

The regression models accounted for approximately 20 to 27 percent of the variance in science achievement scores. For males in both Canada and Norway, the percentage of variance accounted for by the regression model (25 percent for each) was higher than for females (22 and 20 percent, respectively). In the USA, the percentage of variance accounted for by the model was similar for females and males (27 and 26 percent, respectively). The largest difference, 5 percent, was observed between males and females in Norway, indicating that the independent variables

Table 14.1 Descriptive statistics for science literacy score by country and gender.

Variable	Country											
	Canada				USA				Norway			
	Combined sample	Female	Male	Diff.[a]	Combined sample	Female	Male	Diff.[a]	Combined sample	Female	Male	Diff.[a]
Science literacy score												
N	5,205	2,533	2,672		5,807	2,968	2,839		2,518	1,328	1,190	
Mean	533	518	550	−31*	480	468	492	−24*	544	514	574	−60*
Standard deviation	85	81	86		94	88	98		91	79	92	
Percentage participation												
Biology	39	45	34	11*	21	21	21	0	16	22	10	12*
Chemistry	35	34	37	−3*	13	12	14	−2*	17	14	19	−5*
Physics	21	17	26	−9*	25	24	26	−2*	11	5	16	−11*

Notes. * $p < 0.001$.
[a] Diff = female − male.

Table 14.2 Standardized β coefficients from multiple regression analyses.

	Canada		USA		Norway	
	Male	Female	Male	Female	Male	Female
Attitude to science	0.16*	0.10*	0.14*	0.13*	0.13*	0.13*
Confidence in science	0.20*	0.17*	0.10*	0.07*	0.21*	0.16*
Father's expectations	0.07*	0.02*	0.03*	0.03*	0.07*	0.03
Home support for learning	−0.12*	−0.21*	0.17*	0.21*	0.11*	0.13*
Highest parent education level	−0.16*	−0.17*	−0.19*	−0.19*	−0.01	−0.08*
Self expectations	0.17*	0.04*	0.14*	0.12*	0.22*	0.21*
Area student intends to study	−0.03*	−0.06*	−0.01*	0.03*	−0.04*	−0.05*
Teacher's expectations	0.07*	0.01*	0.07*	0.03*	0.06*	−0.04*
Friends' expectations	0.04*	0.12*	0.07*	0.08*	0.08*	0.07*
Mother's expectations	0.01	−0.01	−0.06*	0.00	0.00	0.04*
R^2	0.25	0.22	0.26	0.27	0.25	0.20
Adjusted R^2	0.25	0.22	0.26	0.27	0.24	0.20

Note. *significant at the 0.01 level.

accounted for a larger amount of variance in the science achievement scores for males than females.

In Canada, Home Support for Learning was the strongest predictor for females. Alternatively, the strongest predictor for Canadian males was Confidence in Science, a variable that was also important for females but not to the same extent.

In the USA, Confidence in Science was not a strong predictor for either males or females. Instead, predictors related to SES status proved to be the strongest for both gender groups. For American males the strongest predictor was Parent's Education Level, while for American females the strongest predictor was Home Support for Learning. In Norway, for both males and females, the strongest predictor of science achievement was Self-expectations regarding going to university.

Interesting inter-country comparisons were noted when observing the relative importance of affective variables versus variables that provide proximal measurement of SES factors. It appears that while the science literacy of Norwegian students is most strongly related to self-expectations, the science literacy of American students is most strongly related to either Parent Education Level (males) or Home Support for Learning (females). Canadian students appear to occupy the medial position with Home Support for Learning (females) and Confidence in Science (males) presenting as the strongest predictors, suggesting that both affective and proximal SES measures have a central role in Canada.

Discriminant Function Analysis for Participation in Advanced Science Courses

The results for the discriminant function analyses by country and gender are presented in Table 14.3. This table presents the correlations between each of the independent

Table 14.3 Discriminant function analysis results: correlations with the discriminant function.

| | Canada | | | | | | USA | | | | | | Norway | | | | | |
| | Female | | | Male | | | Female | | | Male | | | Female | | | Male | | |
	B[a]	C[a]	P[a]	B	C	P	B	C	P	B	C	P	B	C	P	B	C	P
Confidence in science	0.34	0.44	0.40	0.25	0.31	0.34	0.33	0.23	0.42	0.48	0.32	0.29	0.39	0.45	0.26	0.24	0.39	0.33
Friend's expectations	0.20	0.44	0.28	0.28	0.48	0.45	0.35	0.51	0.57	0.55	0.66	0.66	0.20	0.24	0.15	0.16	0.32	0.26
Father's expectations	0.10	0.31	0.27	0.28	0.53	0.38	0.63	0.56	0.50	0.43	0.47	0.50	0.28	0.30	0.22	0.36	0.48	0.35
Home support	-0.12	-0.26	-0.24	-0.00	-0.10	-0.18	0.44	0.28	0.43	0.26	0.36	0.36	0.39	0.37	0.08	0.21	0.31	0.25
Like biology	0.84	0.58	0.54	0.89	0.63	0.56	0.44	0.73	0.56	0.61	0.64	0.60	0.84	0.62	0.74	0.55	0.35	0.69
Mother's expectations	0.11	0.34	0.31	0.28	0.52	0.39	0.68	0.56	0.55	0.42	0.46	0.54	0.28	0.29	0.21	0.43	0.46	0.32
Highest parent education level	-0.12	-0.22	-0.26	-0.16	-0.01	-0.22	-0.42	-0.37	-0.60	-0.36	-0.50	-0.56	-0.36	-0.50	-0.42	-0.50	-0.45	-0.29
Self expectations	0.28	0.45	0.38	0.33	0.53	0.40	0.67	0.61	0.54	0.46	0.60	0.62	0.37	0.50	0.27	0.68	0.81	0.52
Area student intends to study	-0.30	-0.60	-0.74	-0.21	-0.35	-0.62	0.02	-0.04	-0.19	-0.39	-0.19	-0.19	-0.11	-0.28	-0.44	0.18	0.00	-0.27
Teacher's expectations	0.11	0.26	0.16	0.27	0.27	0.38	0.36	0.38	0.40	0.57	0.45	0.48	0.08	0.10	0.07	0.35	0.36	0.31
Classification accuracy (%)	84.9	80.0	72.7	78.0	80.8	74.2	78.7	77.4	74.5	72.9	77.8	77.0	77.7	77.8	76.8	80.0	77.9	78.9

Note. [a]B = biology, C = chemistry, P = physics.

variables and the discriminant functions, which distinguish between those who participate in advanced level science courses and those who do not. Interpretation of correlations is similar to those of loadings in a factor analysis. Correlations are preferred over discriminant function coefficients for interpretation due to their stability with small and moderate sample sizes. The last row of each table displays the accuracy of classification of examinees based on the discriminant functions. These classification accuracy percentages indicate the degree of accuracy of predictions of participation of students in advanced level science courses using the independent variables.

Biology

In Canada, the participation prediction accuracies for biology were higher for females than for males (85 versus 78 percent), indicating that the model could classify participation in biology more accurately for females than for males. Next to endorsed Liking of Biology, which was the strongest predictor of participation for both genders, the strongest predictor for females was Confidence in Science, while for males it was Self-Expectations.

In the USA, participation prediction accuracies for biology were again higher for females then for males (79 versus 73 percent). While endorsed Liking of Biology was the strongest predictor for males, it ranked only as the fourth strongest predictor for females. Teacher's Expectations was also a relatively strong predictor for males in the USA. For American females, conversely, Mother's Expectations were the strongest predictor, followed closely by Self-expectations and Father's Expectations.

For students in Norway, participation prediction accuracies for biology were marginally higher for males than for females (80 versus 78 percent). While endorsed Liking of Biology was the strongest predictor for females, Self-expectations were the strongest predictor for males, with endorsed Liking of Biology being the second strongest predictor for males.

Chemistry

For Canadian students, the model accounted for female and male participation in advanced chemistry courses equally, as evidenced by virtually identical classification accuracies (80 versus 81 percent). While endorsed Liking of Chemistry was the strongest predictor for males, the strongest predictor for females was Area Student Intends to Study, followed by endorsed Liking of Chemistry. For males, the second strongest predictor was Father's Expectations.

For students in the USA, participation prediction accuracies for chemistry were similar for males and females (78 and 77 percent respectively), indicating that the model accounts similarly for male and female participation rates. Within the model, the strongest predictor for females was the endorsed Liking of Chemistry, followed by Self-expectations and Mother's Expectations. The strongest predictor for males was Friend's Expectations, followed by endorsed Liking of Chemistry, and Self-expectations.

In Norway, participation prediction accuracies in chemistry were equal for males and females (78 percent for both). Within the model, the endorsed Liking of Chemistry was the strongest predictor for females, while Self-expectations was the strongest predictor for males (much more so than for females).

Physics

For Canadian students, participation prediction accuracies for physics were very similar for males and females (74 and 73 percent). Delineation of the strongest predictors was also similar across gender groups for Canadian students, with the endorsed Liking of Physics being the strongest for both males and females, followed by Area the Student Intends to Study.

In the USA, participation prediction accuracies for physics were slightly higher for males than for females (77 versus 75 percent). While a proximal SES variable, Highest Parent Education Level was the strongest predictor for females, the strongest predictor for males was Friend's Expectations, followed closely by Self-expectations. For females, the second strongest predictor was the endorsed Liking of Physics, followed closely by Mother's Expectations.

For Norwegian students, participation prediction accuracies for physics were slightly higher for males than for females (79 versus 77 percent). The strongest predictor for both males and females was endorsed Liking of Physics, followed by Self-expectations for males, and Area Student Intends to Study for females.

Summary and Discussion

The variables used in the current study can be grouped into three categories. These are: (1) variables related to self; (2) variables related to SES; and (3) expectations. These variables accounted for 20 to 27 percent of variability in achievement in science across the three countries, namely Canada, the USA and Norway. SES was the strongest predictor of achievement in the USA, and expectations regarding going to university was the strongest in Norway. In Canada, self was the strongest predictor for males and SES for females. Within each country there were distinct differences between gender groups regarding strength of predictors for achievement. In Canada and Norway, Confidence in Science was the strongest predictor of science achievement for males, while in the USA Parents' Education Level was the strongest predictor of achievement in science for both females and males. The latter suggests that SES variables may be more influential on science achievement for the American sample. This suggestion is further supported by the observation that another SES related variable, Home Support for Learning, was also a strong predictor of science achievement in the USA for both gender groups. The importance of SES related variables is markedly different in Canada and Norway. Canada appears to occupy a medial position in terms of the degree to which SES is associated with science achievement. Specifically, SES is not as strong a predictor of achievement for the males in the Canadian sample as it is for the females. Norway, on the other hand, seems to occupy the opposite end of the continuum when compared to the USA in terms of the strength of SES as a predictor, with SES having much less importance for predicting science achievement for both males and females in Norway. Overall, it appears that societal factors are more strongly associated with achievement in science in the USA for both males and females, and in Canada for only females, as compared to males in Canada, and males and females in Norway. In Norway, the strongest predictor of achievement in science learning was Self-expectations – namely, whether or not the student expected to go to a four-year university.

Participation in the three advanced level science courses could be predicted with 73 to 81 percent accuracy by the included variables that are not directly related to schooling. The gender differences in prediction of participation in advanced level courses were small, with the largest difference, seven percent, in classification accuracy between the gender groups in Canada for participation in advanced level biology courses. In Canada, even though the variables considered in this study predicted achievement for males better than for females, participation in biology was more accurately predicted by these variables for females.

Some differences between how each of the independent variables were related to participation in advanced level biology, chemistry and physics courses were observed. The strongest predictor of participation in biology courses was whether students liked biology or not for all groups except for USA females and Norway males, for which the strongest predictors were expectations (theirs in Norway males and mother's in the case of USA females) regarding going to a four-year university. Prediction of participation in chemistry courses, however, was more mixed. Whether students liked chemistry or not was the strongest predictor for Canadian males, US females and Norwegian females. However, expectations regarding going to university and the area they intended to study were the strongest predictors for the other groups. Prediction of participation in physics courses was similarly mixed among the three groups. In Canada, the area student intends to study predicted participation in physics courses most strongly both for females and males. In Norway, the strongest predictor of participation was whether the students liked physics or not. In the USA, conversely, the strongest predictor of participation in physics was father's expectations regarding going to university for males and the parents' education level for females.

The participation in advanced level science courses were predicted most accurately by variables related to self (whether students liked the subject or not and the area students wanted to study at university) and expectation regarding going to a four-year university. Unlike the prediction of science achievement, SES related variables, such as Parents' Education Level and Home Support for Learning, were not strong predictors of participation in advanced science courses, except for females in the USA sample.

The findings in this study provide consistent evidence that student personal and home environment variables are strongly associated with science achievement and participation in advanced science courses in all three countries. In the USA, SES is an influential predictor of science achievement for both males and females and in predicting participation rates for females. In Canada, SES is similarly influential for predicting science achievement for females, whereas in Norway SES is not an influential predictor. In order to teach science to children from all socio-economic backgrounds, personal and home environment related variables and how they affect science learning and participation need to be taken into account. However, how to take such associations into account in teaching science is not an easy matter. First, the associations identified in this research are correlational and they are not causal. Therefore, the complex set of factors that might be the sources of associations are not known. Yet consistent evidence based on representative samples from three countries and other similar studies make a strong case for further investigating the associations identified in this research. In order to determine the practical implications of the findings here, in-depth studies that investigate how personal and home environment factors might be affecting learning and participation are needed. In particular, we need to know how students develop positive and negative attitudes towards science, what the role of home environment on the development of these attitudes is, and how these attitudes affect motivation for learning science and participating in advanced level science courses.

References

Butler-Kahle, J. and Lakes, M. (1983) The myth of equality in science classrooms. *Journal of Research in Science Teaching*, 20, 131–40.

Ercikan, K., McCreith, T. and Lapoint, V. (2005) Factors associated with mathematics achievement and participation in advanced mathematics courses: An examination of gender differences from an international perspective. *School Science and Mathematics Journal*, 105, 11–18.

Fleming, M.L. and Malone, M.R. (1983) The relationship of student characteristics and student performance in science as viewed by meta-analysis research. *Journal of Research in Science Teaching*, 20, 481–95.

Gifi, A. (1990) *Nonlinear multivariate analysis*. Chichester: John Wiley and Sons.

Greenfield, T.A. (1996) Gender, ethnicity, science achievement, and attitudes. *Journal of Research in Science Teaching*, 33, 901–33.

Greenfield, T.A. (1997) Gender- and grade-level differences in science interest and participation. *Science Education*, 81, 259–76.

Hendley, D. and Parkinson, J. (1995) Gender differences in pupil attitudes to the national curriculum foundation subjects of English, mathematics, science and technology in key stage three in South Wales. *Educational Studies*, 21, 85–97.

Jones, G., Howe, A. and Rua, M. (2000) Gender differences in students' experiences, interests, and attitudes toward science and scientists. *Science Teacher Education*, 84, 180–92.

Joyce, B. (2000) Young girls in science: Academic ability perceptions and future participation in science. *Roeper Review*, 22, 261–2.

Joyce, B. and Farenga, S. (1999) Informal science experience, attitudes, future interest in science, and gender of high-ability students: An exploratory study. *School Science and Mathematics*, 99, 431–7.

Kagan, D.M. (1992) Implications of research on teacher belief. *Educational Psychologist*, 27, 65–90.

Lee, J. (1998) Which kids can 'become' scientists? Effects of gender, self-concepts, and perceptions of scientists. *Social Psychology Quarterly*, 61, 199–219.

Lorenz, E. and Lupart, J. (2001, May) Gender differences in mathematics, English, and science for grade 7 and 10 students' expectation for success. Paper presented at the annual conferences of the Canadian Society for Studies in Education, Quebec, Canada.

Martin, M.O. and Kelly, D.L. (1996) *Third International Mathematics and Science Study technical report, Volume I: Design and development*. Chestnut Hill, MA: International Association for the Evaluation of Educational Achievement (IEA), Boston College.

Mullis, I.V.S., Martin, M.O., Fierros, E.G., Goldberg, A.L. and Stemler, S.E. (2000) *Gender differences in achievement: IEA's Third International Mathematics and Science Study*. Chestnut Hill, MA: International Study Centre, Lynch School of Education, Boston College.

Osbourne, R. and Whittrock, M. (1983) Learning science: A generative process. *Science Education*, 67, 489–508.

Rennie, L. and Dunne, M. (1994) Gender, ethnicity, and students' perceptions about science and science-related careers in Fiji. *Science Education*, 78, 285–300.

Sayers, M.P. (1988) The relationship of self-efficacy expectations, interests, and academic ability to the selection of science and non-science college majors. *Dissertation Abstracts International*, 48(10-A), 25–44.

Schibeci, R. and Riley, J. (1986) Influence of students' background and perceptions on science attitudes and achievement. *Journal of Research in Science Teaching*, 23, 177–87.

Simpson, R. and Oliver, J. (1985) Attitude toward science and achievement motivation profiles of male and female science students in grades six through ten. *Science Education*, 69, 511–26.

Simpson, R., Koballa, T. Jr., Oliver, J. and Crawley, F. (1994) Research on the affective dimension of science learning. In D.L. Gabel (Ed.), *Handbook of Research on Science Teaching and Learning* (pp. 211–34). Don Mills, ON: Maxwell Macmillan Canada Inc.

Trankina, M. (1993) Gender differences in attitudes toward science. *Psychological Reports*, 73, 123–30.

Vockell, E. and Lobonc, S. (1981) Sex-role stereotyping by high school females in science. *Journal of Research in Science Teaching*, 18, 209–19.

Ware, N. and Lee, V. (1988) Sex differences in choice of college science majors. *American Educational Research Journal*, 25, 593–614.

Weinburgh, M. (1995) Gender differences in student attitudes toward science: A meta-analysis of the literature from 1970 to 1991. *Journal of Research in Science Teaching*, 32, 387–98.

Wentzel, K. (1999) Social-motivational processes and interpersonal relationships: Implications for understanding motivation at school. *Journal of Educational Psychology*, 91, 76–97.

15

Predictors of Student Achievement in TIMSS 1999 and Philippine Results

Vivien M. Talisayon, Celia R. Balbin, and Fe S. De Guzman

Introduction

This chapter presents a Philippine study that determines the predictors of student achievement in science and mathematics in TIMSS 1999 data from the Philippines and a number of other countries.

The Philippines ranked third last (out of 38 countries) in science and mathematics student achievement in TIMSS 1999. Other Asian countries, namely, Chinese Taipei, Singapore, Japan and the Republic of Korea, topped the TIMSS 1999 achievement tests. The Filipino students who took the test were in Grade 7 and were mostly 13 years old. However, at this age, the students in most of the participating countries were generally in Grade 8.

Education officials and teachers in the Philippines are greatly interested in knowing what the predictors of student achievement, discernible from the TIMSS 1999 data, are so that measures can be taken to improve education and, as a consequence, the performance of Filipino students in future TIMSS achievement tests. It is in this context that the study sought to answer the following questions:

1. What are the levels of the student, teacher and school indices for the top five and the bottom five countries in student achievement in TIMSS 1999?
2. Which of the indices are predictors of science and mathematics scores for the all participating countries, using 'country' as the unit of analysis?
3. Are the indices, mathematics score and class size, predictors of the science score in the Philippine data, using the 'student' as unit of analysis?
4. Are the indices and class size predictors of mathematics score in the Philippines data, using the 'student' as the unit of analysis?

Most of the top five countries in the TIMSSS 1999 science and mathematics results are in Asia where the Philippines is located. It is believed that an initial comparison of student, teacher and school indices of the top five countries and the bottom five countries (including the Philippines) would help the Philippines to design educational interventions on indices that are related to achievement and for which the top countries rated high and the bottom countries low.

Following the first analysis of indices, a quantitative determination of predictors of science and mathematics achievement involving all the countries was sought, with 'country' being unit of analysis. This analysis could yield further information for educational interventions. Since 38 countries participated in the study, a maximum of four independent variables or indices could be regressed against student achievement in science or mathematics, the criterion variable.

Using a limited regression analysis, the predictors of science and mathematics achievement could be determined more precisely in the Philippine context if the data of the country (involving more than 6,000 students) were used, with 'student' as the unit of analysis. Class size was included, since the Philippines has a large average class size of 51 students, the largest in TIMSS 1999, as well as class size being a significant predictor in previous TIMSS research (see Martin, Mullis, Gregory *et al.*, 2000: 85).

For the science score, the mathematics score was considered a possible predictor, because mathematics is a language of science. Mathematics skills, such as computational skills, are important in understanding scientific concepts.

Conceptual Framework

The conceptual framework for this research (Figure 15.1) shows the student, teacher and school indices, including class size, as hypothesized predictors. The questionnaire items constituting the indices were selected for their perceived relationship with science and mathematics achievement by the TIMSS Study Center, in consultation with the national research coordinators. The selection was also guided by results of two earlier TIMSS surveys, indicating relationship of certain background variables with achievement (see Martin, Mullis, Gregory *et al.*, 2000).

To have operational definitions of the predictor variables proposed in the framework, the student, teacher and school indices of the TIMSS Study Center (Martin, Mullis, Gonzales *et al.*, 2000 and Mullis *et al.*, 2000) were used. An index is based on responses to between two and five questions. Indices were selected for their likely contribution to the mathematics and science scores in the Philippine setting. Furthermore, indices, rather than individual questionnaire items, were chosen to facilitate comparisons and reduce the number of independent variables.

For the mathematics score, the following student indices were selected (see Figure 15.1): home educational resources (HER), out-of-school study time (OST), students' positive attitudes towards mathematics (PATM) and students' self-concept in mathematics (SCM). The teacher indices were: teacher's confidence in preparation to teach mathematics (CPM), teachers' emphasis on mathematics reasoning and problem-solving (EMRPS), emphasis on use of calculators in mathematics (ECMC) and

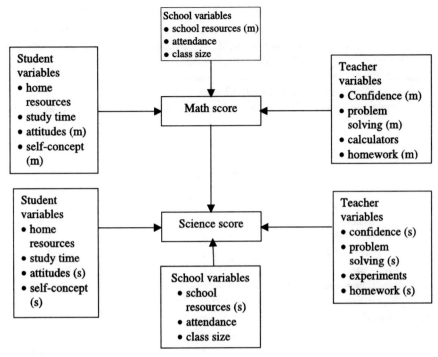

Legend:

- *home resources*: home educational resources (HER)
- *study time*: out-of-school study time (OST)
- *attitudes (m)(s)*: positive attitudes towards math/science (PATM/PATS)
- *self-concept (m)(s)*: self-concept in math/science (SCM/SCS)
- *confidence (m)(s)*: teachers' confidence in preparation to teach math/science (CPM/CPS)
- *problem solving (m)(s)*: teachers' emphasis on mathematics/scientific reasoning and problem solving (EMRPS/ESRPS)
- *calculators*: teachers' emphasis on calculators in math (ECMC)
- *experiments*: teachers' emphasis on conducting experiments in science (ECES)
- *homework (m)(s)*: teachers' emphasis on math/science homework (EMH/ESH)
- *school resources (m)(s)*: availability of school resources for math/science instruction (ASRMI/ASRSI)
- *attendance*: school and class attendance (SCA)

Figure 15.1 Hypothesized student, teacher and school predictors of mathematics and science scores.

teachers' emphasis on mathematics homework (EMH). The school indices were: availability of school resources for mathematics instruction (ASRMI) and school and class attendance (SCA).

For the science score, the school indices: attendance and class size, and the student indices: home resources and study time, are the same as for mathematics. The teacher indices and other student and school indices are similar, but for science instead of for mathematics, having a different, but related set of questionnaire items for science. Thus, the student indices for science are (Figure 15.1) students' positive attitudes towards science (PATS) and students' self-concept in science (SCS).

The teacher indices for science are: teachers' confidence in preparation to teach science (CPS), teachers' emphasis on scientific reasoning and problem-solving

(ESRP), emphasis on conducting experiments in science classes (ECES) and teachers' emphasis on science homework (EMS), the school index, availability of school resources for science instruction (ASRSI). Mathematics score was added as a possible predictor of science achievement, since mathematics concepts and skills are needed in understanding science concepts like force and motion.

The rest of the chapter presents the methodology for the analyses, a discussion of results, and possible implications and recommendations for the Philippine school system.

Methodology

This section presents a qualitative approach[1] (levels of indices) to address the first research question and the quantitative method (multiple regression) for the remaining research questions.

Comparisons of the student, teacher and school indices levels (low, medium or high) were made between the top five and bottom five countries in student achievement in science and mathematics. The selection of a particular index level (for example, medium) for a country was determined by the level with the highest percentage of students, taken from tables in the TIMSS 1999 international reports (Martin, Mullis, Gonzalez *et al.*, 2000 and Mullis *et al.*, 2000).

The top five countries in student achievement in science were Chinese Taipei, Singapore, Hungary, Japan and the Republic of Korea, and the bottom five countries Tunisia, Chile, the Philippines, Morocco and South Africa. In mathematics, the top five countries in student achievement were Singapore, the Republic of Korea, Chinese Taipei, Hong Kong and Japan, and the bottom five countries were Indonesia, Chile, the Philippines, Morocco and South Africa.

In order to determine which of the indices were predictors of mathematics and science scores, stepwise multiple regressions were run, using SPSS (Statistical Package for the Social Sciences) for the 38 participating countries, with the country as the unit of analysis. At most four indices were used as predictor variables at a time (four student indices, four teacher indices and two school indices). The TIMSS 1999 average scale score in science or mathematics was the dependent variable. The contribution of predictor variables is expressed in standardized beta coefficients.

For each country, the percentage of students for each index level (obtained from the international reports) was multiplied by its value (low level = 1, medium level = 2, high level = 3), divided by 100, and totaled for the three levels. The Appendix lists the single-number values of the indices per country. A final multiple regression was conducted with significant predictors from the student, teacher and school indices and added variables.

As a form of validation and focus of interest, similar multiple regressions were done using the Philippine data. The index, emphasis on conducting experiments in science classes (ECES), was not in the Philippine database and, therefore, excluded as a predictor variable. The student was the unit of analysis, and the first plausible values[2] of science and mathematics scores were the dependent variables.

The Philippine sample consisted of more than 3,300 students providing complete data with 150 mathematics teachers and 150 science teachers in 150 schools in 13 regions of the country. The TIMSS 1999 two-stage stratified cluster sample design was used, where the region was the explicit stratum. The first stage was the selection of schools using a probability proportional to size systematic method. At the second stage, intact classrooms were sampled with equal probabilities, one classroom per school.

Findings

The discussion of findings begins with the international results for the first and second research questions and ends with the Philippine results for the third and fourth questions. For the international findings, a discussion of the analysis of the indices levels is presented first, followed by the results of the multiple regression analyses. The international and local predictors of student achievement are presented by subject area, science first and then mathematics. In each case, the results are discussed starting with student predictor variables, followed by teacher variables and, lastly, school variables.

International Predictors of Science Achievement

Levels of indices

Table 15.1 lists the levels of student, teacher and school indices in science. The indices were designed such that the category 'high' level for a country is expected to result in a higher TIMSS 1999 average scale score. For the international average of all indices in science, the highest percentage of students were found in the medium-level category. The meaning of medium level varied for each index on the basis of the contents of the items comprising each index.

In science, four of the high-performing five countries were found to have a medium level for the four student indices, which matched the international average. Similarly, for home educational resources (HER) and self-concept in science (SCS), four of the bottom five countries had a medium level, the international average. Unexpectedly, four of the bottom five countries were found to have attained high levels of out-of-school study time (OST) and positive attitude towards science (PATS), suggesting that more than three hours of study time after school and very positive attitudes towards science do not necessarily result in higher science achievement.

The teachers' index on science homework (ESH) did not differentiate among the top and bottom five countries. All had a medium level, the same as that of the international average. For teachers' confidence in preparation to teach science (CPTS), most of the bottom five countries were found to have a medium level, and more of the top five countries had a low level. It is not clear why teachers in the bottom countries had greater confidence in their preparation to teach science than those in the top countries; this finding should be investigated further.

VIVIEN M. TALISAYON *ET AL.*

Table 15.1 TIMSS 1999 levels of indices of five top and five bottom countries in science.

	Students				Teachers				Schools	
	HER	OST	SCS	PATS	CPTS	ESRPS	ECES	ESH	ASRSI	SCA
Top five countries										
1. Chinese Taipei	Medium	Medium	Medium	Medium	Medium	Low	Medium	Medium	Medium	Medium
2. Singapore	Medium	High	Medium	Medium	Medium	Low	High	Medium	High	Medium
3. Hungary	Medium	Medium	–	–	Low	Medium	–	Medium	Medium	Medium
4. Japan	–	Medium	Medium	Medium	Low	Medium	High	Medium	Medium	Medium
5. Korea, Republic of	Medium	Medium	Medium	Medium	Low	Medium	Medium	Medium	Medium	Medium
Bottom five countries										
34. Tunisia	Medium	High	Medium	High	Low	Medium	High	Medium	Medium	Medium
35. Chile	Medium	Medium	Medium	High	Low	Medium	–	Medium	Medium	Medium
36. Philippines	Medium	High	Medium	High	Medium	Medium	Medium	Medium	Medium	Medium
37. Morocco	Low	High	–	–	Medium	Low	–	Medium	Low	Low
38. South Africa	Medium	High	Medium	High	Medium	Medium	Medium	Medium	Low	Low
International Average	Medium	Medium	Medium	Medium	Medium	Medium	Medium	Medium	Medium	Medium

Note. For legend see Figure 15.1.

In the teachers' emphasis on scientific reasoning and problem-solving (ESRPS), the bottom country group had a slightly higher level than the top group. Four bottom countries had a medium level, while only one bottom country was found to have a low level. Three top countries had a medium level, while the top two countries had a low level.

In contrast, the top group showed only a slightly higher index level than the bottom group in the teachers' emphasis on conducting experiments in science classes. For the top group, two countries had a high level on this index, while the two other countries had a medium level. One of the bottom countries had a high level, and two other bottom countries had a medium level. No level was indicated for one top country and two bottom countries. On average, teachers in top-performing countries conducted rather more experiments in science classes than those in low-performing countries.

Similarly, for the two school indices, the availability of school resources for science instruction (ASRSI) and school and class attendance (SCA), the top group had a slightly higher index level than the bottom group. The top countries all had at least medium level, which was the international average level, indicating that high-performing countries had an average level of resources slightly better than those in low-performing countries. Likewise, attendance of students in top-performing countries' schools was similar to the international norm, slightly higher than those in low-performing countries' schools.

As a whole, the differences in the levels of indices between the top and bottom groups are not marked. The comparisons of the levels of student, teacher and school indices suggest that higher levels do not necessarily appear to lead to higher student achievement in TIMSS 1999. The expected positive relationship between level of index and student achievement, that is, high level leads to high achievement, is not evident from the data.

Multiple regressions on indices

Using the four student indices HER, OST, SCS and PATS as predictors of the science average scale score, only HER and PATS were significant at 0.000 and 0.025 levels, respectively. HER had a positive, greater contribution (0.570), while PATS had a negative contribution (−0.307) to the science score. While the contribution of HER could not be seen in the earlier analysis, the negative contribution of PATS could be gleaned from the index's high level of the bottom performing countries compared to the medium level of the top countries.

The multiple regression model with HER and PATS as significant predictors accounted for about 57 percent of the variation in the science average scale score (with HER, about 51 percent, and with PATS, an increase of about 6 percent, the latter being negative) at the 0.000 significance level.

Of the four teacher indices, only the teachers' ESRPS contributed significantly ($p < 0.005$) to the science average scale score of a country. The model accounted for about 21 percent of the variation in the science score at the 0.005 significance level. However, the ESRPS contribution (−0.488) was negative. This is partly corroborated by the initial analysis where the index of the bottom group was slightly higher than the top group on ESRPS.

With the school indices as predictors, ASRSI was a significant ($p < 0.005$) positive predictor of science score. The regression model accounted for about 18 percent of the variation in the science score at the 0.005 significance level. However, SCA was a non-significant predictor.

The indices HER, PATS, ESRPS and ASRSI, were significant in separate multiple regressions runs. HER and ASRSI were positive predictors, while PATS and ESRPS were negative predictors. When put together in one multiple regression run, only HER and ESRPS remained significant predictors. PATS and ASRSI became non-significant predictors. The contribution of HER (0.630) was still positive and greater than that of ESRPS (–0.290). The contribution of HER increased when combined with other student indices. ESRPS' contribution remained negative but its contribution decreased when combined with other teacher indices. The model can account for about 56 percent of the variation in the science average scale score at the 0.000 significance level.

Mathematics score and class size

In view of the large unexplained variance of about 44 percent in the science average scale score, the contribution of mathematics average scale score and class size were explored based upon the fact that mathematics is fundamental to science as both a tool and as the language of science and, secondly, given the large class size in the Philippines (national average of 50 students compared to the international average of 31), which may have some effect on its low student performance.

In a multiple regression of HER, ESRPS, mathematics average scale score and class size, HER and ESRPS were no longer significant predictors. The mathematics average scale score was the single most important predictor ($p < 0.000$) of the science average scale score with a contribution of 0.910. It is important to note that this could also be a reflection of the scholastic aptitude or cognitive ability. Class size was a negative significant ($p < 0.012$) predictor having a contribution of –0.132. The model accounted for about 93 percent of the variation in the science score at the 0.000 significance level.

A comparison of the class size of the top and bottom groups of countries (Table 15.2) showed that the top countries and two of the bottom countries had a large class size (36 or more students), while two others in the bottom group had 21–35 students per class, which was the international average. Class size did not appear to be a predictor in the initial analysis in determining the top performing countries. However, when all countries were included in the quantitative analysis, class size was a significant predictor; meaning the larger the class, the lower was the science achievement.

Predictors of Mathematics Achievement

Levels of indices

As for science, the international average for all the indices was the medium level, with the exception of the teacher's CPTM, which was found to exhibit a high level

Table 15.2 Class size of top and bottom countries in science and mathematics.

Country	Rank in science	Rank in mathematics	Class size[a]
Top			
Singapore	2	1	36 or more
Korea, Republic of	5	2	36 or more
Chinese Taipei	1	3	36 or more
Japan	4	5	36 or more
Bottom			
Chile	35	35	21–35
Philippines	36	37	36 or more
Morocco	37	37	21–35
South Africa	38	38	36 or more

Note. [a]Refer to the Appendix for actual class size.

(Table 15.3). As with the science indices, high-level indices were expected to be associated with higher mathematics scores.

Regarding the student indices, the top five and bottom five countries in student achievement all had a medium level in SCM. In OST and PATM, the bottom countries on the whole had a higher level than the top countries. The HER level of the top group (all had a medium level) was only slightly higher than that of the bottom group (four countries had a medium level and one had a low level).

For the teacher indices, there was no difference between the top and bottom groups in EMH, most countries had a medium level. Teacher's EMRPS had a similar pattern; all bottom countries had a medium level, while the top countries effectively had a medium level, with the high level of one country offsetting the low level of another. For teacher's confidence in their CPTM and teacher's ECMC, the bottom countries as a group had a higher level than the top countries taken as a whole.

Regarding the school indices, the SRMI and SCA, the top group was only slightly higher than the bottom group. In ASRMI, four top countries had a medium level, and one top country had a high level. Four bottom countries also had a medium level, while one other country had a medium-low level.

As in science, the overall relationship of the student, teacher and school indices with students' mathematics achievement was not apparent. The top countries were expected to have a higher level of indices than the bottom countries, but this expectation is not supported by the data.

Multiple regressions on indices

When for 35 countries the student indices were regressed against the average scale score for mathematics, only HER and PATM were significant ($p < 0.000$ and $p < 0.036$, respectively) predictors of the mathematics score. They accounted for about 47 and 6 percent, respectively, of the variation in the mathematics score at a 0.000 significance level. PATM was a negative predictor with a contribution of –0.298, while HER had a positive contribution (0.550) to the mathematics score.

Table 15.3 TIMSS 1999 levels of indices of the five top and five bottom countries in mathematics.

	Students				Teachers				Schools	
	HER	OST	SCM	PATM	CPTM	EMRPS	ECMC	EMH	ASRMI	SCA
Top five countries										
1. Singapore	Medium	High	Medium	Medium	Medium	Medium Low	Medium Low	High	High	Medium
2. Korea, Republic of	Medium	Medium	Medium	Medium	Medium	Medium	Medium	Medium	Medium	Medium
3. Chinese Taipei	Medium	Medium	Medium	Medium	Medium	Medium	Medium	Medium	Medium	Medium
4. Hong Kong, SAR	Medium	Medium Low	Medium	Medium	Medium	Medium	Medium	Medium	Medium	Medium
5. Japan	Medium	Medium	Medium	Medium	Low	Low	Low	Medium	Medium	Medium
Bottom five countries										
34. Indonesia	Medium	High	Medium	High	High	Medium	Medium	High	Medium	Medium
35. Chile	Medium	Medium	Medium	Medium	Low	Medium	Medium	Medium	Medium	Medium
36. Philippines	Medium	High	Medium	High	Medium	Medium	Medium	Medium	Medium	Medium
37. Morocco	Low	High	Medium	High	High	Medium	Medium	Medium	Medium	Medium
38. South Africa	Medium	High	Medium	High	High	Medium	High	Medium	Medium Low	Low
International Average	Medium	Medium	Medium	Medium	High	Medium	Medium	Medium	Medium	Medium

Note. For legend see Figure 15.1.

None of the teacher indices were significant predictors of mathematics score. Only the school index, ASRMI, was a significant ($p < 0.008$) positive predictor of mathematics score with a contribution of 0.432. The model accounted for about 16 percent of the variance in the mathematics score at a 0.008 significance level.

When all significant predictors of mathematics score were combined, the ASRMI was no longer a significant predictor. HER and PATM remained significant predictors. When class size was entered as a possible predictor, it was non-significant. HER and PATM continued to be significant predictors of the mathematics score.

Science and Mathematics Patterns

In the initial analysis of the indices, EMH/ESH and SCS/SCM did not differentiate between the top and bottom groups (Table 15.4). As a group, the top countries had a slightly higher level than the bottom countries in HER, ASRSI/ASMRI and SCA. The difference in level was more marked for OST and PATS/PATM, in favor of the bottom countries. Overall, the bottom group had a slightly higher teacher's confidence in their preparation to teach science/mathematics (CPTS/CPTM) level than the top group.

The results were mixed for teacher's ESRPS/EMRPS, ECES and ECMC. EMRPS did not differentiate between the two groups, while ESRPS was slightly higher for the bottom group. Compared to the top group, the overall level of the bottom group was slightly higher for the index ECMC and a little lower in ECES.

Of the 17 indices, the top countries were found to have slightly higher levels for four indices than the bottom countries (Table 15.4), namely, HER, teacher's ECES, ASRMI/ASRSI and SCA. The two groups had the same level in four indices. The top group was slightly lower than the bottom group in four indices and lower in three indices. Thus, from this initial analysis of the indices in relation to both the science

Table 15.4 Comparison of levels of indices of top (T) and bottom (B) countries.

	Index	Comparison[a]
Student	Home Resources	T > B
	Study Time	T << B
	Self-concept (s/m)	T ~ B
	Attitudes (s/m)	T << B
Teacher	Confidence (s/m)	T < B
	Problem Solving (m)	T ~ B
	Problem Solving (s)	T < B
	Calculators	T < B
	Experiments	T > B
	EMH/ESH	T ~ B
School	ASRMI/ASRSI	T > B
	SCA	T > B

Note. [a] > slightly higher; ~ about the same; < slightly lower; and << lower.
For legend see Figure 15.1.

and mathematics scores, there appeared to be no general pattern associating level of indices with student achievement.

In the quantitative analysis, HER and PATM were overall significant predictors of the mathematics score (Figure 15.2).

Both the mathematics score and class size were overall significant predictors of the science score. PATM and class size were negative predictors, a logical result for class size but perhaps not for PATM. Figure 15.2 illustrates the relative magnitude of the contributions (standardized beta coefficients) of the predictors of science and mathematics scores. The significance levels (in parentheses) of the regression coefficients are also shown.

The mathematics score is the single most important predictor of science score. However, this could also be associated with other variables such as cognitive ability. Students with high mathematics score are also likely to score high in science and this could be a result of the fact that many of the science items needed mathematical thinking and computational skills. That HER was the biggest predictor of mathematics score suggests the need for schools to compensate the lack of home educational resources for disadvantaged children.

Philippine Predictors of Student Achievement
Once the cross-national data had been analyzed, multiple regression analysis was applied specifically to the Philippine data. For the science score, the multiple regression of all the student, teacher and school indices, mathematics score and class size predictors showed that mathematics score, as in the inter-country analysis, was the single most important predictor for science, accounting for about 48 percent of the

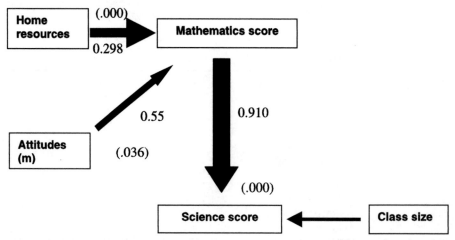

Figure 15.2 Overall significant predictors of mathematics and science scores of participating countries (unit of analysis: country).

variation in the science score (Figure 15.3). However, class size was a non-significant predictor.

Unlike in the inter-country analysis, the student indices, SCS, PATS, HER and OST, negatively, turned out to be significant predictors, accounting for 3 percent of the variation in the science score. Some teacher and school indices, teacher's CPTS, ESH, SCA, ASRSI, were significant predictors, but accounting for only 0.2, 0.2 and 0.1 percent, respectively, of the variation in the science score. Teacher's ECES was not included in the Philippine database and therefore not included in this analysis. The regression model accounted for only about 52 percent of the variation in the science score.

The model for mathematics achievement, including all the predictors, accounted for much less (about 16 percent) of variation in mathematics score (Figure 15.3). All four student indices (not only HER and PATM as in the international data) but also SCM and OST were significant predictors of the mathematics score, accounting for about 12 percent of the variation in the score. PATM had a positive contribution (0.156) in the Philippine data but a negative predictor in the international results.

Unlike in science, OST was a positive predictor in the Philippine data for mathematics. The school index ASRMI, unlike in the inter-country analysis, had a significant but small contribution to the mathematics score, explaining 2 percent of the variation in the score. In addition to the teacher indices, EMRPS and EMH, the index CPM was also a significant predictor. The teacher indices, however, explained only 1.7 percent of the variation in the score.

In sum, the Philippine data supported the inter-country regression results, illustrating that mathematics score was the single, largest significant predictor of the science score and that HER and PATM were significant predictors of mathematics score, although the PATM contribution was positive for the Philippines and negative for the international sample.

The TIMSS 1995 regression results (Mullis *et al.*, 2000) found different sets of predictors for each country included in the analysis of top one-third versus bottom one-third of the schools per country. Only the predictors, HER, PATM, and class size in this study concurred with the TIMSS 1995 regression studies for 14 countries. Home educational resources had a significant positive influence in the majority of the countries. Attitude to science was a significant ($p < 0.1$) positive predictor in only three countries, while attitude to mathematics was a significant positive predictor in only four countries. Class size was a significant positive predictor in only two countries for science and only in three countries for mathematics.

Conclusion

The initial 'qualitative' comparisons of student, teacher and school indices indicated no apparent trend that higher levels of indices would result in higher student achievement in science and mathematics. However, the quantitative analysis revealed significant predictors of science and mathematics scores.

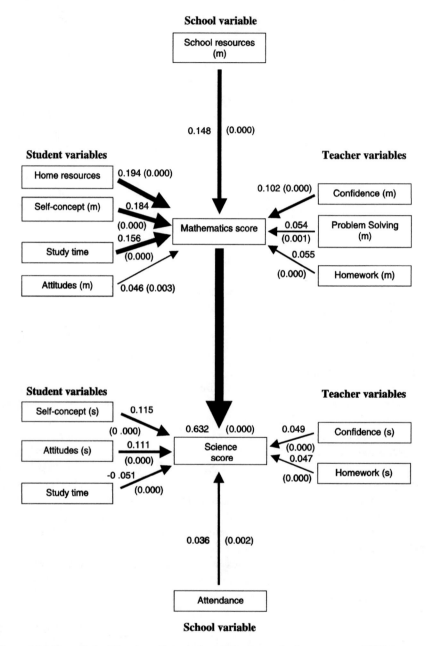

Figure 15.3 Overall significant predictors of mathematics and science scores of Filipino students (unit of analysis: student).

Since mathematics score is the single most important predictor of the science score in the international and Philippine data, the teaching of mathematics needs to be given at least the same time allotment as the teaching of science in Philippine schools. Efforts to improve mathematics achievement at the national level appear to lead to increasing science achievement as well. However, as mentioned previously, the association between mathematics and science could also be the result of other variables such as scholastic and cognitive ability, which also needs to be considered.

The fact that Philippine schools have a much higher average class size than the international average should be given serious consideration by local authorities, in as much as class size is a significant predictor of science achievement in the international results. This recommendation is given notwithstanding the national result of class size as a non-significant predictor of science achievement. Perhaps, training of Filipino teachers can also focus on large class management, particularly, in doing student experiments in science.

The international and national results of the home educational resources index as a significant positive predictor of mathematics achievement can be addressed by schools to compensate for inadequate home resources of some students. School administrators can also consider the Philippine result of the availability of school resources for mathematics instruction and its positive contribution to mathematics achievement.

The index of positive attitude towards mathematics was a positive significant predictor of the mathematics score in the national data but a negative significant predictor in the international results. One implication seems to be that enhancing students' positive attitude towards mathematics contributes to local mathematics achievement, although one has to be aware of the difficulty of determining whether this is because of or as a consequence of good achievement in mathematics, but that it does not necessarily improve the international ranking of the Philippines in mathematics achievement.

Filipino teachers, educators, curriculum developers and education officials can work together to reflect upon the significant effects on student achievement of the other indices on student achievement, such as teacher's self-concept in science and mathematics and confidence in preparation to teach mathematics/science, and out of school study time, based on national data, and to translate these to concrete teaching and administrative strategies.

This study dealt only with questionnaire indices, class size, and (for science) mathematics score as predictors of science and mathematics achievement. Multiple regression analysis can be carried out on the questionnaire items not covered by the indices. Since a limited number of non-randomly selected countries were involved in the regression runs, the results can be further validated with regression analyses per country using 'school' as the unit of analysis.

In future studies, the usefulness of the student, teacher and school questionnaire data in predicting the science and mathematics scores needs to be reexamined. Validation of the questionnaire responses with interview and observation data may improve the predictive ability of the questionnaire variables.

Appendix 15.1 TIMSS 1999 scores, class size, and student, teacher and school indices.

Country	Average scale score (science)	Average scale score (math)	Class size	Student Indices						Teacher Indices								School Indices		
				HER	OST	SCS	SCM	PATS	PATM	CPTS	CPTM	ESRPS	EMRPS	ECES	ECMC	ESH	EMH	ASRSI	ASRMI	SCA
Chinese Taipei	569	585	39	2.00	1.88	2.25	1.97	2.36	2.39	1.50	2.57	1.82	1.84	2.35	1.52	1.80	2.46	2.17	1.90	2.17
Singapore	568	604	37	1.97	2.54	1.89	2.04	2.39	2.51	1.74	2.56	1.55	1.62	2.16	2.85	2.06	2.66	1.88	2.46	2.27
Hungary	552	532	21	2.12	2.32	2.29	2.17	2.47	2.36	nd	2.20	1.54	2.06	2.57	2.09	2.18	2.17	2.22	2.29	2.06
Japan	550	579	36	nd	1.84	2.04	1.94	2.34	2.40	1.44	1.40	1.60	2.44	2.29	1.21	1.71	1.77	1.90	2.33	1.61
Korea, Rep. of	549	587	42	2.07	1.75	2.00	2.05	2.51	2.34	1.80	2.27	1.45	2.08	2.56	1.29	2.26	2.13	2.52	1.90	2.24
Netherlands	545	540	25	2.07	2.12	2.20	2.19	2.47	2.37	2.00	2.72	1.60	1.84	2.46	2.95	1.89	2.10	2.22	2.40	2.06
Australia	540	525	27	2.19	1.95	2.25	2.17	2.39	2.47	nd	2.69	nd	1.68	nd	2.80	nd	2.09	1.90	2.27	2.04
Czech Republic	539	520	24	2.09	1.94	2.21	2.04	2.48	2.38	1.79	2.84	1.90	2.15	2.48	2.30	2.00	1.89	2.22	2.50	2.28
England	538	496	99	nd	nd	2.11	2.21	2.33	2.47	1.56	nd	1.57	1.72	2.21	2.79	1.91	2.27	1.63	2.24	nd
Belgium (Flemish)	535	558	19	2.02	2.34	2.39	2.12	2.37	2.43	1.82	2.62	1.45	1.42	2.45	2.32	1.92	1.93	2.32	2.54	2.49
Slovak Republic	535	534	25	2.06	2.12	2.27	2.02	2.38	2.35	2.27	2.85	1.59	2.06	2.35	2.36	1.28	2.00	2.43	2.01	2.19
Finland	535	520	19	nd	2.00	1.98	2.20	2.32	2.38	1.52	2.68	1.45	1.76	2.81	2.38	1.97	2.10	2.11	2.22	1.97
Canada	533	531	27	2.25	2.08	2.23	2.26	2.38	2.48	1.82	2.63	1.62	1.89	2.49	2.76	1.95	2.15	2.21	2.26	2.09
Slovenia	533	530	22	2.06	2.19	2.22	2.11	2.35	2.39	nd	2.00	nd	2.14	nd	1.83	nd	2.17	1.97	2.02	2.37
Hongkong, SAR	530	582	37	1.84	1.74	2.05	1.98	2.40	2.32	1.21	2.50	2.01	1.68	2.55	2.75	1.61	2.39	2.25	2.10	2.18
Russian Federation	529	526	24	2.05	2.42	2.21	2.34	2.43	2.39	1.84	nd	1.29	1.96	2.44	2.19	1.41	2.57	2.60	1.49	1.90
Bulgaria	518	511	22	2.07	2.30	2.14	1.95	2.49	2.49	1.88	2.37	1.91	2.13	2.31	1.84	1.60	2.50	1.66	1.63	2.08
United States	515	502	26	2.16	2.01	2.13	2.20	2.44	2.46	2.12	2.85	1.55	1.92	2.52	2.62	1.91	2.26	2.36	2.33	2.06

Country																				
New Zealand	510	491	25	2.12	1.97	2.24	2.13	2.42	2.44	1.80	2.80	1.98	1.58	2.27	2.73	2.50	2.01	2.18	2.30	1.99
Latvia (LSS)	503	505	22	2.04	2.34	2.30	1.99	2.46	2.33	2.09	2.56	1.83	1.76	2.33	1.85	2.07	2.21	2.28	1.62	1.85
Italy	493	479	20	1.94	2.52	2.05	2.11	2.35	2.42	nd	2.47	1.78	2.18	nd	2.41	1.98	2.80	1.83	2.22	2.24
Malaysia	492	519	38	1.82	2.60	2.56	2.16	2.39	2.75	nd	2.70	1.76	2.01	2.23	1.37	2.31	2.71	1.47	2.13	0.61
Lithuania	488	482	23	1.97	2.27	2.15	2.05	2.75	2.35	1.48	nd	2.04	1.84	2.54	nd	2.43	2.25	2.16	1.83	1.80
Thailand	482	467	42	1.51	2.37	2.01	1.83	2.44	2.42	2.04	1.61	1.88	1.70	2.46	1.43	2.10	2.73	2.10	1.51	2.01
Romania	472	472	24	1.83	2.43	1.88	1.81	2.39	2.39	2.09	2.78	1.92	2.16	2.22	1.45	1.72	2.76	1.76	1.78	1.86
Israel	468	466	26	2.18	2.23	2.16	2.17	2.33	2.50	1.70	2.72	1.76	1.98	2.19	2.65	2.39	2.52	1.30	2.26	1.65
Cyprus	460	476	29	2.06	2.21	1.77	2.00	2.42	2.44	1.57	2.87	1.80	1.94	2.53	1.84	2.53	2.36	1.45	2.15	1.92
Moldova	459	469	26	1.84	2.42	2.27	1.93	2.50	2.33	1.90	2.31	1.85	2.05	2.29	2.07	2.01	2.57	2.31	1.33	1.64
Macedonia	458	447	28	1.81	2.49	2.02	1.95	2.62	2.59	2.06	2.93	1.81	2.09	2.48	1.75	2.07	2.38	1.70	1.63	1.14
Jordan	450	428	36	1.79	2.48	1.89	1.90	2.50	2.62	2.10	2.83	1.87	1.85	2.25	1.82	2.48	2.32	2.10	1.74	2.34
Iran, Islamic Rep. of	448	422	33	1.47	2.65	2.14	1.99	2.48	2.60	2.19	2.69	2.01	1.77	.84	1.46	2.11	2.91	1.56	1.83	2.35
Indonesia	435	403	45	1.59	2.38	2.27	1.91	2.67	2.50	1.33	2.65	2.10	1.79	2.51	1.72	1.72	2.61	1.77	2.12	1.77
Turkey	433	429	43	1.53	2.52	2.23	1.97	2.60	2.48	1.80	2.71	1.65	2.15	2.18	1.48	2.54	2.46	1.80	1.72	2.14
Tunisia	430	448	34	1.65	2.50	2.21	1.97	2.59	2.67	2.41	1.93	1.84	1.74	2.18	1.67	1.79	2.28	1.62	1.85	1.91
Chile	420	392	34	1.68	2.09	2.05	1.90	2.52	2.52	1.41	1.79	1.81	1.78	2.19	1.91	1.98	2.01	2.08	2.12	2.01
Philippines	345	345	50	1.73	2.41	1.83	1.85	2.65	2.65	1.73	2.25	2.14	1.95	2.44	1.60	2.16	2.12	1.77	1.83	1.88
Morocco	323	337	33	1.38	2.36	1.91	1.84	2.58	2.58	1.99	2.52	1.71	1.65	nd	1.94	1.86	2.11	1.80	1.82	1.64
South Africa	243	275	50	1.58	2.29	1.82	1.81	2.62	2.62	1.96	2.42	1.81	1.90	2.28	2.43	2.14	2.25	1.59	1.62	1.50

Note. nd = no data.

Notes

1. Due to the fact that the levels of the indices were categorized originally by Martin, Mullis, Gonzalez *et al*. (2000), based upon contents of the items rather than upon a statistical procedure, this approach is seen as qualitative and is further discussed as such within this section.
2. TIMSS used Item Response Theory (IRT) to generate country means based upon plausible values. The IRT scaling procedures produced five inputed scores or plausible values for each student (see Martin, Gregory and Stemler, 2000).

References

Martin, M., Mullis, I., Gonzales, E. Gregory, K., Smith, T., Chrostowski, S., Garden, R. and O'Connor, K. (2000) *TIMSS 1999 international science report*. Boston, MA: International Study Center, Lynch School of Education, Boston College.

Martin, M., Mullis, I., Gregory, K., Hoyle, C. and Shen, C. (2000) Effective Schools in Science and Mathematics. IEA's *Third International Mathematics and Science Study*. Boston, MA: International Study Center, Lynch School of Education, Boston College.

Martin, M., Gregory, K. and Stemler, S. (2000) *TIMSS 1999 Technical Report*. Boston, MA: International Study Center, Lynch School of Education, Boston College.

Mullis, I., Martin, M., Gonzales, E., Gregory, K., Garden, R., O'Connor, K., Chrostowski, S. and Smith, T. (2000) *TIMSS 1999 International Mathematics Report*. Boston, MA: International Study Center, Lynch School of Education, Boston College.

Part 4

Curriculum Related Variables and Achievement

16

Exploring the Change in Bulgarian Eighth-grade Mathematics Performance from TIMSS 1995 to TIMSS 1999

Kelvin D. Gregory and Kiril Bankov

Introduction

Historically, Bulgarians have viewed education as being of distinct social value. The education system has been characterized as progressive, stable, inclusive and offering a high quality of education (UNESCO, 2000). In 1995, Bulgaria took part in the Third International Mathematics and Science Study, subsequently renamed Trends in International Mathematics and Science Study, or TIMSS. On that assessment, Bulgarian eighth-grade students had a rescaled mean achievement of 527, statistically higher than the international mean of 500 (Beaton *et al.*, 1996; Yamamoto and Kulick, 2000).[1] However, in the 1999 follow-up study, the Bulgarian mathematics achievement fell to 511 scale score points (Mullis, Martin and Stemler, 2000). While still significantly above the international average, the decrease in achievement was both substantial and significant. However, policy-related interpretations of the decrease should be made only after two major sources of error, sampling and measurement, have been thoroughly explored. This chapter examines issues related to sampling across the two studies, briefly investigates the psychometric model used to summarize the Bulgarian achievement data, compares the performance of Bulgarian students across the content areas for the two assessments, and then examines student performance within the context of opportunity to learn, paying special attention to when that opportunity to learn was most likely to occur.

Sampling Issues for Bulgaria

The TIMSS studies were designed to be fair, comparative assessments. Each participating country was expected to meet or exceed a number of standards designed to
ensure comparability of the achievement results in accordance with the International
Association for the Evaluation of Educational Achievement (IEA) standards (Martin
et al., 1999). As part of these requirements, each country was expected to test a statistically representative sample from the upper of the two grades that contained the most
13-year-old students.

Most TIMSS samples were designed as two-stage stratified cluster samples (Foy,
1997; Foy and Joncas, 2000a). The first stage consisted of sampling individual
schools in which 13-year-olds were enrolled in the school. Schools were generally
sampled systematically with a probability proportional to the estimated number of eligible (13-year-old) students enrolled. A minimum of 150 schools was selected in each
country (where that number existed), although the requirement for national analyses
often called for a somewhat larger sample. As the schools were sampled, two replacement schools were simultaneously identified in case a sampled school chose not to
participate. The second stage of the sampling design involved randomly selecting a
Grade 8 mathematics class from the participating school.

Bulgaria selected the same target population in both the TIMSS 1995 and 1999
assessments, that is, the Bulgarian eighth grade at the time of the first study. In the
early 1990s, the usual age for entry in Grade 1 of the Bulgarian system was six years.
In the years following the TIMSS 1995 assessment, this rose to seven years because
of the loose interpretation of the age entrance policies and parents decided to keep
their children home for an extra year. The knock-on effects of these changes can be
seen in the 1999 target population, with the 1995 mean age being 14.01 years compared with the 1999 mean age of 14.78 years. However, the number of years of
schooling remained the same over the two assessments.

Data quality standards in TIMSS required minimum participation rates for schools
as well as for students (Foy and Joncas, 2000b; Martin *et al.*, 1999). These standards
were established to minimize response bias. In the case of countries meeting these
standards, it is likely that any bias resulting from non-response will be negligible, that
is, typically smaller than the sampling error. The allocation of the school sample for
Bulgaria for TIMSS 1995 and 1999 is shown in Table 16.1. In 1995, Bulgaria's minimum required sample consisted of 167 schools. Of the attained sample, 114 schools
were from among those first sampled, four more were replacement schools, and a total
of 49 schools did not participate. Most of the non-participating schools actually completed the assessment but simply did not return the completed tests to the National
Research Center in time to be included in the assessment. Since over 30 percent of the
schools failed to submit the required data in time, meant that Bulgaria did not meet
the sampling standards for TIMSS 1995.

For TIMSS 1999, Bulgaria's minimum school sample allocation was 172. Of the
sampled schools, 163 participated. The unweighted school-level participation rate of
95 percent, together with a similar student response rate, meant that Bulgaria met the
sampling standard in TIMSS 1999. Given that Bulgaria did not have a representative

Table 16.1 Bulgarian school samples for TIMSS 1995 and 1999.

Year	Total sampled schools	Ineligible schools	Participating schools		Non-participating schools
			Sampled	Replacement schools	
1995[a]	167	–	114	4	49
1999	172	3	163	0	6

Note. [a]Includes schools with Grade 7 classes.
Source. 1995 data Foy (1997); 1999 data Martin *et al.* (1999).

sample in TIMSS 1995 but did have one in TIMSS 1999, it is feasible that the difference in national mathematics achievement between the two assessments can be attributed, at least partly, to a non-response bias in the 1995 study. However, this supposition is based on the premise that the TIMSS achievement estimates appropriately summarize the student responses.

The TIMSS Psychometric Models and Bulgarian Achievement

TIMSS relies on a sophisticated form of psychometric scaling, known as item response theory (IRT), to combine student responses in a way that provides accurate estimates of achievement (Yamamoto and Kulick, 2000). TIMSS used two- and three-parameter IRT models to summarize the achievement data from the two assessments and place the data onto common scales. For mathematics, there was one overall scale (the mathematics scale) and five sub-scales defined by mathematics content domains (fractions and number sense, measurement, data representation, geometry, and algebra). However, the usefulness of these scales is predicated on the degree to which they accurately characterize how students within any country actually responded to the TIMSS items.

TIMSS used a cluster-based, rotated design that resulted in 277 mathematics and science items being distributed across eight test booklets (Garden and Smith, 2000). The first cluster in each test booklet, the A cluster, consisted of six mathematics and six science multiple choice items. The A cluster was used in both the TIMSS 1995 and 1999 and is one of several item clusters used to link the two assessments. As a brief investigation of the accuracy of the IRT models, the predicted and empirical probability of obtaining each of the six mathematics items was explored.

The probability of obtaining a correct response on each mathematics item in cluster A was computed using the imputed or plausible achievement values of the students and the item characteristics (Yamamoto and Kulick, 2000). The average country-level probability of getting a mathematics item correct was then computed for the two assessments using these model-based probabilities. Similarly, the average country-level empirical probability of getting an item correct was computed using students' actual responses to the multiple choice items. The results of these analyses are presented in Table 16.2.

Table 16.2 Bulgarian empirical and model-based probablities of correct scores (item block A).

	TIMSS 1995				TIMSS 1999			
	Empirical probability		Model-based probability		Empirical probability		Model-based probability	
M012001	0.62	(0.02)	0.63	(0.01)	0.66	(0.02)	0.61	(0.02)
M012002	0.87	(0.01)	0.79	$(0.01)^b$	0.86	(0.01)	0.77	$(0.01)^b$
M012003	0.43	(0.02)	0.68	$(0.01)^a$	0.40	(0.02)	0.63	$(0.01)^a$
M012004	0.57	(0.01)	0.58	(0.01)	0.55	(0.02)	0.54	(0.01)
M012005	0.80	(0.01)	0.65	$(0.01)^b$	0.80	(0.02)	0.66	$(0.01)^b$
M012006	0.87	(0.01)	0.80	$(0.01)^b$	0.85	(0.01)	0.78	$(0.01)^b$

Notes. [a]Indicates that the model-based probability is statistically higher than the empirical probability at the $p < 0.05$ level, two-tailed when adjusted for multiple comparisons.
[b]Indicates that the empirical probability is statistically higher than the model-based probability at the $p < 0.05$ level, two-tailed when adjusted using Dunn (Bonferroni) correction for multiple comparisons.

Several points are noteworthy from this table. First, the empirical probabilities for each item across the two assessments are very similar. For example, the empirical probability of getting item M012006 correct is 0.87 in TIMSS 1995 and 0.85 in TIMSS 1999. Second, the model-based probabilities are also similar across the two assessments. For example, the model-based probability of getting M012001 correct is 0.63 in 1995 and 0.61 in 1999. Third, the empirical and model-based probabilities are statistically different, after adjusting for multiple comparisons, for four of the six items for both assessments ($p < 0.05$). In one case, M012003, the model-based probability is significantly higher than the empirical probability in both assessments. The item is from the measurement domain. For three items, M012002, M012005 and M012006, Bulgarian students performed statistically better than the IRT model predicted. These items are from the algebra, geometry and data representation domains, respectively.

If the drop in achievement was attributable to changes in sampling over the two assessments, then it would be reasonable to expect the empirical probability for each item to reflect the decrease. It is apparent from Table 16.2 that the expected decrease does not occur in cluster A. The empirical achievement on the cluster A mathematics items is essentially the same over the two assessments.

Expanding the analyses to include all 47 mathematics items[2] that link the 1995 and 1999 assessments suggests that the discrepancy between the empirical and model-based probabilities is somewhat persuasive. Table 16.3 shows the number of linking items for each domain, the number of these items for which the empirical probability of obtaining a correct response was significantly higher or lower than the model-based probability for the two assessments, and the number of items where the empirical probabilities are statistically different across the two assessments. The empirical or actual probability of a correct response is statistically higher, after applying a conser-

Table 16.3 Number of statistically significant differences between Bulgarian empirical and model-based probabilities for TIMSS linking Items.[a]

Domain	Number of linking items	TIMSS 1995		TIMSS 1999		Difference between 1995 and 1999 empirical probabilities	
		Model-based probability larger than empirical probability	Empirical probability larger than model-based probability	Model-based probability larger than empirical probability	Empirical probability larger than model-based probability	1999 greater than 1995	1995 greater than 1999
Data	8	0 (0)	3 (3)	2 (2)	2 (2)	3 (3)	4 (4)
Algebra	11	2 (0)	7 (5)	2 (0)	4 (4)	2 (2)	6 (5)
Fractions	17	4 (3)	3 (1)	5 (4)	2 (0)	3 (2)	11 (7)
Geometry	6	0 (0)	5 (5)	0 (0)	4 (4)	2 (1)	1 (1)
Measurement	6	1 (1)	3 (2)	1 (1)	2 (2)	1 (1)	4 (2)
All Items	47	7 (4)	21 (16)	10 (7)	14 (12)	11 (9)	26 (19)

Note. [a]Numbers in parentheses indicate number of statistically different items when adjusted using Dunn (Bonferroni) correction for multiple comparisons, over all items. All calculations use $p < 0.05$, two-tailed tests.

vative adjustment for multiple comparisons, than the model-based probability for 16 items in TIMSS 1995 and 12 items in TIMSS 1999. The model-based probability of a correct response is higher than the empirical probability for four of the linking items in TIMSS 1995 and seven items in TIMSS 1999. Taken together, the suggestion is that the psychometric models do not capture the Bulgarian achievement data for over 40 percent of the linking items. This suggests that the IRT models used to report the Bulgarian achievement data are unsatisfactory.

The final two columns in Table 16.3 show the number of linking items that had a significant change in empirical probability of a correct response. After applying the adjustment for multiple comparisons, the Bulgarian students showed a decreased performance on 19 items and an increased performance on nine items. Thus, it may be concluded that the Bulgarian performance on the TIMSS assessments did decrease over the four year period, but this decrease was not evenly spread over the items.

While sampling may be at the heart of the change in Bulgarian achievement and cannot be ruled out, there is at least a suggestion that the reason for the decrease may involve the measurement model used to summarize the achievement data. Further, the fact that student performance increased on some items over the two assessments and decreased on others suggests that there may be specific curriculum, instructional or other effects at play. Harnisch and Linn (1981), in their analysis of item response patterns, found that schools in different parts of the USA had very different response patterns to a standardized test. They took this to be indicative of localized mismatches between the test and curricula. Similarly, Tatsuoka and Tatsuoka (1982) found that aberrant responses related to differences in instruction. The next section explores these notions in more detail by examining the changes in achievement of the Bulgarian school types participating in both TIMSS 1995 and 1999.

Bulgarian Profile and Non-profile Schools

Bulgaria follows a 4:4:4 model of education (Bankov, 1997): primary school, for Grades 1–4, is the first level; junior secondary, for Grades 5–8, makes up the second level and is the end of compulsory education, although the minimum age for leaving school is 16; and the third level, for Grades 9–12, comprises technical, vocational and senior secondary education. The TIMSS assessments discussed in this chapter were administered to students in grade 8, the grade that marks the end of junior secondary education. This grade also serves as the beginning grade for some Bulgarian schools.

Bulgaria has six main school types:

- Primary schools (Grades 1–4)
- Basic schools (Grades 1 to 7 or 8)
- Gymnasium (Grades 9–12)
- Profile schools (Grades 8–12)
- Secondary general schools (Grades 1–12)
- Special schools

Bulgarian students participating in the TIMSS assessments at the eighth grade were sampled from basic schools, secondary schools and profile schools. Profile schools have competitive entry requirements, with students required to take entrance examinations in Bulgarian or mathematics, or both subjects. Other school types do not have these entry requirements. For purposes of this discussion, all of the sampled schools in the Bulgarian sample will be referred to either as profile (selective, academic gymnasium and some vocational) schools or non-profile (non-selective) schools. The non-profile school category includes rural schools, urban schools, various types of vocational schools, basic schools and secondary schools (Bankov, 2002).

In 1995, 104 non-profile schools and 11 profile schools participated in the eighth-grade assessment. In contrast, in 1999, 135 non-profile schools and 28 profile schools participated at the eighth-grade level. This change reflects the large increase in profile schools in Bulgaria over this time period. In 1995 there were less than 50 profile schools in Bulgaria, and competition to gain entry into these schools was very high. By 1999 the number of profile schools had increased to over 150, and consequently the schools were at least marginally less selective.[3]

Average Mathematics Raw Score for Bulgarian Profile and Non-Profile Schools

The TIMSS 1995 and 1999 studies used a multiple-test-book design. In this design, a large number of mathematics and science test items are distributed across eight test books. The large number of items allowed for the assessment of student achievement across the broad content areas described in the TIMSS Mathematics and Science Curriculum Framework (Robitaille *et al.*, 1993). Each student received only a subset of the test items, thereby lowering the amount of testing time for each student.

The TIMSS 1999 assessment was designed to replicate the TIMSS 1995 assessment as far as possible. Since approximately two-thirds of the test items were made public after the 1995 study, the TIMSS 1999 test books contained some items that were used in TIMSS 1995 and a substantial number of replacement items. TIMSS 1999 developed 'clone items' – items meant to be copies of the original, but released, items with the goal of making the 1999 test books look like the 1995 test books. The replacement items assessed the same content areas and performance expectations and, as near as possible, matched the difficulty level of the 1995 items (Garden and Smith, 2000; Robitaille and Beaton, 2002). However, taking into account a desire to increase content coverage and the fact that students in TIMSS 1995 had ample time to complete the test, some books used in TIMSS 1999 also included a few additional items (Garden and Smith, 2000).

In an effort to better understand the change in performance of the Bulgarian students, the average percent correct for each test book was computed for the 1995 and 1999 assessments for the school types. Although the 1999 assessment was designed to replicate the 1995 assessment, the two assessments were not identical. The differences in the number and difficulty of items between the two assessments are reflected in the raw scores obtained for the two assessments. A linear equating method

(Kolen and Brennan, 1995) was used to equate the TIMSS 1999 raw scores to TIMSS 1995 raw scores for each book at the total and content area levels. Random samples of one thousand students were drawn from each assessment for each country that participated in both assessments that met the sampling standards. The samples were then pooled and linear equating constants computed. Linear equating methods, like other equating methods, are sensitive to the size of the tests being equated. Thus, it would be expected that errors associated with equating the overall tests would be substantially lower than those of the various content areas. Although not reported here, the method of drawing random samples enables the calculation of standard errors of equating.

Table 16.4 shows the total number of raw score points for each book for the two assessments shown, the mean percent correct, and the equated mean percent correct for each Bulgarian school type. Several differences in the average percent correct between the 1995 and 1999 assessments are worth noting. First, non-profile schools tended to perform at or about the same level for five of the test books (1, 2, 3, 5, 7 and 8) over the two assessments. Second, the performance of non-profile schools on test books 4 and 6 was significantly lower in 1999 compared with 1995. Third, the performance of profile school students decreased significantly across all test books between the two assessments, with an average decrease of 20 percentage points over the test books. The profile school students performed exceptionally well in TIMSS 1995. In TIMSS 1999 they still performed better than their non-profile colleagues, but significantly less well then their 1995 counterparts. The decrease in profile student performance may be attributable to the fact that there were approximately three times as many profile schools in 1999 compared with 1995, and thus these schools were less selective. However, since the test books are distributed evenly across the schools, and the 1999 test books were designed to be very similar to the 1995 test books, these results also support the notion that there are specific items that may explain the poorer performance of non-profile students on some test books. Analyses designed to check for differential item functioning would enable the testing of this hypothesis but such analyses are beyond the scope of this chapter. The next section explores the notion that differential performance on specific content areas may help explain these changes in performance.

TIMSS Mathematics Curriculum Framework and the Bulgarian Curriculum

TIMSS 1995 and 1999 shared a common curriculum framework (Robitaille et al., 1993). This framework had three central aspects: content, performance and perspectives. The content aspect breaks mathematics subject matter into sub-areas. The main content categories for TIMSS 1995 and 1999 were:

- Numbers (reported as Fractions and Number Sense),
- Proportionality (reported with Fractions and Number Sense),
- Measurement,
- Geometry,

Table 16.4 Number of mathematics score points and average percent correct for each test book for Bulgarian non-profile and profile school students in TIMSS 1995 and 1999.

Test book	Maximum points		Non-profile schools			Profile schools		
	1995	1999	1995 Percent correct (SE)	1999 Percent correct (SE)		1995 Percent correct (SE)	1999 Percent correct (SE)	
				Actual	Equated[d]		Actual	Equated[d]
1	43	47	53 (3.5)	48 (1.5)	51 (1.5)	87 (1.7)	66 (4.4)[b]	68 (4.3)[b]
2	33	33	64 (3.4)	61 (1.4)	60 (1.3)	96 (2.2)	76 (3.1)[b]	75 (3.1)[b]
3	40	46	51 (3.1)	51 (1.7)[a]	50 (1.6)	90 (3.6)	70 (3.2)[b]	68 (3.0)[b]
4[c]	34	34	64 (2.6)	56 (1.7)[a]	58 (1.6)[a]	89 (1.3)	77 (2.3)[b]	76 (2.1)[b]
5	40	47	48 (2.9)	49 (1.6)	49 (1.5)	88 (2.0)	68 (3.5)[b]	67 (3.3)[b]
6[c]	33	33	66 (2.7)	58 (1.6)[a]	58 (1.6)[a]	93 (1.5)	74 (2.8)[b]	72 (2.8)[b]
7[c]	39	41	58 (3.2)	54 (1.7)	54 (1.6)	88 (0.1)	67 (6.0)[b]	66 (5.9)[b]
8	41	41	58 (3.2)	53 (1.7)	55 (1.7)	95 (1.1)	69 (4.2)[b]	71 (4.2)[b]

Notes. [a]Decrease was significant at the 0.05, two-tailed, level.
[b]Decrease was significant at the 0.05, two-tailed, level when adjusted using Dunn (Bonferroni) correction for multiple comparisons within the school type for that specific assessment.
[c]One item, M012043, was mistranslated in the TIMSS 1999 Bulgarian assessment. The maximum score points for books 4, 6, and 7 for Bulgaria in TIMSS 1999 is one less than that shown in the table.
[d]Equated values were obtained using linear equating parameters obtained from pooled samples drawn from countries that participated in both assessments that met the sampling standards.

- Functions, relations, and equations (reported as Algebra), and
- Data representation, probability, and statistics.

The performance aspect describes the kinds of performances that students are expected to demonstrate while engaged with the content. The mathematics perform-ance expectations used for the TIMSS 1995 and 1999 studies were known, using routine procedures, investigating and problem solving, mathematical reasoning and communicating. The TIMSS reports do not contain results using these categories. The perspectives aspect was intended to depict curricula goals that focus on the develop-ment of students' attitudes, interests and motivations in mathematics education.

The number of score points allocated to each content area for each test book in the 1999 assessments is shown in Table 16.5. A similar distribution of score points was used in the TIMSS 1995 assessment, with the actual number of score points for any particular content area, for any book, usually being within two points of those pre-sented in Table 16.5.

To investigate whether or not the changes in performance noted in Table 16.4 were evenly spread across the content areas, the change attributable to each content area was calculated for each test book. This statistic was computed by calculating, for each test book, the difference in the equated proportion correct for each content area across the two assessments. The formula is as follows:

$$\Delta X_{i,95} = X_{i,99}^e - X_{i,95}$$

where i is the ith content area; $X_{i,99}^e$ is the equated mean score point for the content area for TIMSS 1999; $X_{i,95}$ is the mean score point for the content area for TIMSS 1995; and $\Delta X_{i,95}$ is the change in mathematics score points for the content area across the two assessments, expressed in 1995 score points.

According to this formula, a value of zero would be expected if there were no change in the performance level across the two assessments, a negative number would indicate a decrease in score points, while a positive number would reflect an increase in performance.

Table 16.5 Maximum number of mathematics score points in each test book by subject matter content category – TIMSS 1999.

Content category	Test book							
	1	2	3	4	5	6	7	8
Fractions and number sense	16	12	16	12	16	12	14	18
Measurement	9	5	11	4	9	4	3	4
Data representation, probability, and statistics	5	4	4	6	8	6	8	5
Geometry	5	6	6	3	5	4	5	5
Algebra	12	6	9	9	9	7	11	9
Total	47	33	46	34	47	33	41	41

Source.   Garden and Smith (2000: 66).

Table 16.6 presents a summary of the changes in terms of the TIMSS 1995 item score points allocated to each content area, and the percent of the decline that can be attributed to each content area. For non-profile school students, four of the changes in performance on algebra items, two of the changes on measurement items, three of the changes for the data representation, probability and statistics content area, one of the changes in geometry performance, were statistically significant. No change for the fractions and number sense area was statistically significant. For all content domains, and across all books, the change was less than one score point.

In contrast, profile school student performance declined markedly in the fractions and number sense content area for all test books. In this content area, students lost approximately double the number of points compared with other content areas on six of the eight test books. That is, they lost two or more score points from fractions and number sense items in five of the eight books. Statistically significant decreases in performance were recorded over all test books, even after adjusting for multiple comparisons, for the fractions and number sense content area. Smaller, but significant, declines were also evident in six of the eight test books for the other four content areas.

If changes in performance were evenly distributed across a test book, then the expected change in any particular content area would be equal to the overall change multiplied by the proportion of score points assigned to that content area:

$$\Delta X_{i,95}^{w} = p_i \Delta X_{95}$$

where p_i is proportion of mathematics score points allocated to content area i; ΔX_{95} is the overall change in mathematics score points across the two assessments, expressed in 1995 score points; and $\Delta X_{i,95}^{w}$ is the change in mathematics score points for content area i, expressed as a proportion of the change in 1995 score points.

Applied to each category and for each school type, the expectation is that there will no difference between the actual change and the predicted change if the change was spread proportionally across the domains. TIMSS 1999 non-profile school students tended to lose disproportionately more points on the algebra items. For example, on test books 4, 5 and 8, they lost 0.30 or more score points on algebra items than expected. These students also tended to lose fewer points than expected on the factions and number sense items. With the exception of test book 7, where they lost 0.27 more score points then expected, these students performed disproportionately better on this domain by 0.21 to 0.63 points. These students did slightly better than expected in the data representation domain on test book 4 and worse than expected on test book 7. The students performed roughly as expected on the other two domains.

In contrast, TIMSS 1999 profile score students always lost more points than expected on the factions and number sense domain. The losses ranged from a low of 0.12 to a high of 1.75, with the loss being greater than a score point on half of the test books. For test books 2 and 4, these students also lost more than expected algebra points (0.22 and 0.35, respectively). However, on the other test books, the profile students performed disproportionately better on the algebra items, and on test books 1, 2, 4 and 6, their performance was 0.5 points or more better than expected. Profile students performed disproportionately better in the measurement domain on test books 3 and 5 but worse than expected on test books 6 and 8. They lost more points than

Table 16.6 Change in TIMSS 1995 raw score points for each reporting category from TIMSS 1995 to 1999.[c]

School type	Test book	Algebra	Data representation, probability and statistics	Fractions and number sense	Geometry	Measurement
Non-profile schools	1	-0.22 (0.43)	-0.09 (0.17)	0.16 (0.67)	-0.23 (0.20)	-0.25 (0.38)
	2	-0.29 (0.23)	-0.40 (0.13)[b]	0.08 (0.59)	-0.44 (0.23)	-0.38 (0.20)
	3	-0.32 (0.41)	-0.13 (0.12)	0.05 (0.61)	-0.39 (0.23)	-0.04 (0.33)
	4	-0.91 (0.27)[b]	-0.11 (0.21)	-0.25 (0.46)	-0.24 (0.15)	-0.53 (0.14)[b]
	5	-0.48 (0.19)[a]	-0.06 (0.28)	0.73 (0.60)	0.30 (0.28)	-0.21 (0.39)
	6	-0.75 (0.23)[b]	-0.67 (0.21)[b]	-0.84 (0.51)	-0.31 (0.15)[a]	-0.32 (0.16)[a]
	7	-0.58 (0.21)[b]	-0.22 (0.36)	-0.98 (0.59)	-0.08 (0.20)	-0.21 (0.18)
	8	-0.57 (0.34)	-0.49 (0.18)[b]	0.02 (0.72)	-0.08 (0.23)	-0.04 (0.16)
Profile schools	1	-0.69 (0.72)	-0.77 (0.32)[a]	-3.73 (0.74)[b]	-0.93 (0.28)[b]	-1.62 (0.42)[b]
	2	-1.43 (0.18)[b]	-0.58 (0.13)[b]	-2.54 (0.47)[b]	-1.02 (0.37)[b]	-1.08 (0.28)[b]
	3	-1.26 (0.31)[b]	-1.31 (0.25)[b]	-5.05 (0.71)[b]	-0.81 (0.10)[b]	-1.45 (0.70)[a]
	4	-1.36 (0.35)[b]	-0.36 (0.24)	-1.49 (0.48)[b]	-0.01 (0.08)	-0.58 (0.23)[b]
	5	-0.47 (0.20)[a]	-0.86 (0.23)[b]	-3.39 (1.06)[b]	-0.47 (0.36)	-0.32 (0.34)
	6	-1.27 (0.31)[b]	-1.50 (0.28)[b]	-1.98 (0.59)[b]	-0.50 (0.14)[b]	-1.17 (0.18)[b]
	7	-1.61 (0.43)[b]	-1.15 (0.42)[b]	-4.91 (1.18)[b]	-0.96 (0.17)[b]	-0.62 (0.14)[b]
	8	-1.75 (0.31)[b]	-0.87 (0.27)[b]	-3.44 (0.83)[b]	-1.39 (0.36)[b]	-1.30 (0.12)

Note. [a]Change is significant at the 0.05 level, without adjustments for multiple comparisons.
[b]Change is significant at the 0.05 level after adjusting for multiple comparisons using the Dunn (Bonferroni) method within content areas for each school type.
[c]TIMSS 1999 values were first equated to TIMSS 1995 values.

Table 16.7 Expected change in TIMSS 1995 raw score points for each reporting category from TIMSS 1995 to 1999.

School type	Test book	Algebra	Data representation, probability and statistics	Fractions and number sense	Geometry	Measurement
Non-profile schools	1	0.35	0.14	−0.25	0.37	0.40
	2	0.20	0.28	−0.06	0.31	0.27
	3	0.39	0.16	−0.06	0.47	0.05
	4	0.45	0.05	0.12	0.12	0.26
	5	−1.71	−0.21	2.61	1.07	−0.75
	6	0.26	0.23	0.29	0.11	0.11
	7	0.28	0.11	0.47	0.04	0.10
	8	0.49	0.42	−0.02	0.07	0.03
Profile schools	1	0.09	0.10	0.48	0.12	0.21
	2	0.22	0.09	0.38	0.15	0.16
	3	0.13	0.13	0.51	0.08	0.15
	4	0.36	0.09	0.39	0.00	0.15
	5	0.09	0.16	0.62	0.09	0.06
	6	0.20	0.23	0.31	0.08	0.18
	7	0.17	0.12	0.53	0.10	0.07
	8	0.20	0.10	0.39	0.16	0.15

expected on the data representation domain on test books 3 and 6, but lost less points than expected on test books 4 and 7. In geometry, the profile students generally lost fewer points than expected, with the exception of test book 8.

Taken together, these analyses suggest that curriculum differentiation may at least partially explain the changes in performance. Given that all students follow a national curriculum, and are taught the same topics each year, this differentiation would most likely be evident in the implementation of the curriculum.

Opportunity to Learn and Mathematics Performance of Bulgarian Students

Bulgarian education is under the control of the Ministry of Education. The Ministry sets policies related to curriculum, textbooks and accreditation of schools. The intended mathematics curriculum, published by the Ministry, describes the content to be taught by each grade level, accompanied by a statement of course goals and the intended learning outcomes (Bankov, 1997). Bulgaria introduced a revised national curriculum in 1997 (Mullis, Martin, Gonzalez et al., 2000). The changes mostly affected the lower grades (up to Grade 5), and Grades 10, 11 and 12. The number of hours for mathematics education was reduced as was some course content for these grade levels. The only significant impact on the Grade 8 mathematics curriculum was to slightly reduce the number of topics covered in geometry.

All students study the national curriculum, with teachers adapting the curriculum to suit particular local needs. The curriculum is supported through a variety of

methods including pre-service teacher education, in-service teacher education, mandated textbooks and a system of school inspections. The curriculum provides a map
of students' opportunity to learn specific mathematics topics, with students in the two
school types covering the same topics, but at varying degrees of difficulty. Profile students typically have between two and four times more mathematics lessons than their
non-profile student colleagues. For most lessons, the profile students go into curriculum topics at a much greater depth, or engage in exercises designed to enable mastery
of advanced mathematical techniques. However, at the start of Grade 8, all Bulgarian
students have been exposed to the same mathematical curriculum content and have
similar formal mathematics experiences.

Opportunity to learn (OTL) was originally intended as a measure of 'whether or
not students have had an opportunity to study a particular topic or learn how to solve
a particular type of problem presented by the test' (Husen, 1967: 162). The Second
International Mathematics Study (SIMS) refined this concept and operationalized it as
the implemented curriculum or implemented coverage (Travers *et al.*, 1989). More
recent work has suggested that OTL should not be limited to purely curriculum coverage, but should include broader dimensions, taking into account factors known to
affect learning. OTL is now generally considered to be a multifaceted concept, and
incorporates many factors known to affect learning, including teacher preparedness,
student readiness to learn and adequacy of facilities (Nodding, 1997). Even given the
more developed understanding we now have of opportunity to learn (or as it is now
widely called, educational opportunity), whether or not students have been at least
exposed to the topic lies at the core of the concept.

OTL, in the narrow sense, has been a fundamental concept underpinning recent
International Association for the Evaluation of Educational Achievement (IEA) mathematics and science assessments (see Robitaille and Beaton, 2002). Early IEA assessments operationalized OTL very narrowly as whether particular tested items were
taught beforehand to students who took the test (Husen, 1967). Since teachers sometimes misinterpreted the items, researchers cautioned that the approach was 'too
bound to the form of the specific item and more representative of teachers' judgments
of items rather than the content categories on which the item is an example' (Schmidt
and McKnight, 1995: 345). TIMSS 1999 operationalized OTL through a series of content specific topic questions based primarily upon the TIMSS curriculum framework
(Mullis, Martin and Stemler, 2000).

The analysis of teachers' reports of students' opportunity to learn the specific
mathematics topics assessed by TIMSS, corroborated with the Bulgarian mathematics curriculum, suggest that most students have had a chance to learn most of the
topics assessed by TIMSS. However, much of this learning has taken place one to
four years before the eighth grade. This suggestion is supported by the Test-
Curriculum Matching Analysis (TCMA) conducted by TIMSS. That analysis rested
largely upon country-based panels deciding whether or not an item was appropriate,
i.e. 'had probably been taught to them' (Howsen, 2002: 96; see also Beaton and
Gonzalez, 1997). Seventy-three percent of the TIMSS 1995 mathematics items were
deemed suitable for both Grades 7 and 8 Bulgarian students (Beaton and Gonzalez,
1997). Since there is a negligible amount of spiraling present in the Bulgarian math-

ematics curriculum, the TCMA results suggest that all of the items deemed suitable for Grade 8 students must have been taught either during or before Grade 7. Thus, it is reasonable to conclude that Grade 8 students in Bulgaria are using knowledge acquired in previous years to answer the TIMSS questions. The TCMA results, in agreement with the curriculum analysis, suggest that many, if not all, of the recent mathematics learning experiences of these Grade 8 students have little direct relationship to the TIMSS assessment. Put another way, the TIMSS mathematics assessment does not assess what the Bulgarian students have been taught during the eighth grade. It may be argued that the TIMSS mathematics items assess foundational mathematical concepts that are essential components of the advanced concepts taught to the Bulgarian students. However, this assumes that students acquire a hierarchy of skills sequentially – something that is at odds with the literature (see Rothstein, 2004).

As early as 1885, psychologists have reported that recall of an event declines over time (for example, Ebbinghaus, 1885; Bahrick and Phelps, 1987). Typically, a student forgets a substantial amount of learned material shortly after the learning cycle, and the rate of forgetting slows down over time. Research in the USA, aimed at examining the effect of the three-month long summer vacation on student achievement, has shown that summer loss is dramatic in mathematics related areas (Cooper *et al.*, 1996). The fractions and number sense domain accounted for approximately 35 percent of the mathematics score points available in each test book in TIMSS 1999 (Garden and Smith, 2000). Most of the items used by TIMSS to assess this domain map onto the Grades 5 and 6 Bulgarian mathematics curriculum. None of the items map onto the Grade 8 curriculum.

One possible explanation for the marked decline in the performance of profile school students could be that these students, like their non-profile student colleagues, were exposed to the TIMSS content many years before the eighth grade. However, unlike their non-profile student colleagues, this exposure occurred not only at another time but in another school. Since these profile school students focus primarily upon more abstract algebra and geometry concepts than their non-profile school colleagues, they typically do not have chance to revisit and revise past topics. Unless a spiral approach to learning the topics is used, in which students are continually exposed to past learning experiences, and have a chance to retrieve, relearn and revise these concepts, it is likely that a significant amount of this earlier learning will not be remembered at the time of testing. A further complicating reason could be the reluctance of profile students (and their teachers) to revisit essentially concrete concepts from a much earlier grade at the expense of learning more challenging, abstract concepts.

Implications of the TIMSS Assessments for Bulgarian Student Achievement

The philosophy underlying TIMSS assumes the 'validity and legitimacy of mapping the achievement outcomes of the world's school systems onto a common scale'

(Clarke, 2003: 147). Under such an assumption, an idealized international curriculum is treated as unproblematic and lies behind the entire TIMSS endeavor (Clarke, 2003; see also Keitel and Kilpatrick, 1999). Given that the TIMSS mathematics assessment covers a broad range of topics, it is reasonable to expect that the assessment might favor countries which have an integrated, spiral curriculum model. Such a model is found in Australia, England, Japan and Singapore (Ruddock, 1998). Further, the TIMSS mathematics assessment is very similar to the US National Assessment of Educational Progress (NAEP) (National Assessment Governing Board, 2002; see also McLaughlin *et al.*, 1998). While the TIMSS and NAEP assessments are both based upon an integrated mathematics assessment framework, an integrated approach to mathematics education in the USA is relatively new. In this country, the tradition has been to teach arithmetic for the first eight years, followed by a year each of algebra and geometry (Ruddock, 1998). The Bulgarian mathematics curriculum has many similarities with other Eastern European countries and relatively few similarities with curricula like those used in England and the USA.

For example, the geometry concepts of congruence and similarity are often taught within the same year under the British system since the two concepts are analogous. In Bulgaria and other Eastern European countries, congruence topics are generalized on the basis that the congruent figures have the same shape and size. Similarity deals with the upper level of this generalization since similar figures have the same shape. Clearly, congruent figures are similar as well, but the reverse is not necessarily true. Based on this distinction, in Eastern European countries, congruence tends to be taught earlier than similarity. For example, in Bulgaria, congruence is taught in Grade 7 while similarity is taught in Grade 9. Given that Bulgaria, and some other former Eastern Bloc countries, use a very different curriculum model, it remains to be seen whether it is valid to map student achievement from those countries onto a common international scale.

Freudenthal (1975) has argued that a country's success on the international test was an artifact of the degree of alignment between the test instrument and the mathematics curriculum of the particular country. Furthermore, and in addition to whatever stakes countries may attach to achievement on the TIMSS assessments, there is a temporal dimension to the degree of alignment. TIMSS is designed to assess concepts that the vast majority of Grade 8 students have had an opportunity to learn. However, this design incorporates the assumption that the time between the learning experience and assessment is irrelevant. While Bulgarian students have been taught much of what TIMSS assesses, the fact that in many cases the key learning experiences are one to four years before the assessment date may serve as a crucial mediating factor.

Conclusion

TIMSS was designed to provide a fair, comparative assessment of mathematics achievement at specific target ages. Part of this requirement is that representative samples of students be drawn from each participating country. Bulgaria met the IEA's sampling standards for TIMSS 1999 but did not do so for TIMSS 1995. It is possible that the change in Bulgaria's mathematics achievement over these two assessments represents nothing more than sampling errors.

However, closer examination of the Bulgarian achievement suggests that there is a substantial amount of measurement error in the Bulgarian results. The root cause of this error may be the lack of alignment between the TIMSS assessments and the Bulgarian eighth grade mathematics curriculum. Given that many topics assessed in the TIMSS 1995 and 1999 mathematics assessments are taught much earlier than the eighth grade in Bulgaria, it seems reasonable to expect that students' knowledge of these content areas may have faded somewhat. That being the case, the TIMSS assessments appear to have a structural bias against Bulgaria and any similar countries that do not share the TIMSS mathematics curriculum framework.

Notes

1. Bulgarian mean of 527 is a rescaled mean obtained when the 1995 and 1999 assessments were placed on the same scale. The Bulgarian mean of 540 reported in TIMSS 1995 is on a different scale.
2. One mathematics items, M012043, was mistranslated in the Bulgarian 1999 assessment. This item is deleted from the analyses presented here.
3. Data is not available to allocate the 49 schools that did not participate in TIMSS 1995 to either the profile or non-profile category.

References

Bahrick, H.P. and Phelphs, E. (1987) Retention of Spanish vocabulary over eight years. *Journal of Experimental Psychology: Learning, Memory, and Cognition*, 13, 344–49.

Bankov, K. (1997) Bulgaria. In D.F. Robitaille, *National contexts for mathematics and science education: An encyclopedia of the education systems participating in TIMSS*. Vancouver, Canada: Pacific Education Press.

Bankov, K. (2002) *Reasoning in school mathematics in Bulgaria. Reasoning, explanation and proof in school mathematics and their place in the intended curriculum* (pp. 178–85). Proceedings of the QCA International Seminar, 4–6 October, Cambridge, England.

Beaton, A.E. and Gonzales, E.J. (1997) TIMSS test-curriculum matching analysis. In M.O. Martin and D.L. Kelly (eds.), *TIMSS technical report volume II: Implementation and analysis*. Chestnut Hill, MA: Boston College.

Beaton, A.L., Mullis, I.V.E., Martin, M.O., Gonzalez, E.J., Kelly, D.L. and Smith, T.A. (1996) *Mathematics achievement in the middle school years*. Chestnut Hill, MA: Boston College, TIMSS International Study Center.

Clarke, D. (2003) International comparative research in mathematics education. In A.L. Bishop, M.A. Clements, C. Keitel, J. Kilpatrick and F.K.S. Leung (eds.),

Second international handbook of mathematics education, part one, pp. 143–84. London: Kluwer.

Cooper, H., Nye, B., Charlton, K., Lindsay, J. and Greathouse, S. (1996) The effects of summer vacation on achievement test scores: A narrative and meta-analysis review. *Review of Educational Research*, 66(3), 227–68.

Ebbinghaus, H. (1885) *Uber das Gedachthis*. Leipzig: Dunckes and Humblot.

Foy, P. (1997) Implementation of the TIMSS sample design. In M.O. Martin and D.L. Kelly (eds.), *Third International Mathematics and Science Study technical report, volume II: Implementation and analysis, primary and middle school years (population 1 and population 2)*. Chestnut Hill, MA: Boston College.

Foy, P. and Joncas, M. (2000a) TIMSS sample design. In M.O. Martin, K.D. Gregory and S.E. Stemler, *TIMSS 1999 technical report*, pp. 29–48. Chestnut Hill, MA: Boston College.

Foy, P. and Joncas, M. (2000b) Implementation of the sample design. In M.O. Martin, K.D. Gregory and S.E. Stemler, *TIMSS 1999 technical report*, pp. 157–70. Chestnut Hill, MA: Boston College.

Freudenthal, H. (1975) Pupils' achievements internationally compared – the IEA. *Educational Studies in Mathematics*, 6, 127–86.

Garden, R.A. and Smith, T.A. (2000) TIMSS test development. In M.O. Martin, K.D. Gregory and S.E. Stemler, *TIMSS 1999 technical report*, pp. 49–70. Chestnut Hill, MA: Boston College.

Harnisch, D.L. and Linn, R.L. (1981) Analysis of item response patterns: Questionable test data and dissimilar curriculum practices. *Journal of Educational Measurement*, 18(3), 133–46.

Howsen, G. (2002) TIMSS, common sense, and the curriculum. In D.F. Robitaille and A.E. Beaton, *Secondary analysis of the TIMSS data*, 95–111. Dordrecht: Kluwer.

Husen, T. (ed.) (1967) *International study of achievement in mathematics: A comparison of twelve systems*. New York: Wiley.

Keitel, C. and Kilpatrick, J. (1999) The rationality and irrationality of international comparative studies. In G. Kaiser, E. Luna and I. Huntley (eds.), *International comparisons in mathematics education*, pp. 241–56. London: Falmer Press.

Kolen, M.J. and Brennan, R.L. (1995) *Test equating methods and practices*. New York: Springer.

Martin, M.O., Rust, K. and Adams, R.J. (1999) *Technical standards for IEA studies*. Delft, Netherlands: International Association for the Evaluation of Educational Achievement.

McLaughlin, D., Dossey, J. and Stancavage, F. (1998) Validation studies of the linkage between NAEP and TIMSS eighth grade mathematics assessments. In E. Johnson (ed.), *Linking the National Assessment of Educational Progress (NAEP and the Third International Mathematics and Science Study (TIMSS): A technical report*. Washington, DC: National Center for Education Statistics.

Mullis, I.V.S., Martin, M.O. and Stemler, S.E. (2000) TIMSS questionnaire development. In M.O. Martin, K.D. Gregory and S.E. Stemler, *TIMSS 1999 technical report*, pp. 71–88. Chestnut Hill, MA: Boston College.

Mullis, I.V.S., Martin, M.O., Gonzalez, E.J., Gregory, K.D., Garden, R.A., O'Connor, K.M., Chrostowski, S.J. and Smith, T.A. (2000) *TIMSS 1999 International mathematics report: Findings from IEA's Repeat of the Third International Mathematics and Science Study at the eighth grade.* Chestnut Hill, MA: Boston College.

National Assessment Governing Board (2002) *Mathematics framework for the 2003 National Assessment of Educational Progress.* Washington, DC: National Assessment Governing Board (available from: www.nagb.org).

Nodding, N. (1997) Thinking about standards. *Phi Delta Kappan* 79(3), 184–9.

Robitaille, D.F. and Beaton, A.E. (2002) TIMSS: A brief overview of the study. In D.F. Robitaille and A.E. Beaton, *Secondary analysis of the TIMSS data*, pp. 11–20. Dordrecht: Kluwer.

Robitaille, D.F., Schmidt, W.H., Raizen, S., McKnight, C., Britton, E. and Nicol, C. (1993) *Curriculum frameworks for mathematics and science: TIMSS monograph no. 1.* Vancouver, Canada: Pacific Education Press.

Rothstein, R. (2004) A wider lens on the black–white achievement gap. *Phi Delta Kappan*, 86(2), 105–10.

Ruddock, G. (1998) *Mathematics in the school curriculum: an international perspective.* London: NFER.

Schmidt, W.H. and McKnight, C.C. (1995) Surveying educational opportunity in mathematics and science: An international perspective. *Educational Evaluation and Policy Analysis*, 17(3), 337–53.

Tatsuoka, K.K. and Tatsuoka, M.M. (1982) Detection of aberrant response patterns and their effect on dimensionality. *Journal of Educational Statistics*, 7(3), 215–31.

Travers, K.J., Garden, R.A. and Rosier, M. (1989) Introduction to the study. In D.F. Robitaille and R.A. Garden (eds.), *The IEA study of mathematics II: Contexts and outcomes of school mathematics*, pp. 1–16. New York: Pergamon.

UNESCO (2000) *Bulgaria: report: part I: Descriptive section*, http://www2.unesco.org/wef/countryreports/bulgaria/rapport_1.html

Yamamoto, K. and Kulick, E. (2000) Scaling methodology and procedures for the TIMSS mathematics and science scales. TIMSS questionnaire development. In M.O. Martin, K.D. Gregory and S.E. Stemler, *TIMSS 1999 technical report*, pp. 237–63. Chestnut Hill, MA: Boston College.

17

Curriculum Coherence: Does the Logic Underlying the Organization of Subject Matter Matter?

Richard S. Prawat and William H. Schmidt

Introduction

Curriculum coherence has emerged as a key variable in accounting for students' test performance in mathematics in international studies, most notably TIMSS. This chapter examines one aspect of coherence – the logic behind the ordering of content. Specifically, we examine the possibility that more child-centered and thus more process-oriented countries such as the US (and Norway) may opt for a 'content in service of process' orientation, while high performing countries such as Japan or the Czech Republic employ a subtly different approach, best characterized as 'process in service of content'. These two differing logics, we hypothesize, account for the now well-known differences in curricular profiles between the US and the top performing countries.

This issue has special currency for educators in the US in light of the emphasis most of the professional organizations in mathematics and science place on so-called 'constructivist' teaching and learning. The latter can also be defined as 'process-oriented' instruction. The National Research Council's well-regarded science standards are an example; they highlight inquiry as the 'central strategy for teaching science' (Keys and Bryan, 2001: 631). The issue in this chapter is whether or not TIMSS has anything to say on this matter. Educational policy makers in the US have learned from TIMSS. This study has resulted in much more attention being devoted to the important issue of curriculum design and delivery. In fact it has broadened our understanding of one important concept related to curriculum – that of 'coherence'.

Prior to release of the TIMSS 1995 data, this concept had emerged as being a particularly powerful one in many aspects of school reform in the US. Studies done in the early 1990s viewed this construct through an organizational lens, which emphasizes the importance of administrators developing a shared vision of the educational process among members of the school community – teachers, students and even parents. The ability to develop a shared vision, and thus a common purpose in teaching and learning, became one of the important standards advocated by the Interstate School Leaders Licensure Consortium, one of the largest school administrators' groups in the US.

Other studies done prior to release of the TIMSS findings in 1997 looked at coherence across schools primarily from an 'alignment' perspective: Are the schools' instructional programs aligned with external standards and policies? Do they map onto tests or other assessments? Often the focus in this work has been on the relationship between the content taught and the content tested. Standards have an important mediating role to play in this regard. It is not enough to increase teacher accountability, which one can do by simply aligning the content taught with the content tested; teachers need to know in a more explicit way *what* they are being held accountable for, which is where external standards come in. Historically, states and the federal government in the US have failed to recognize the important role that standards play in this regard. The failure to specify outcome criteria for tests was less of an issue in the US because of the way standardized tests are used, which, unlike in other countries, is more for college or job selection than as a way to certify students' mastery of the curriculum.

Coherence as Logical Organization

TIMSS, as suggested, has opened the eyes of US educators to another important aspect of coherence as it relates to education. The validity of this third way of thinking about the concept is supported by dictionary definitions of the term. Thus, in addition to the organizational definition (that is, a common and widely shared vision of what the learning outcomes should be in a particular disciplinary domain), and the policy definition (that is, a set of instruments – standards, textbooks, and tests – that send a consistent, mutually reinforcing message), it suggests a third alternative that puts a slightly different spin on the other two: something (for example, a mathematics curriculum) is coherent if it is systematically or logically organized or connected (Merriam-Webster, 1984).

This is the definition adopted in the present chapter. Content standards, we argue, are coherent in the aggregate if they reflect a logical ordering of content (Schmidt *et al.*, in press). The problem with this definition, of course, is that it does not attempt to delineate what the word 'logical' means. At the risk of over-simplifying this issue, we argue that there is an important difference of opinion between educators in the US and in its six top scoring competitors on the TIMSS population 2 (eighth grade in most TIMSS countries) mathematics achievement test as to how one should define the term 'logical ordering of content'. These six countries are hereafter referred to as the 'A+ countries'. It might be argued that these countries do well *not* because they put content first but because the assessments have this orientation. However, if one compares the process-oriented domains on the eighth grade mathematics tests (for example, 'estimating quantity and size', 'rounding' and 'estimating computations') with those that are more content oriented, one finds that the pattern of high performance for the A+ countries holds (Schmidt *et al.*, 1999: 122–3).

Based on an examination of the most widely cited content standards document in mathematics in the US, the National Council of Teachers of Mathematics (NCTM) (1989) *Curriculum and Evaluation Standards for School Mathematics*, many educators in that country have adopted an organizational scheme that is consistent with their avowed goal of assigning equal importance to content *and* process standards. The latter, which highlights outcomes like problem solving, communication, reasoning and proof, eschews a hierarchical model of standards presentation in favor of one that relies on process to constantly revisit and deepen understanding of a set of recurring 'themes' (that is, number and operations, algebra, geometry, measurement, and data analysis and probability) that apply across the K-12 'grade bands'.

The key notion here is that processes, like students developing the ability to clarify and justify their ideas or to learn how to function in a collaborative problem-solving environment, are important ends in the mathematics curriculum in and of themselves. Highlighting process, it should be pointed out, is consistent with an approach advocated by Jerome Bruner (1966) some 40 years ago: The notion that, if students could bring to bear increasingly sophisticated inquiry skills from one year to the next, they could gain a deep understanding of content by 'spiraling back' on it throughout the elementary school years. The top performing A+ countries appear not to have bought into this argument, opting instead for a hierarchical ordering of content that minimizes redundancy of topic coverage from one grade to the next.

Greeno's distinction between two kinds of reasoning processes is relevant here. There is a 'thinking with the basics' approach, he states, and a 'thinking is basic' approach:

> According to the 'thinking with the basics,' the job of classroom learning is to provide basic scientific or mathematical knowledge that students can then use in thinking mathematically or scientifically after they have learned enough and if they are sufficiently talented and motivated. According to the 'thinking is basic' [which clearly is Greeno's preference], learning to think scientifically and mathematically should be a major focus of classroom activity from the beginning. (Kouba *et al.*, 1998: 3)

Our point in quoting Greeno is not to claim that either approach – the content first, process second approach or the converse of that – is inherently better but simply to assert that the second alternative plays a more dominant role in the NCTM standards, while the first may be more consistent with the approach taken by top scoring TIMSS countries like Japan, Belgium or the Czech Republic.

The authors of the report from which the Greeno quote is taken agree with the first assertion. They did a careful examination of the 1989 NCTM mathematics document and concluded that nearly half of the standards that deal with reasoning and communication processes espouse a 'thinking is basic' perspective; these processes, they state, 'form an umbrella structure with which to make sense of and interpret the more content oriented components of the standards'.

Two Differing Logics for Curriculum Organization

The data presented in this chapter contrast the NCTM logic with the logic that apparently underlies content standards in countries that clearly outperform the US. The NCTM logic might be characterized as that of a 'content in service of process', while the logic evident in the high performing countries is apparently that of 'process in service of content'. In these countries, as our analysis reveals, there is little doubt about what exactly in the way of mathematics *content* ought to be emphasized at each and every grade level.

Mathematical processes like problem solving, reasoning and proof, and the like, we will argue, are not unimportant in these countries but they do play a subsidiary role in the sequencing of content across the primary and lower secondary years. Perhaps it would be helpful to provide an example. The content is so well defined in countries such as Japan, Stevenson and Stigler (1992) argue in the their well known book *The Learning Gap*, that observers can visit classrooms from one end of the country to the other on the same day and be reasonably assured of encountering exactly the same content taught, more often than not, in very similar ways.

Stevenson and Stigler compare the substance of the curriculum in Japan to that of a script or score. The teacher's task, like that of the actor or musician, is to operate as effectively and creatively as possible within the constraints laid down by the script or score. Not surprisingly, teachers in Japanese schools, meeting in study groups, spend a good deal of time and energy in exploring ways to 'tweak' and improve the script, subtly changing the wording of a problem here or coming up with an alternative way of representing an important concept there. Being absolutely certain about the scope and sequence of the important concepts in mathematics at each and every grade level is what allows for this carefully planned 'variation'.

Again, it may be helpful to contrast the content scaffolding that one finds in top performing countries like Japan with the more free-wheeling approach being pushed in the NCTM standards. Thus, the teacher's role in 'orchestrating discourse' is defined as that of 'listening carefully to students' ideas', 'deciding what to pursue in depth from among the ideas that students bring up during discussion', 'deciding when and how to attach mathematical notation and language to student ideas', and so forth (NCTM, 1991: 35). In reading these instructions for interactively negotiating the development of student thinking about mathematics, one has the distinct impression that, as Greeno suggests, 'thinking is basic' and takes precedence over 'thinking with the basics'.

This chapter tests this hypothesis by examining the patterns that apparently underlie content standards recommendations in the US and in the six top performing competitor nations examined as part of TIMSS 1995. One of the two types of standards documents examined for the US is that of the NCTM Mathematics Standards, the 1989 version, which was the one subjected to analysis. The second, described more thoroughly below, was a set of documents that detail the mathematics content standards for a representative sample of 21 states in the US.[1] The 1989 version of the NCTM standards, rather than the 2002 version, was analyzed because it was more likely to have influenced the state mathematics standards. Thus, the pattern of content coverage from K to 12 is compared and contrasted across *two* efforts in the US – one national (NCTM), and one carried out at the state level – and across *six* high performing countries on the international scene.

We have reason to believe that the pattern of content coverage in the US will assume what might be characterized as a relatively 'flat' profile, while that of the international competitor nations will resemble a hierarchy. The former, if true, is consistent with an underlying logic that is primarily pedagogical in origin – which is to say, it places a premium on students' constructivist working and reworking of certain content 'themes' throughout the elementary, middle and high school years. The latter, on the other hand, is unabashedly content oriented in focus. The intent here is to be clear about content focus first, and then to try to figure out how to engage students in more active ways with that content.

Design and Methodology

The same content analysis scheme and coding procedure was used to examine primary and lower secondary school mathematics standards for the US and for the six high scoring international competitor nations (Singapore, Korea, Japan, Hong Kong, Belgium and the Czech Republic). The scheme and procedures are described in depth in Appendix D of *Many Visions, Many Aims, Volume 1* (Schmidt *et al.*, 1997). The TIMSS Curriculum Framework was designed to capture the vast majority of the different kinds of possible topics taught at the three grade levels tested in mathematics (and science). The TIMSS assessment framework sampled topics for testing from the more inclusive framework. For that reason, the latter is more appropriate for our purposes than the former. For the international sample, data relating to content standards were obtained from education officials, typically curriculum specialists in the national ministry, in each nation. These individuals, utilizing their national content standards or an aggregate of regional standards, indicated for each grade level whether a content topic was intended or not.

The procedure for the content analysis includes dividing the content standards into small segments called *blocks*. After defining the blocks the actual instructional material in each block is described using categories from the TIMSS curriculum frameworks (see 1997 reference above; in this approach coders identified each block's content as to the topic[s] involved [44 different topic codes for mathematics]). More complex standards can be identified with more than one topic as appropriate. The result was a map reflecting the grade level coverage of each topic for each country. Topic trace maps were available for each of the six A+ countries. While none were identical they all bore strong similarities. The following procedures were followed to develop a composite map: First the mean number of intended topics at each grade level was determined across the A+ countries. Next the topics were ordered at each grade level based on the percent of the A+ countries that included it in their curriculum. Those topics with the greatest percentage of coverage across the top achieving countries were chosen first. The number selected equaled the mean number of intended topics at that grade level across A+ countries.

Essentially the same procedure as that described above was used to analyze content standards in the US. Two data sources were utilized for this purpose: The first is the mathematics standards document developed by NCTM, the leading mathematics education professional organization in the US. This document, as indicated, is the 1989 version; this was used because it was the one in effect when the TIMSS study took place. The second data source is the standards documents from the 21 states referred to earlier (Schmidt *et al.*, in press).

Results

NCTM Standards

Figure 17.1 illustrates the intended coverage for the original NCTM (1989) standards. As will become evident, the mathematics topics listed are the same, and are displayed in the same order, as the topics listed for the A+ countries. One thing to keep in mind in interpreting Figure 17.1 is the design of the NCTM standards. It employs what can best be characterized as a 'cluster' organization as regards grade level: Content standards are distributed in clusters of grades, including Grade 1 through Grade 4, Grades 5–8 and Grades 9–12. As our analysis focuses on grade eight mathematics, we consider only the grades up to and including that grade.

This structure results in blocks of topics being assigned to each cluster – in particular, to the Grade 1–4 cluster and the Grade 5–8 cluster, the two that are relevant for the grade range under consideration. Obviously, this grade-level placement ambiguity has to be resolved in some fashion. The NCTM standards do not contribute to the resolution of this issue since, according to their own statement (1989: 252), 'The standards is a framework for curriculum development. However, it contains neither a scope-and-sequence chart nor a listing of topics by specific grade level'.

Based on additional data on how these standards have been used by states and districts in developing their standards, and given the content of US textbooks, we have taken the standards at face value. For example (1989: 48), when they state 'In grades K-4, the mathematics curriculum should include two-and-three-dimensional geometry so that students can ...', we have made the assumption that first grade teachers, reading this, will assume that they are to cover these topics and so will second-, third- and fourth-grade teachers. The TIMSS data on teacher coverage of topics supports this conjecture.

As Figure 17.1 shows, there are noticeable blank areas for grades one through four. This is true for a number of important topics which apparently are not included in the US elementary school mathematics curriculum: properties of common and decimal

Topic	Grade 1	Grade 2	Grade 3	Grade 4	Grade 5	Grade 6	Grade 7	Grade 8
Whole Number Meaning	•	•	•	•	•	•	•	•
Whole Number Operations	•	•	•	•	•	•	•	•
Measurement Units	•	•	•	•	•	•	•	•
Common Fractions	•	•	•	•	•	•	•	•
Equations & Formulas	•	•	•	•	•	•	•	•
Data Representation & Analysis	•	•	•	•	•	•	•	•
2-D Geometry: Basics	•	•	•	•	•	•	•	•
Polygons & Circles	•	•	•	•	•	•	•	•
Perimeter, Area & Volume	•	•	•	•	•	•	•	•
Rounding & Significant Figures	•	•	•	•	•	•	•	•
Estimating Computations	•	•	•	•	•	•	•	•
Properties of Whole Number Operations	•	•	•	•	•	•	•	•
Estimating Quantity & Size	•	•	•	•	•	•	•	•
Decimal Fractions	•	•	•	•	•	•	•	•
Relationship of Common & Decimal Fractions	•	•	•	•	•	•	•	•
Properties of Common & Decimal Fractions					•	•	•	•
Percentages						•	•	•
Proportionality Concepts						•	•	•
Proportionality Problems						•	•	•
2-D Coordinate Geometry	•	•	•	•	•	•	•	•
Geometry: Transformations	•	•	•	•	•	•	•	•
Negative Numbers, Integers & Their Properties					•	•	•	•
Number Theory	•	•	•	•	•	•	•	•
Exponents, Roots & Radicals					•	•	•	•
Exponents & Orders of Magnitude	•	•	•	•	•	•	•	•
Measurement Estimation & Errors	•	•	•	•	•	•	•	•
Constructions w/ Straightedge & Compass	•	•	•	•	•	•	•	•
3-D Geometry	•	•	•	•	•	•	•	•
Congruence & Similarity	•	•	•	•	•	•	•	•
Rational Numbers & Their Properties					•	•	•	•
Patterns, Relations & Functions	•	•	•	•	•	•	•	•
Slope & Trigonometry					•	•	•	•

Figure 17.1 Mathematics topics intended at each grade by the 1989 NCTM standards.

fractions, percentages, proportionality concepts, proportionality problems, negative numbers, integers, and their properties, exponents, roots and radicals, rational numbers and their properties, and slope and trigonometry. As regards the other 24 of 32 mathematics topics, the intent of the NCTM standards is clear – that they be covered across the first eight grades. Specific information about how long a topic will remain in the curriculum is ambiguous as a consequence of the clustering of grades. However, taking the standards at face value would suggest that the answer to this question ranges from four to eight grades, with the preponderance of topics being included in all eight grades.

The 32 topics from the NCTM standards listed in Figure 17.1 include all of the topics intended for coverage by a majority of the top-achieving countries (see below). It should be pointed out that NCTM assigns a high priority in grades one through eight to more topics than are included in this figure – seven more to be exact. The TIMSS mathematics framework, used to analyze the NCTM standards, also contains more topics than are listed in Figure 17.1; in this case, 12 more. Interestingly, there is a strong overlap between the list of seven high priority NCTM topics that are *not* part of the composite curriculum for A+ countries and the set of twelve TIMSS topics that are in the same category (that is, not part of the A+ curriculum but part of the overall framework). In fact, all seven of the NCTM non-A+ country topics are included in the list of 12 non-A+ country TIMSS framework topics. Thus, 39 of the 44 TIMSS mathematics topics represent high priority content for grades one through eight according to the NCTM standards. One cannot claim any systematic bias in the framework used to analyze the NCTM standards. Nor, by the way, is there any bias in the framework as regards the A+ countries; only two additional TIMSS topics need to be added to the 32 in Figure 17.1 to describe the composite curriculum for the high performing countries.

State Standards

Figure 17.2 illustrates the results of applying a similar statistical technique to the mathematics content standards of the representative sample of 21 states. A perusal of Figure 17.2 shows that the content coverage pattern across states more closely resembles that of the NCTM standards than it does that of the A+ countries (see Figure 17.3; the silhouette of Figure 17.3 has been superimposed on Figure 17.2 to outline the A+ pattern and thus facilitate comparison). The similarity between NCTM standards and state standards is not surprising since the former were used as a model in the development of many state standards. Clearly, the state standards do not reflect the three-tier structure described below. In other words, the majority of the 32 mathematics topics identified for the grade one through eight students in the A+ countries are likely to be taught to American students *repeatedly* throughout the eight years of elementary and middle school. In fact, the average duration of a topic in state standards was almost six years – twice as long as for the A+ countries. The added duration is achieved by increasing the time of coverage for topics at the earlier grades. If one were to make a summary statement, it is fair to say that, for the individual states represented in Figure 17.2, most mathematical topics are supposed to be taught to all students at all grades.

Curriculum Structure of the Top Achieving Countries

Figure 17.3 portrays the set of topics for Grades 1–8 that represents only those common topics that were intended by a majority of the A+ countries. As indicated, the data suggest a three-tier, hierarchical pattern of increasing mathematical complexity. The first tier, covered in primary Grades 1–5, emphasizes primarily arithmetic, including whole number concepts and computation, common and decimal fractions, estimation and rounding. The third tier, covered in Grades 7 and 8, consists primarily of advanced number topics, including exponents, roots, radicals, orders of magnitude, and the properties of rational numbers; algebra, including functions, and slope; and geometry, including congruence and similarity and three-dimensional geometry.

Topic	Grade 1	Grade 2	Grade 3	Grade 4	Grade 5	Grade 6	Grade 7	Grade 8
Whole Number Meaning	•	•	•	•	•	•		
Whole Number Operations	•	•	•	•	•	•		
Measurement Units	•	•	•	•	•	•	•	•
Common Fractions	•	•	•	•	•	•	•	
Equations & Formulas		•	•	•	•	•	•	•
Data Representation & Analysis	•	•	•	•	•	•	•	•
2-D Geometry: Basics	•	•	•	•	•	•	•	•
Polygons & Circles	•	•	•	•	•	•	•	
Perimeter, Area & Volume		•	•	•	•	•	•	•
Rounding & Significant Figures								
Estimating Computations	•	•	•	•		•		•
Properties of Whole Number Operations	•	•	•	•				
Estimating Quantity & Size			•					
Decimal Fractions			•	•	•	•		•
Relationship of Common & Decimal Fractions				•	•	•		
Properties of Common & Decimal Fractions								
Percentages					•	•	•	
Proportionality Concepts						•	•	
Proportionality Problems						•	•	
2-D Coordinate Geometry			•	•		•	•	
Geometry: Transformations	•	•	•		•	•	•	
Negative Numbers, Integers & Their Properties						•	•	
Number Theory					•	•	•	•
Exponents, Roots & Radicals						•	•	•
Exponents & Orders of Magnitude							•	•
Measurement Estimation & Errors	•	•	•	•	•	•	•	•
Constructions w/ Straightedge & Compass								
3-D Geometry	•	•	•	•	•	•		•
Congruence & Similarity					•	•	•	•
Rational Numbers & Their Properties						•	•	•
Patterns, Relations & Functions	•	•	•	•	•	•	•	•
Slope & Trigonometry								

Figure 17.2 Mathematics topics intended at each grade by a majority of the 21 states. *Note.* The area outlined by the dark-lined box represents the pattern of intended topics among A+ countries as found in Figure 17.3 superimposed on Figure 17.2.

Grades 5 and 6 appear to serve as an overlapping transition or middle tier with continuing attention to arithmetic topics (especially fractions, decimals, estimation and rounding), but with an introduction to the topics of percentages, negative numbers, integers and their properties, proportional concepts and problems, two-dimensional coordinate geometry, and geometric transformations, all of which, except for percentages, were also topics found in the third stage. Thus, Grades 5 and 6 serve as a point of transition, where attention to topics such as proportionality and coordinate geometry led to the formal treatment of algebra and geometry, characteristic of the third stage (Schmidt *et al.*, in press).

The 'upper triangular' appearance of the display in Figure 17.3 implies a hierarchical sequencing of the topics in the top achieving countries over the first eight grades. As discussed in the preceding paragraphs, the sequencing moves from

Topic	Grade 1	Grade 2	Grade 3	Grade 4	Grade 5	Grade 6	Grade 7	Grade 8
Whole Number Meaning	•	•	•	•	•			
Whole Number Operations	•	•	•	•	•			
Measurement Units	•	•	•	•	•	•	•	
Common Fractions			•	•	•	•		
Equations & Formulas			•	•	•	•	•	•
Data Representation & Analysis			•	•	•	•		•
2-D Geometry: Basics			•	•	•	•		•
Polygons & Circles				•	•	•	•	•
Perimeter, Area & Volume				•	•	•	•	•
Rounding & Significant Figures				•	•			
Estimating Computations				•	•	•		
Properties of Whole Number Operations				•	•			
Estimating Quantity & Size				•	•			
Decimal Fractions				•	•	•		
Relationship of Common & Decimal Fractions				•	•	•		
Properties of Common & Decimal Fractions					•	•		
Percentages					•	•		
Proportionality Concepts					•	•	•	•
Proportionality Problems					•	•	•	•
2-D Coordinate Geometry					•	•	•	•
Geometry: Transformations						•	•	•
Negative Numbers, Integers & Their Properties						•	•	
Number Theory							•	•
Exponents, Roots & Radicals							•	•
Exponents & Orders of Magnitude							•	•
Measurement Estimation & Errors							•	
Constructions w/ Straightedge & Compass							•	•
3-D Geometry							•	•
Congruence & Similarity								•
Rational Numbers & Their Properties								•
Patterns, Relations & Functions								•
Slope & Trigonometry								•

Figure 17.3 Mathematics topics intended at each grade by a majority of A+ countries.

elementary to more advanced topics in a way that is consistent with a content-based as opposed to a content *and* process based structuring principle.

Conclusion

A two-part argument was made in the introduction to this chapter. First, it was thought that different logics might be in evidence when one compares the way curriculum specialists and, more specifically, mathematics curriculum specialists, organize content standards in the US and how that same task is carried out in top-performing nations around the world. Second, it was argued that these different logics likely reflect the fact that educators in the US – especially educators who have taken the lead in formulating mathematics standards like the NCTM standards – place great value on student ('constructivist') activity as a prime mechanism, not only for engaging students in deeper levels of understanding, but also as a structuring principle for the theme oriented sequencing of content across grade level. Content is seen as grist for this activity in an approach that Greeno characterizes as 'thinking is basic'. One can conclude from this that educators in the six top performing countries, while they do not slight student reasoning and problem solving processes, are much more insistent than their US counterparts about the need to lead with content. Content comes first; student thinking processes in mathematics are a means to the end of increased under-standing of the content.

The data presented in this chapter lend support to the 'alternative logic' argu-ment. Although conjecture, one can also make a case for the fact that it is the NCTM's more process-oriented or 'horizontal' approach to content coverage (that is, calling for the constant revisiting of a few themes like number across all grade levels) that accounts for the sequencing of content across grades at the state level in the US. States, we know, took the NCTM document very seriously as they were for-mulating their own grade by grade content standards. It is reasonable to assume some directionality here.

The second part of the argument, the connection between what might account for the non-hierarchical ordering of content observed in the NCTM and state standards, is less amenable to test given the TIMSS data set. It is likely that the overall approach reflects important cultural norms. The US, rightly or wrongly, has often been cast as one of the most child-centered of the 50 some nations participating in TIMSS. The TIMSS country that comes closest to mirroring this orientation may be Norway. This, at least, is one of the lessons one derives from a case study of primary education (Grades 1–6) in that country (Schmidt *et al.*, 1996). Examples supporting the child-centered claim include the fact that primary children are never differentiated by ability or achievement, and that the same group of children typically is taught by the same teacher throughout the primary grades.[2]

If the child-centered argument has validity (that is, if this orientation also translates into an honoring of child 'process' in the educational system), one might expect that the pattern of content sequencing in Norway will closely resemble that of the US. In fact, it does. Thus, superimposing the hierarchical silhouette of content sequencing in the A+ countries on the pattern for Norway (see Figure 17.2 for example), reveals that fully 40 percent of all possible topics fall outside its boundary; this is comparable to the 43 percent that falls outside the silhouette in the 21 state analysis depicted in Figure 17.2. Cultural norms may indeed dictate the kind of logic that guides content selection and sequencing in the US and in the six top performing countries examined in this chapter. This hypothesis is certainly worthy of further study.

Notes

1. The 21 states included the following: California, Colorado, Delaware, Florida, Georgia, Illinois, Indiana, Kentucky, Maryland, Massachusetts, Michigan, New Hampshire, New Jersey, New York, North Carolina, Ohio, Oregon, South Carolina, Vermont, Washington and Wisconsin.

2. Based on his 21 country study, Sjøberg (2000) concludes that Norway's science curriculum, at least, is a model of what a 'child-centered' curriculum should look like.

References

Bruner, J.S. (1966) *The process of education*. Cambridge, MA: Harvard University Press.

Keys, C.W. and Bryan, L.A. (2001) Co-constructing inquiry-based science with teachers: Essential research for lasting reform. *Journal of Research in Science Teaching*, 38(6), 631–45.

Kouba, V.L., Champagne, A.B., Piscitelli, M., Havasy, M., White, K. and Hurley, M. (1998) *Literacy in the national standards: Communication and reasoning*. Albany, NY: National Research Center on English Learning and Achievement.

Merriam-Webster, A. (1984) *Webster's ninth new collegiate dictionary*. Springfield, MA: Merriam-Webster, Inc.

National Council of Teachers of Mathematics (1989) *Curriculum and evaluation standards for school mathematics*. Reston, VA: Author.

National Council of Teachers of Mathematics (1991). *Professional standards for teaching mathematics*. Reston, VA: Author.

Schmidt, W.H., Jorde, D., Cogan, L.S., Gonzalo, I., Moser, U., Shimizu, K., Sawada, T. Valverde, G., McKnight, C., Prawat, R., Wiley, D.E., Raizen, S., Britton, E.D. and Wolfe, R.G. (1996) *Characterizing pedagogical flow: An investigation of mathematics and science teaching in six countries*. Dordrecht: Kluwer.

Schmidt, W.H., McKnight, C.C., Valverde, G.A., Houang, R.T. and Wiley, D.E. (1997) *Many visions, many aims, volume 1: A cross-national investigation of curricular intentions in school mathematics*. Dordrecht: Kluwer.

Schmidt, W.H., McKnight, C.C., Cogan, L.S., Jakwerth, P.M. and Houang, R.T. (1999) *Facing the consequences. Using TIMSS for a closer look at US mathematics and science education*. Dordrecht: Kluwer.

Schmidt, W.H., Wang, H.C. and McKnight, C.C. (in press). Curriculum coherence: An examination of US mathematics and science content standards from an international perspective. *Journal of Curriculum Studies*.

Sjøberg, S. (2000) *Science and scientists: The SAS-study. Cross-cultural evidence and perspectives on pupils' interests, experiences and perceptions background, development and selected results*. Oslo: Oslo University, Department of Teacher Education and School Development.

Stevenson, H.W. and Stigler, J. (1992) *The learning gap*. New York: Summit Books.

18

Curricular Effects in Patterns of Student Responses to TIMSS Science Items

Carl Angell, Marit Kjærnsli and Svein Lie

Introduction

This chapter deals with cultural patterns in students' understanding of science based on data from TIMSS 1995 (Beaton *et al*, 1996). An analysis of Grade 8 students' responses from selected countries on a sample of science items was carried out. Based on analyses of overall similarities and differences between countries concerning relative strengths and weaknesses in science, a set of countries was selected to represent maximum cultural diversity. Moreover, the countries were chosen to represent quite similar overall proficiency in science achievement. Thus, country differences in response patterns will be easier to relate to curricular factors. For that purpose, the characteristic country differences were compared with Opportunity To Learn (OTL) data, that is, data from the science teacher questionnaire on the number of lessons devoted to particular science domains. This was to see how well the variation between countries might be explained by this measure of OTL.

Science items for analysis were selected from two TIMSS science domains: 'Earth in the Universe' and 'The particle structure of matter'. Within each of the two domains the results from three different items that complement each other in probing students' understanding of the basic concepts are discussed. By choosing three items (out of five or six possible) we were able to cover some of the basic concepts in each domain without too much variety in detailed topics.

In two earlier papers (Angell *et al*., 2000; Kjærnsli *et al*., 2002) we discussed students' conceptual understanding of science based on their responses to TIMSS items. In our latest work (Kjærnsli *et al*., 2002) we focused on notable differences between groups of countries concerning response patterns to individual items. We concluded that there are curricular impacts within some areas, accounting for significant differences between countries that cannot be understood by cultural or geographical factors alone. In order to further investigate the differences in student response patterns across countries, it would be necessary to carry out detailed curricular comparisons between countries.

Our present approach represents an attempt to take up this challenge and set the next step along that road by addressing the following research question: To what extent can curricular and OTL differences between countries explain the differences in students' conceptual understanding that are demonstrated in the responses to selected TIMSS items? On our way to a response, we have addressed the fact that both curricular and other cultural factors lie behind the pattern of p-values from item to item for a particular country. In order to mainly focus on the curriculum factor we have selected countries for our analysis that are clearly different in a broader cultural sense. The similarities and differences between the TIMSS countries were investigated by comparing patterns of strengths and weaknesses across science items. In the present study 'cultural' factors are defined as factors that contribute to setting up similarities between certain countries concerning these patterns of science understanding. These factors have not been measured explicitly in our study.

Science Domains and Items for Analysis

The two science domains were selected because they represent important and fundamental aspects of science. The items within each domain, both individually and as a group, probe basic understanding of the essential principles involved. Furthermore, OTL data are available for these domains. The items focus on the following topics, each of which will be separately discussed later:

- Earth in the Universe: The earth rotating around its axis and, together with the moon as its close neighbor circling around the sun, with the nearest star extremely far away.
- The particle structure of matter: The role of particles (atoms and molecules) as building blocks of matter in all substances, including living things.

These domains obviously represent major bases of scientific knowledge, both from a perspective of philosophy and history of science. By selecting three related (and available) items within each domain we were able to study how the interplay between them could provide deeper insight into students' fundamental conceptions.

Grouping and Selection of Countries

As explained above, the sample of countries was based on diversity in item-by-item achievement combined with similarity in overall performance in science in TIMSS 1995 (Grade 8). For that purpose we first carried out a cluster analysis of countries based on what we will call a 'p-value residual' method.

The TIMSS project is a typical quantitative study where one of the main goals is to establish valid and reliable estimates of student achievement. Regardless of the measurement model (classical or IRT models), the main idea behind a test score is that responses to individual items are not important beyond their contribution to the overall score. From a psychometrical point of view, the details about how students from different countries respond to individual items, often called 'item-by-country interactions', is regarded as a sort of random noise, or 'error variance'. Seen from a different perspective, however, the details of this interaction represent something very interesting, namely a guide into strong and weak areas for each country. The item-by-item sets of percent correct responses (p-values) establish highly interesting country-specific educational 'fingerprints'. Earlier analyses (Zabulionis, 2001; Kjærnsli and Lie, 2002) have shown that meaningful groups of countries can be established based on similarity among patterns of responses. From these results it is evident that both language and other cultural, and even geographical factors, contribute to group countries together. For simplicity we use the term 'cultural' for this combined effect.

As mentioned above, our starting point was a matrix of p-values by item by country. The cell residuals were calculated by subtracting from each cell value the average over countries for the actual item and the average over items for the actual country. Thus we were left with a residual matrix, where each cell tells how much better or worse that particular country scored on that particular item, compared to what could be expected based on the country's overall achievement and the international difficulty of the item. The fact that some countries scored higher than others and that some items were harder than others no longer showed up in the matrix cells. As an example, a cell residual of 0.05 for country A and item q means that the p-value for the item in country A was 5 percentage points higher than expected, based on the general achievement of country A and the international difficulty for item q. Thus the value 0.05 shows a moderate positive strength for this country on this particular item.

Cluster analysis allowed us to find a pattern of clustering of countries based on similarities between the sets of p-value residuals. The magnitudes of the correlations were taken as measures of similarities. High (positive) correlation is a measure of short 'distance', that is, high degree of similarity, between the p-value residuals. The next decision was to establish a rule to decide which countries or country clusters should be combined, according to their 'closeness', at each clustering step. At the first step, the two 'closest' countries define a group. At the next step, two other countries group together, or one country links to the already established group. In this process it makes a difference how distances from one country to a country group are measured. Here we have used the average distance to all members of the group. Similarly, the distance between two clusters is defined as the average of the distances between all pairs of cases in which one member of the pair is from each of the clusters.

Figure 18.1 is a dendrogram that displays the results of the clustering process. It shows how and at what distance countries link together into clusters. When we move from left to right, that is, to longer distances, we go from high to low and further to negative correlations. The horizontal distance scale runs linearly from a correlation of 0.80 to the far left (the highest, between England and Scotland, amounts to 0.75) to −0.10 to the far right. The exact magnitudes, however, are not important here.

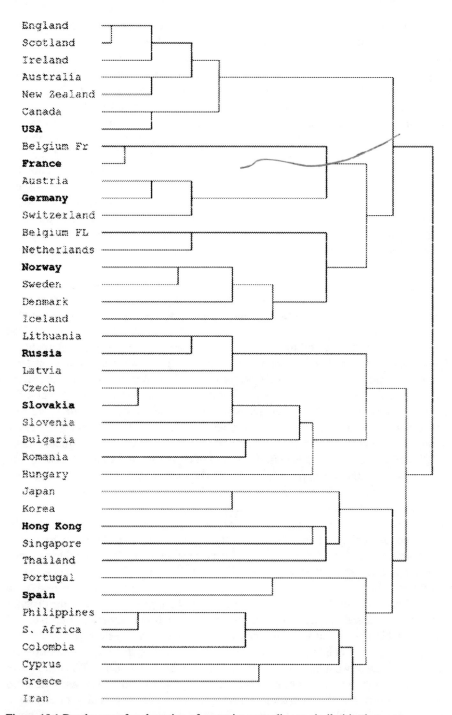

Figure 18.1 Dendrogram for clustering of countries according to similarities between countries in patterns across science items.

For example, the dendrogram clearly displays the clustering of the English-speaking countries, which in turn seem to be divided into a European, a southern, and an American subgroup. Next from the top is a West European continental group with a French-speaking and a German-speaking subgroup. Similarly, a strong Nordic group is clearly visible. A lot of further interesting features emerge from the diagram, some of which are discussed elsewhere (Grønmo *et al.*, 2004). Our main purpose here was to select countries that were well spread out across the figure. Taking into account also the requirement of similar overall achievement in science we ended up with the following eight countries: France, Germany, Hong Kong, Norway, Russia, Slovak Republic, Spain and USA. To facilitate the impression of diversity, these countries have been bold-faced in Figure 18.1.

Opportunity to Learn (OTL)

As explained earlier, we wanted to make use of information about students' opportunity to learn the subject matter topics discussed here. In the science teacher questionnaire there was a set of questions concerning time spent on various topics from the framework for the TIMSS project (Robitaille *et al.*, 1993). For our purpose the two domains selected were the following (with number labels according to the framework):

1.1.3 **Earth in the Universe** ('Interaction between sun, earth and moon; planets and the solar system; things beyond the solar system; evolution of the universe')

1.3.2 **Structure of matter** ('Atoms, ions, molecules, macromolecules, crystals')

For the OTL questions, teachers were asked to indicate for how many lessons the topic had been taught, or that it would be taught later the same year (usually one or two months remaining of the school year). They could also check a box to indicate that the topic would not be covered that year, or that it had been covered in earlier years. For our purpose we concentrated on the number of lessons taught and calculated an average number for each of the two topics and for each country. In Russia, the questions were not included in the teacher questionnaire, but 'correct' data according the very strict requirements in the official curriculum guide was included. Accordingly, to some extent our OTL data is partly a mix between what are called the intended and the implemented curriculum.

Population 2 consisted of two grades, in most countries corresponding to Grades 7 and 8. Our focus grade is Grade 8, but for our OTL measures we combined the two grades by adding the number of lessons reported. Accordingly, our OTL data provide average estimates of how many lessons students in a country have been taught the actual topic during their last two school years. Given the fact that countries have been selected to have similar overall proficiency, one would expect a pronounced positive relation for each domain between OTL and the average scores on the items.

The first phase of TIMSS also contained detailed investigations of curricular intentions by close analysis of curriculum guides and of samples of textbooks (Schmidt *et al.*, 1997). For each subcategory in the TIMSS framework (Robitaille *et al.*, 1993) it was recorded *whether or not* that topic was included in the national curriculum guide or in the analyzed textbooks for Grade 8. However, there is no reported information on *how much* attention was devoted to the subtopics (Schmidt *et al.*, 1997: 116).

In Table 18.1 these data for our two domains are shown together with the OTL data. On the rows labeled 'Taught' the number of lessons for the two grades together are given. These sums will be correlated with achievement data in the following sections. On the rows labeled 'Curriculum' and 'Textbook' we have given the number of relevant subtopics that are included in the curriculum and textbooks respectively. It is important to note that, whereas the rows 'Taught' refer to data from two grades combined, the other two rows include data from Grade 8 only. The number of relevant subtopics for each domain is given in parentheses (see column 'Domain').

Earth in the Universe

The three selected items, shown in Figures 18.2 through 18.4, all deal with our understanding of the Universe; from the Moon and the Sun to the great distances between stars. The topic represents broad aspects of our image of space and even our world view in a quite literal meaning.

Figure 18.5 displays the percent of correct responses for each item for each of the selected countries. Roughly speaking, Item O14 (Figure 18.3) is fairly easy; Q16 (Figure 18.2), more difficult; and Q11 (Figure 18.4), something between the two others. However, data from USA and Hong Kong show a somewhat different pattern. For USA the free response item about the Moon and the Sun seems to be relatively more difficult than the other two. For Hong Kong the percent correct response for Item Q16 is among the lowest, while it is the highest for item Q11.

Despite the selected countries' relatively similar overall results, the variation on single items within this domain is large and significant; typically between 20 and 30 percentage points. At the item level, differences between countries turn out to be

Table 18.1 Lessons taught (grades 7 and 8) and coverage in curriculum/textbooks (grade 8 only) for the domains at hand.

Domain	Source	France	Germany	Hong Kong	Norway	Russia	Slovak Rep.	Spain	USA
Earth in	Taught	2.5	2.8	0.3	10.1	0	5.6	7.9	10.9
Universe	curriculum	3	0	0	3	0	0	3	3
(3 topics)	textbook	1	1	0	3	0	2	3	3
Structure	Taught	3.6	4.5	5.3	5.6	8.3	10.7	10.2	13.4
of matter	curriculum	1	1	1	1	0	1	1	1
(1 topic)	textbook	1	0	1	1	1	1	1	1

Q 16 How long does it take light from the nearest star other than the Sun to reach Earth?

A. Less than 1 second

B. About 1 hour

C. About 1 month

*D. About 4 years

Figure 18.2 Item Q16.

O14 The Sun is bigger than the Moon, but they appear to be about the same size when you

look at them from the Earth. Why is this?

(Correct answer: Referring to the Sun's larger distance from the Earth)

Figure 18.3 Item O14.

Q11 Which statement explains why daylight and darkness occur on Earth?

*A. The Earth rotates on its axis.

B. The Sun rotates on its axis.

C. The Earth's axis is tilted.

D. The Earth revolves around the Sun.

Figure 18.4 Item Q11.

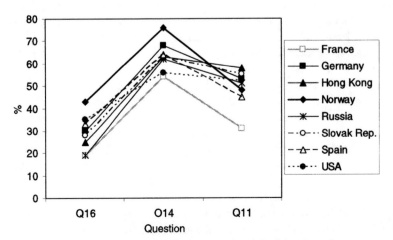

Figure 18.5 Percentage correct responses for items on 'earth in the universe'.

significant when differences in percent correct amount to around 5 percentage points. This also applies to all other items discussed in this chapter. Especially France had weaker results, and Norwegian scores were clearly better than expected on the basis of overall results on the test as a whole.

The Earth orbiting the Sun and at the same time rotating around its own axis causing day and night (Item Q11) is a complex movement. Only 31 percent of the French students got this item correct while 58 percent of the Norwegians provided a correct answer. Alternative D (the Earth revolves around the Sun), a 'correct' but inadequate response, attracted more than 30 percent of the students in each of the eight countries.

More than 50 percent of the students in all the countries could explain that the Sun and Moon seem to be of the same size because the Sun is more remote from the Earth (O14). The percentage of correct responses differs, however, from 54 percent (France) to 76 percent (Norway).

As mentioned above, the most difficult of these three items was the one about how long it takes light from the nearest star other than the Sun to reach the Earth (Q16). In almost all countries the most popular response was the incorrect alternative A. In five countries more than 40 percent of the students seemed to believe that it takes the light less than a second to reach us. This could well indicate that they had an idea of light moving momentarily, an idea supported by all observations on Earth. Furthermore, understanding the enormous and almost empty space and the huge distances between stars requires a lot of both knowledge and imagination. Maybe this complexity can help explain why relatively few chose the correct alternative on Item Q16. Alternative B (about 1 hour) may have been selected by some students on the basis of light taking about 8 minutes to travel from the Sun to the Earth. Consequently, it may be reasonable to assume an hour from the nearest star.

It is easy to underestimate the difficulty for students of understanding the models for both our solar system and the enormous Universe. It might also be argued that there is a need for greater emphasis on the teaching of the role and purpose of the concept of scientific models in science (Treagust *et al.*, 2002).

USA and Norway reported much more teaching time for the domain 'Universe' than any of the other countries selected here (see Table 18.1). And these two countries also had the best results on the most demanding question. A little surprising then is that the percent correct answers on the item about the Moon and the Sun (O14) for USA is not higher. Hong Kong, reporting very little teaching time for this domain, shows a notable response pattern. Weak result on Q16, average on O14 and best of the selected countries on Q11. Thus, relatively many Hong Kong students could explain day and night even with almost no teaching time appropriate to this phenomenon. As mentioned above, France had unconvincing results for these items. From Table 18.1 we can see that France reported clearly fewer lessons taught than Norway and USA.

As explained in the introduction, we have computed the correlation between average score and the time the subject has been taught in the different countries. If we look at these three items together, the correlation is 0.45. That means that about 20 percent ($r^2 = 0.45^2 \approx 0.20$) of the variance between countries can be explained by our data on teachers' reports on what is taught in the classrooms. However, if we look at each of the three items, we find that the correlation between lessons taught and single item results vary substantially. The correlation with item Q16 is 0.86, with O14 it is 0.21 and with item Q11 close to zero (−0.05). Of the three items, Q16 is probably the most 'school (or teaching) dependent' in the sense that both the size of the Sun and the Moon and why it is day and night are more well-known topics, both from outside school and previous learning in school, than the enormous distances between stars. Only by reasoning or by everyday conversation it might not be easy to get to the right answer for this task.

Particle Structure of Matter

The three items within this domain, shown in Figures 18.6 through 18.8, address aspects of the fundamental particle structure of nature; from atoms to cells.

Figure 18.9 displays the percentage of correct responses for each item for each country. Also for this topic we find a sort of common pattern, however not as apparent as for the topic about the Universe. Item G10 seems to be somewhat less difficult than the two others. For Item J3, Germany was on the bottom and Russia on the top with 21 percent and 53 percent correct responses, respectively. For J6, the percents correct range from 14 percent in France to 42 percent in USA. And for G10, the variation is from 44 percent in Hong Kong to 68 percent in USA and Russia. We also observe that USA, Germany and Hong Kong exhibited a somewhat different pattern than the other countries. Both USA and Germany had weak results on J3 and better on G10, while USA had a particularly good result on Item J6.

J3 The words *cloth, thread,* and *fiber* can be used in the following sentence:

cloth consists of *threads* which are made of *fiber.*

Use the words *molecules, atoms,* and *cells* to complete the following sentence:

_____consist of _____ which are made

of _____ .

Figure 18.6 Item J3.

G10. If you took all of the atoms out of a chair, what would be left?

    A. The chair would still be there, but it would weigh less.

    B. The chair would be exactly the same as it was before.

    *C. There would be nothing left of the chair.

    D. Only a pool of liquid would be left on the floor.

Figure 18.7 Item G10.

J6 Animals are made up of many atoms. What happens to the atoms after an animal has

died?

    A. The atoms stop moving.

    *B. The atoms recycle back into the environment.

    C. The atoms split into simpler parts and then combine to form other atoms.

    D. The atoms no longer exist once the animal has decomposed.

Figure 18.8 Item J6.

    Items J6 and G10 may appear reasonably similar. They both deal with atoms as fundamental and 'never disappearing' particles. The students gave very different answers to these two questions. On average 54 percent gave a correct answer to item G10, but only 26 percent gave a correct answer to item J6. For J6 the most frequent answer (not correct) was alternative D (the atoms no longer exist). However, students may have chosen the correct alternative (nothing left of the chair) on G10, even though they had the misconception that atoms can disappear into nothing, since the item focuses on what is happening to the chair and not to the atoms. Furthermore, for many students it might be hard to accept that the 'dead' chair and the living animal both are build up of atoms, and nothing else. On the other hand, the wording of the correct response to J6, *recycle back into the environment*, might have caused a problem for a lot of students. Probably they were led to think about what happens exactly at the time the animal died, whereas recycling takes time. Some students may also have thought about the environment more as the 'air' around.

    In a study with 25 students at the same age as population 2, this question (J6) was given in the context of an interview (Schoultz, 2000). There seemed to be no obvious connection between atoms and a dead animal; and, in discussing the issue, 21 out of

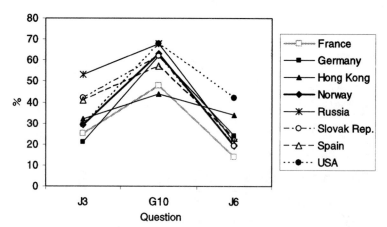

Figure 18.9 Percentage correct responses for items on 'particle structure of matter'.

the 25 students argued that the atoms will 'fall into the ground', 'disappear into the ground' or something similar.

Item J3 is about structure of matter from atoms to cells. In Figure 18.10 we have put together some of the codes from the scoring guide in order to emphasize some interesting findings. As we see from the figure, relatively few students give a correct answer to this item. The average is 31 percent for the selected countries. However, the differences between countries are substantial. Correct responses differ from 21 percent in Germany to 53 percent in Russia. Furthermore, the distribution of codes for incorrect answers is quite different from country to country.

The results indicate that many students did not have much understanding of the particle structure of nature. They apparently mixed up the concepts of atoms, molecules and cells. Codes 71 and 73 both include the notion of cells being the smallest particles. About 16 percent of the students' answers were classified as one of these two. France and Germany had the highest percentage with 20 and 23 percent, respectively. We may well argue that the most important distinction is between cells (small, but not atom size) on the one side and atoms/molecules on the other. Responses being coded as 70 could even be regarded as 'correct' if we ignore the difference between atoms and molecules, both being micro-particles. The average correct responses would then increase from 31 to 47 percent, mainly due to the large 'contribution' from Hong Kong and USA. However, there are still about 50 percent of the students who did not seem to be aware of the fact that cells are enormous compared to molecules and atoms, of which the cells are composed.

As seen in Table 18.1, USA, Spain, Slovak Republic and to some extent Russia, reported more teaching time than the other countries on this domain. USA reported the most teaching time, but in spite if this only 29 percent of the students gave a correct answer to Item J3. Germany in particular, and also France, had high percentages for the conception of cells as the smallest particles. Both countries also reported less teaching time than most of the other countries.

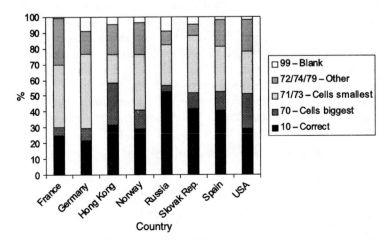

Figure 18.10 Distribution of responses to item J3 as a percentage.

Taken together as a domain, the correlation between average percentage correct and time taught is 0.81. That means that as much as 65 percent of the variance between countries can be explained by what is taught in classrooms. The particle structure of nature is not apparently or immediately observable, and for many students the only way to get any knowledge about it is from school lessons. Looking at the correlations between time taught and single items within this domain, we find relatively small variations (J3: 0.46, G10: 0.57 and J6: 0.50).

Discussion

Our analyses have focused on two main issues; on similarities between countries as regards student responses to science items, and on possible curricular explanations for these patterns. Let us first try to sum up the findings on country differences across items within domains. The data on *p*-values presented revealed large differences across the countries. The differences between the highest and lowest percentages for an item typically amount to 30 percentage points. This is high considering our selection of countries with similar overall science scores. On average, for all science items, the highest-scoring country (Slovak Republic) scored only 5 percentage points higher than the lowest (France).

Figures 18.5 and 18.9 clearly show that for both domains there is a tendency towards homogeneity between items (that is, parallel lines in the two figures), but also that there are substantial disorders (crossing of lines). Furthermore, in both cases the three items differed in how much of the country variance could be explained by the OTL data from the teacher questionnaires. Table 18.2 sums up these features. For each domain and for the item average, correlations with teacher OTL data are shown.

The amount of variance that can be explained by OTL varies substantially from item to item, but there are a number of items with remarkably high OTL values. To some extent these items may also be regarded as the most 'school-relevant' across countries, in the sense that these represent issues that most likely are focused on when the topics are actually taught. Items that represent not so 'relevant' problems at Grade 7 or 8 may not be 'covered', even if the topic as such receives substantial attention. The teaching that will enable students to actually handle the challenge in the item may happen at earlier or later grade levels.

We have not related *p*-values to the curriculum and textbook data in Table 18.1, because of the low variance and the correspondingly low explanatory power for these sets of data. There is, however, some general agreement between these data and data

Table 18.2 Correlations between *p*-values and teacher OTL data for items and domain averages.

Domain	Item 1	Item 2	Item 3	Average
Earth in the Universe	0.86 (Q16)	0.21 (O14)	−0.05 (Q11)	0.45
Particle structure of matter	0.57 (G10)	0.50 (J6)	0.46 (J3)	0.81

on teachers' coverage of topics ('Taught'). The few notable discrepancies are mainly due to the fact that the teacher data are based on two grades, while the data on textbooks and curricula are for Grade 8 only. The importance of applying two years' data can be demonstrated by applying only Grade 8 teacher data in a correlation analysis similar to that in Table 18.2. By doing so, the average correlations (last column) are then reduced to 0.26 and 0.58, respectively.

As already mentioned, in an international study like TIMSS, the main purpose of individual items is to contribute to the overall achievement scale. This between-country variance for individual items is, from a psychometrical point of view, regarded as 'error'. However, we have demonstrated that the 'item by country interactions' provide a valuable starting point for interesting analyses of cultural similarities and differences. Outcomes like dendrograms can give fascinating images of the clustering of countries and thus provide visions of cultural linkages and influences. They cannot alone, however, identify causal factors behind the clusters. Obviously curriculum effects, language similarities, and other cultural factors interplay to produce the observed item-by-item patterns that are basis for the country clusters. In our study we have tried to estimate the importance of the former effect. Our finding is that curricular effects (measured as teacher coverage of the 'topic') seem to be able to explain between 25 and 50 percent of the variance between countries for individual items. It is important, however, to bear in mind that our countries were selected to represent maximum diversity.

Conclusion

The present chapter has investigated the relationships between the Opportunity To Learn (OTL) measures and student understanding within two science domains. On one hand, the study has clearly demonstrated the rather obvious fact that instruction helps for better understanding. Further, and more important, it seems that teacher coverage measure provides a more useful measure than whether or not a topic is mentioned in the curriculum guide. It has also been demonstrated that, by adding data also from the year before the actual, OTL can strongly improve it's potential for explaining understanding. Finally, we also have shown how item-by-item data from large-scale international studies can be investigated to throw light on similarities and differences between countries concerning strengths and weaknesses in students' conceptual understanding.

References

Angell, C., Kjærnsli, M. and Lie, S. (2000) Exploring students' responses on free-response science items in TIMSS. In D. Shorrocks-Taylor and E.W. Jenkins (eds.), *Learning from others. International comparisons in education*. Science & Technology Education Library, vol. 8, pp. 159–88. Dordrecht: Kluwer Academic Publishers.

Beaton, A., Martin, M.O., Mullis I.V.S., Gonzales, E.J., Smith, T.A. and Kelly, D.A. (1996) *Science achievement in the middle school years. IEA's Third International Mathematics and Science Study* (TIMSS). Chestnut Hill, MA: Boston College.

Grønmo, L.S., Kjærnsli, M. and Lie, S. (2004) Looking for cultural and geographical factors in patterns of responses to TIMSS items. In *Proceedings of the IRC-2004. The 1st IEA International Research Conference*, Vol. 1, pp. 99–112. University of Cyprus.

Kjærnsli, M. and Lie, S. (2002) TIMSS science results seen from a Nordic perspective. In D.F. Robitaille and A.E. Beaton (eds.), *Secondary analysis of the TIMSS data*, pp. 193–208. Dordrecht: Kluwer Academic Publishers.

Kjærnsli, M., Angell, C. and Lie, S. (2002) Exploring population 2 students' ideas about science. In D.F. Robitaille and A.E. Beaton (eds.), *Secondary analysis of the TIMSS data*, pp. 127–144. Dordrecht: Kluwer Academic Publishers.

Robitaille, D.F., Schmidt, W.H., Raizen, S., Mc Knight, C., Britton, E. and Nicol, C. (1993) *Curriculum frameworks for mathematics and science*. TIMSS monograph no. 1. Vancouver: Pacific Educational Press.

Schmidt, W.H., Raizen, S., Britton, E., Bianchi, L.J. and Wolfe, R.G. (1997) *Many visions, many aims, volume 2. A cross-national investigation of curricular intentions in school science*. Dordrecht: Kluwer Academic Publishers.

Schoultz, J. (2000) Conceptual knowledge in talk and text: What does it take to understand a science question? In *Att samtala om/i naturvetenskap. Kommunikation, context och artefakt*. PhD thesis (partly in English), University of Linkoping.

Treagust, D., Chittleborough, G. and Mamiala, T. (2002) Students' understanding of the role of scientific models in learning science. *International Journal of Science Education*, 24(4), 357–68.

Zabulionis, A. (2001) Similarity of mathematics and science achievement of various nations. *Educational Policy Analysis Archives*, 9(33).

19

Towards a Science Achievement Framework: The Case of TIMSS 1999

Min Li, Maria Araceli Ruiz-Primo and Richard J. Shavelson

Introduction

Increasingly testing has become an essential element in educational reforms. Testing is used to gauge both students' and teachers' performance and to improve students' learning (for example, Baker *et al*, 2002; Linn, 2000; Pellegrino *et al*., 2001). While the debates about testing and related issues have been heated and political, it is recognized that testing can be far too powerful an instrument to be used without very careful scrutiny of what is measured by the tests. Motivated by this belief, we present a secondary analysis that addresses the validity of TIMSS 1999 test score interpretation as a measure of students' science achievement.

In this chapter, we describe and implement a framework for assessing science achievement. The framework conceptualizes science achievement as four types of knowledge that are characteristic of competency or achievement in science. We apply this framework to the TIMSS 1999 science items and scores by logically analyzing test items and statistically modeling the underlying patterns of item scores. We use a coding system to analyze and map test items into the type(s) of knowledge that they most likely measure. We also report results from factor analyses to determine whether the knowledge-factor model accounts for the underlying pattern of item scores better than alternative models. Lastly, we examine the mosaic of instructional factors that influence students' performance in terms of knowledge types.

Theoretical Framework of Science Achievement

Defining Science Achievement

Parallel to the old definition of intelligence as what intelligence test measures, today we can say that science achievement is conventionally considered as what multiple-choice tests measure due to the pervasive and accepted use of the multiple-choice technology. Such a narrow perspective for measuring science achievement not only provides limited and even inaccurate information about students' performance but undermines desirable outcomes of students' learning (for example, Kohn, 2000; Shavelson *et al.*, 1990). In contrast, our knowledge-based conception of science achievement is intended to broaden the notion of science achievement.

In a series of papers (for example, Shavelson and Ruiz-Primo, 1999; Ruiz-Primo, 2002; Ruiz-Primo *et al.*, 2002; Li *et al.*, in press), we have developed and described a framework for science achievement drawing upon a body of related research (for example, Alexander and Judy, 1988; Bybee, 1996, 1997; Chi, Feltovich and Glaser, 1981; Chi *et al.*, 1988; Bennett and Ward, 1993; de Jong and Ferguson-Hessler, 1996; OECD/PISA, 2001; Pellegrino *et al.*, 2001; Sadler, 1998; White, 1999). We believe that such a framework should fully address the connections between standards, science curriculum, student learning, instruction, and educational measurement.

Specifically, we conceptualize science achievement as four interdependent types of knowledge:[1]

1. *Declarative knowledge: knowing that.* Such knowledge includes scientific terms, definitions, facts, or statements, such as 'CO2 is carbon dioxide' or 'green plants manufacture glucose from carbon dioxide and water'. Declarative knowledge can be learned and held in memory as words, pictures, or sensory representations. Indeed, representing concepts in a variety of forms is considered essential to the learning of science (for example, National Research Council [NRC], 1996).
2. *Procedural knowledge: knowing how.* Procedural knowledge takes the form of if-then production rules or a sequence of steps, such as applying algorithms to balance chemical equations, reading data tables, or designing experiments to identify components of a mixture. With practice, individuals can convert their procedural knowledge into automatic responses. Such automaticity is one of the characteristics of expertise (for example, Anderson, 1983; Dreyfus and Dreyfus, 1986; Dutton and Jackson, 1987; Randel and Pugh, 1996; Rasmussen, 1986).
3. *Schematic knowledge: knowing why.* Schematic knowledge typically entails the application of scientific principles or explanatory models, such as explaining why the moon generally rises about 50 minutes later each day or how a particular virus functions. Schematic knowledge can also be used to guide actions, to troubleshoot systems, and to predict the effect that changes in some phenomenon will have on others (De Kleer and Brown, 1983; Gentner and Stevens, 1983).
4. *Strategic knowledge: knowing when, where, and how to apply knowledge.* This knowledge includes domain-specific strategies, such as ways to represent a problem or strategies to deal with certain types of tasks. It also entails general

monitoring performance or planning strategies, such as dividing a task into sub-tasks, or integrating the three other types of knowledge in an efficient manner. Experts, for example, are reported to interpret problems based on relatively absolute principles compared to the novices who just group problems using superficial features such as springs or inclined planes (for example, Chi *et al.*, 1981; Zeitz, 1997).

Linking Assessments to the Knowledge Framework

This achievement framework can be used to specify important aspects of an achievement test. By synthesizing the research of cognitive psychology and assessment (for example, Glaser *et al.*, 1992; Hamilton, *et al.*, 1997; Li *et al.*, in press; Martinez, 1999; Messick, 1993; Shavelson *et al.*, 2002; Shavelson and Ruiz-Primo, 1999; Snow, 1989; Sugrue, 1993, 1995; White, 1999), we argue that some types of science knowledge may be well tapped by certain assessment items because of particular affordances and constraints provided by those items. Accordingly, in order to assess science achievement accurately and adequately, assessors should employ tests varying in their methods for eliciting students' performance regarding knowledge types.

To this end, we take one step further to map assessment items to these knowledge types. Note that it is not a straightforward task. Martinez (1999), for example, has demonstrated that testing method alone does not determine what type of knowledge an item measures. Instead, test items vary in many other characteristics, including the nature of the task demands placed on test takers or degree of task openness. Therefore, all these item characteristics should, if possible, be taken into account during the process of mapping assessment items to science knowledge.

Research Questions

Having sketched our framework for defining and measuring science achievement, in what follows we report a study using the TIMSS 1999 items and scores of the USA sample to examine the viability of this framework. For the sake of simplicity, we only report the results on TIMSS 1999 Booklet 8 items since similar findings were observed from the analysis of the other seven booklets.[2] This study purports to validate the framework, which allows us to measure students' science achievement and interpret scores in a meaningful way rather than using total scores. We ask two research questions.

First, can the science achievement measured by the TIMSS 1999 items be decomposed as four types of knowledge? Research has shown that science achievement measured by large-scale tests can be decomposed as different aspects (for example, Hamilton, *et al.*, 1997). In the logical analysis, two coders independently coded and classified the items into types of knowledge. Then, a set of confirmatory factor analyses (CFA), using the Booklet 8 item scores, was performed to ascertain whether the items measured the types of knowledge as hypothesized and to compare against the competing models for science achievement.

Second, do students' instructional experiences affect their science achievement? Instruction is conceived of as a vital factor that contributes to the variations in students' learning outcomes, such as test scores (NRC, 1996). The structural equation modeling was conducted after controlling other related variables to examine how the instructional variables are associated with students' use of science knowledge.

Students' Science Achievement Measured by TIMSS 1999 Items

In this section, we apply the achievement framework to examine the TIMSS 1999 science items and scores. We first present the logical analysis in which we used a coding system to identify item characteristics and classify the items into knowledge-types. Following this, we report a set of CFA that were executed to verify the item classifications.

Coding System and Example Items

We relied on the research work done by us and other researchers to develop a coding system to analyze item characteristics (Ayala *et al.*, 2000; Baxter and Glaser, 1998; Li, 2001; Quellmalz, 2002). We identified four major categories, listed from most important to least important: task demands; cognitive demands; item openness; and complexity (examples of those coding decisions are presented in Table 19.1). Task and cognitive demands are both critical cues for establishing the item-knowledge link whereas the latter two categories provide additional information to modify or revise the link.

Through generating and weighting the item characteristics, we were able to map the TIMSS 1999 test items into types of knowledge. Below we present several TIMSS 1999 items to illustrate the utility of the coding system to classify test items.

Item D05 was classified as an item tapping declarative knowledge (Figure 19.1). It asks students to identify the organ that carries the sensorial information. The coding led us to conclude that it mainly taps declarative knowledge. First, the response is expected to be in the form of terms (for example, names of the human organs). This item asks for a specific content question (that is, a specific fact), leaving students little opportunity to provide relations between concepts or to apply principles. Second, the cognition evoked is likely to be recalling facts. The cognitive process to answer the item is to directly retrieve information or do a minimum of scientific reasoning to sort out the relevant information. Third, speaking item openness, Item D05 is restricted because of the multiple-choice format and the unfinished stem that forces students to select options instead of providing responses prior to reading options. Being restricted in turn reinforces the task and cognitive demands placed on students. Finally, the coding for complexity supports the declarative link as well. For example, the item is similar to school-type problems. Consequently, when answering the question students may recall exactly what they had been taught. Therefore, weighing the four categories, we classified Item D05 as a declarative-knowledge item.

Item P01 was coded as testing procedural knowledge (Figure 19.2). It provides students with a graph showing a distance that a car has traveled for 75 minutes and asks

Table 19.1 Coding system.

Coding category	Description	Selected examples of coding sub-categories	Possible links with knowledge type[a]
Task demands	Task(s) that students are asked to perform	• Defining/using terms • Providing explanations • Designing an experiment • Mathematical calculation	• D • D/Sc • P • P
Cognitive demands	Inferred cognitive processes students likely act upon to provide responses	• Recalling facts • Reading figures • Reasoning with models • Comparing and selecting strategies	• D • P • Sc • St
Item openness	Degree of freedom students are allowed to have in shaping responses to an item	• Selecting vs. generating responses • Stem presented as a complete question vs. an incomplete stem • One vs. multiple correct answers • Recipe directions	• Selecting good for D • Incomplete stem good for D, but not good for Sc/P/St • One good for D vs. multiple good for Sc/St • Not good for St
Complexity	Domain-general factors that influence item difficulty	• Linguistic challenges • Textbook vs. novel tasks • Inclusion of irrelevant information • Use of everyday knowledge	• Not good for any type • Textbook good for D vs. novel good for St • Good for St • Not good for any type

Notes. [a]D = declarative knowledge, P = procedural knowledge, Sc = schematic knowledge, and St = strategic knowledge. These abbreviations are applied hereafter.

D05. Sensory messages are taken to the brain by
A. arteries and veins
B. arteries and hormones
C. nerves and hormones
D. muscles and veins

Figure 19.1 Example of a declarative-knowledge item.

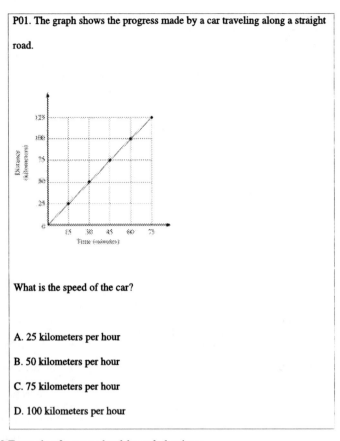

P01. The graph shows the progress made by a car traveling along a straight road.

What is the speed of the car?

A. 25 kilometers per hour

B. 50 kilometers per hour

C. 75 kilometers per hour

D. 100 kilometers per hour

Figure 19.2 Example of a procedural-knowledge item.

them to select the speed of that car. The item requires students to respond with an algorithm to calculate speed and/or to interpret the diagram directly, both of which are procedural knowledge as we defined. Second, the cognitive process students probably undertake can be either applying a calculation algorithm for speed by dividing distance with time, or locating the point corresponding to 60 minutes on the line in the graph (that is, an hour) and simply reading the distance on the vertical axis. However, the item's lack of openness to some degree limits students applying procedural knowledge. Although this item allows students to generate their own responses before reading the options, students can arrive at the correct answer by simply working backwards from options or even merely guessing. Finally, the analysis of complexity does not provide additional information for modifying the coding decision. Considering all the codes, Item P01 was classified as a procedural item though it is not a perfect candidate.

Very different from Item D05 and P01, our logical analysis revealed that Item P02 tended to test students' schematic knowledge. This short-answer item asks students to provide an explanation for why the same, or less, amount of light shines on the wall from two meters away (Figure 19.3). The four item characteristics support students to use schematic knowledge. First, the item is intended to assess students' knowledge about factors that determine the amount of light. It invites students to make a prediction and provide conceptual explanations by using the phrase 'explain your answer' in the stem. The knowledge that students use can be models or theories about light in words or pictures. Second, the dominant cognitive process is reasoning with theories. Very likely students have to develop responses by explaining the key features about light and its movement. Third, Item P02 is moderately open, requiring students to reason and organize their responses. Finally, the low complexity (for example, no heavy reading or irrelevant information) strengthens the posited link to schematic knowledge by reducing construct-irrelevant variances.

Lastly and unfortunately, there are no example items for strategic knowledge that can be selected from TIMSS 1999. Strategic items are expected to have certain item characteristics, such as a complex, novel task that entails students interpreting the problem and selecting domain-specific strategies, an open item structure that supports the use of strategic knowledge, and a scoring that directly captures the differences in students' use of strategic knowledge (for example, types of strategies, efficiency, patterns of monitoring). However, no TIMSS 1999 items satisfied the criteria for being strategic knowledge items.

Logical Analysis

Two coders independently reviewed and classified the TIMSS 1999 science items into knowledge types following the same procedures illustrated in the previous section. The inter-coder agreement was 89 percent.[3] A final item classification was achieved as follows. For those items on which the two coders agreed, the agreed upon classifi-

---

P02. James turns on a flashlight in his bedroom and shines it on his wall one meter

away to produce a small circle of light. He then shines the flashlight on his ceiling

two meter away to produce a larger circle of light.

a) Does more light reach the ceiling than the wall?

(Check one)

Yes ☐

No ☐

b) Explain your answer.

Figure 19.3 Example of a schematic-knowledge item.

cation is reported. For the disagreed items, we consulted the findings of Pine and colleagues (AIR, 2000) and resolved the difference based on their rating.

In some cases where the item might tap a second knowledge type, coders were instructed then to provide a secondary code for the item. This was expected as the TIMSS 1999 items were designed to tap more than one performance expectation domain. Due to ambiguity and items drawing on everyday experience, we found that the knowledge codes were not applicable for a few items (then, those items were labeled as NA; see Table 19.2). This difficulty was not unique. In a recent study analyzing the same items, Pine and his colleagues (AIR, 2000) experienced similar difficulty.

Item classifications are presented in Table 19.2. Given the fact that the Booklet 8 science items varied a great deal in their item characteristics, we anticipated that those items would require the use of different knowledge types. Consistent with this prediction, we observed that the booklet measured three of the four proposed types of knowledge, while – as stated before – none of the items was found to assess the strategic knowledge. The Booklet 8 science test is heavily loaded on declarative knowledge (approximately 50 percent) with balance of procedural and schematic knowledge items about equally distributed. The finding that declarative knowledge is overwhelmingly tested is also reported at other studies that examined large-scale tests at various grade levels (Britton *et al.*, 1996; Lomax *et al.*, 1995).

As expected, three items (A08, P03 and R05) were double coded, which were ambiguous to an extent that students could use either of two types of knowledge. We also classified Item Q16 as NA because students could use many different ways, even general reasoning, to figure out the correct answer. The item itself did not have sufficient affordances for triggering a certain type of knowledge. Rather, the knowledge to be used is more determined by an individual student's preference and other factors.

We also examined item format and item difficulty indexed by p-values in the table. There was little association between item format and the type of knowledge items are intended to test (adjusted $\chi^2 = 0.68, p = 0.71$). This is not surprising. As we explained, the knowledge that one brings to work on an item is elicited by a complicated interaction of his cognition and *a set of* item characteristics, including but not limited to format. Further, a comparison of item difficulties revealed that relatively easier items were used for measuring the declarative knowledge rather than other types of knowledge. On average about 15 percent more students were able to provide correct answers to declarative items than other items. Putting those pieces together, one disturbing conclusion is that the Booklet 8 test largely measured students' declarative knowledge with less challenging items.

Collectively, we found that raters could apply the coding system to consistently classify the TIMSS 1999 items into types of knowledge. The coding showed that the Booklet 8 test was heavily loaded on declarative knowledge while procedural or schematic knowledge was measured by relatively fewer items.

Factor Analysis
The link between construct and assessment does not stop with the logical analysis. A variety of analyses can be performed to verify the posited links, such as examining

Table 19.2 Classifications of TIMSS 1999 booklet 8 science items with the USA booklet 8 sample.

Item label	Description	Format[a]	p-value[b]	D	P	Sc	St	NA
			Classification[c]					
A07	Organ not in abdomen	MC	0.73	✓				
A08	Stored energy in two springs	MC	0.80	✓	✓			
A09	Fanning a wood fire	MC	0.78	✓				
A10	Seeing person in a dark room	MC	0.76			✓		
A11	Overgrazing by livestock	MC	0.62	✓				
A12	Changes in river shape/speed	MC	0.52		✓			
B01	Layers in earth	MC	0.91	✓				
B02	Energy released from car engine	MC	0.64	✓				
B03	Greatest density from mass/volume table	MC	0.18		✓			
B04	Pulse/breathing rate after exercise	MC	0.92	✓				
B05	Elevation diagram of wind/temperature	MC	0.49		✓			
B06	Color reflecting most light	MC	0.82	✓				
P01	Determination of speed from graph	MC	0.47		✓			
P02	Amount of light on wall and ceiling	SA	0.29				✓	
P03	Tree growth	SA	0.41	✓			✓	
P04	Hibernating animal	MC	0.44	✓				
P05	Two reasons for famine	SA*	0.32	✓				
P06	Digestion in stomach	SA	0.42	✓				
P07	Replication of measurement	MC	0.37		✓			
Q11	Earth year	MC	0.63	✓				
Q12	Water level in U-tube	SA	0.26		✓			
Q13	Alcohol/glass expansion in thermometer	MC	0.44			✓		
Q14	Sugar dissolving in water	MC	0.35			✓		
Q15	NOT a chemical change	MC	0.39			✓		
Q16	Light from sun and moon	SA	0.59					✓
Q17	Function of chlorophyll	MC	0.53	✓				
Q18	Mass of freezing water	SA	0.30				✓	
R01	Bacteria/mold experience	MC	0.46		✓			
R02	Appearance of red dress in green light	MC	0.45	✓				
R03	Two outcomes of introducing new species	SA*	0.39	✓				
R04	Atmospheric conditions in jets	SA	0.15	✓				
R05	Small pieces of wood burn faster	SA	0.16	✓	✓			
R06	Results of global warming	MC	0.28	✓				

Notes. [a]MC = multiple-choice, SA = short-answer worth 1 point, and SA * = short-answer worth 2 points. [b]The p-value reported is defined as the proportion of students who got completely correct answers out of those who took the Booklet 8 test. [c]NA indicates that the items cannot be classified into certain knowledge-type(s) since items are too ambiguous or not dealing with science knowledge.

students' responses to the test items, collecting and analyzing students' think-aloud verbalizations while responding to the test items, or statistically modeling the underlying patterns. In this chapter we empirically evaluate the posited link by examining the patterns of how item scores covariate with each other. We performed the CFA using software Mplus (Muthén and Muthén, 1999) to ascertain whether the knowledge model could account for the underlying covariance pattern. We built a statistical model to test the links between the items and the coded knowledge types, which exactly correspond to the item classifications resulting from the logical analysis in Table 19.2 after removing Item Q16.

Mplus reports χ^2 to indicate the model-goodness-fit of the data to the hypothesized model. The statistics of our model ($\chi^2 = 325.49$, df = 294, $p = 0.10$) and RMSEA (0.01) indicate that the fit of the item scores to our model was adequately good and was in the bounds of statistical acceptability. That is, the CFA confirmed the item classifications from the logical analysis: 17 items as declarative-knowledge; eight items as schematic-knowledge; and seven as procedural-knowledge. The results completely supported these classifications. Most factor loadings were statistically significant and generally high, which was also true for at least one loading with those double-coded items. The fit and the factor loadings support the feasibility to use our framework of science achievement to analyze the TIMSS 1999 science items.

Further, the estimate of standard error ratios ranging from 3.73 to 11.04 indicated that the relation between each pair of the three hypothetical knowledge factors was significant, too. We speculate that the strong associations presented in the knowledge types may be due to several reasons: (1) different types of knowledge, in nature, are interdependent aspects of science understanding that one can achieve, therefore knowledge types should be moderately correlated; (2) inevitably an individual's general ability more or less affects her performance on each item, subsequently the knowledge types; and (3) items are unevenly distributed across the format and content, for example, 70 percent of items in multiple-choice format and more than half in life science and physics. If replicating the analysis with appropriately sampled items, we should expect to observe diluted correlations after reducing the influences from format or content.

Another part of our CFA work was to compare the knowledge model with competing models. We built alternative models based on different theoretical claims, such as one factor as general ability, two factors as format (multiple-choice and short-answer), or three factors using the performance expectation framework by TIMSS 1999[4] (Table 19.3). Using the fit measure and other information, we found that the knowledge model had a much better fit than those alternative models. All the models

Table 19.3 Comparison between the knowledge-type model and alternative models.

Tested model	Description of factors	Fit of model	
		χ^2	Probability
One factor model	General reasoning ability	343.16	0.03 (df = 297)
Knowledge model	Declarative, procedural, and schematic	325.49	0.10 (df = 294)
Format model	Multiple-choice and short-answer	339.20	0.04 (df = 296)
Content model	Life science, physics, environmental issues, chemistry, earth science and nature of science	333.89	0.04 (df = 290)
TIMSS 1999 performance expectation model	Simple information, complex information, applying, theorizing and solving problems	341.79	0.04 (df = 296)

did fit well the item scores. Yet, only the knowledge model was significantly superior to the one factor model whereas other models just slightly improved the fit. The results suggest that a knowledge model better reflects the underlying pattern among item scores than using the total score as the general science achievement or using alternative structures. Overall, regardless of the domain, format, or thinking processes the TIMSS 1999 items have, they clustered into three knowledge factors.

Linking Instructional Experiences to Students' Achievement

In order to provide additional evidence bearing on the construct validity, we attend to the relationship between students' achievement and their instructional experiences. Ample research has addressed the relation between learning outcomes and instructional practice based on the very premise that students' learning is influenced by classroom experiences (for example, Hamilton *et al.*, 2003; Wenglinsky, 2002). One challenge that such studies must face is what learning outcomes are measured (Glass, 1990; Haertel, 1986; Linn, 2000). This exactly relates to the validity concerns that we pointed out at the beginning of this chapter. Use of invalid measures of science achievement not only results in misleading interpretations and decisions but eventually undermines the teaching and learning practice.

With this concern, we continue validating our achievement framework by relating students' classroom experiences with their learning defined as distinct types of knowledge. We expect that the associations would vary in their strength depending on the types of knowledge, which renders appropriate interpretations about students' achievement. In contrast to using total scores, this partition approach provides more accurate, diagnostic information in tracing the effects of instructional variables. A similar approach used to conceptualize and decompose the science achievement, but on the basis of reasoning dimensions, has been reported more valuable than using total scores in a series of studies on National Educational Longitudinal Study of 1988 achievement test (for example, Hamilton, *et al.*, 1997; Kupermintz and Snow, 1997).

In summary, this section continues to apply the proposed knowledge framework to examine how instructional variables were associated with different types of knowledge demonstrated by the students. We anticipate that the findings are meaningfully different enough to justify the use of the knowledge framework rather than total scores.

Measures of Instructional Experiences

TIMSS 1999 administered questionnaires to both students and their teachers to collect information about the classroom instructional experiences that students have been exposed to. We limit the scope of our analysis by examining: (1) the instructional variables available in the student questionnaires (Martin *et al.*, 2000); (2) literature on the impact of instructions (for example, De Jong and van Joolingen, 1998; Duit and Treagust, 1998; Hamilton *et al.*, 2003; Mayer, *et al.*, 2000; Monfils *et al.*, 2000; National Center for Education Statistics, 2001; Papanastasiou *et al.*, 2004; Ruiz-

Primo *et al.*, 2002; Smerdon *et al.*, 1999; Tytler, 2002; Von Secker, 2002; Wenglinsky, 2002); and (3) instructional methods that teachers are recommended in science standard documents (for example, NRC, 1996; New Standards, 1997). Specifically, we selected to focus our analysis on eight instructional methods, grouped as textbook-based and hands-on. Textbook-based methods require students to copy notes from the board, to take quizzes and tests, or to work from worksheets or textbook questions. In contrast, hands-on methods are those recommended by national and state standards to teach science through authentic activities and inquiry experience. While the textbook-based methods tend to concentrate on literally rote learning of the materials from the texts, the hands-on methods attempt to help students make sense of what is learned.

Table 19.4 summarizes the means of the instructional variables self-reported by the TIMSS 1999 USA Booklet 8 sample. Generally, students were often exposed to all those instructional methods, indicated by the range of these variables. The textbook-based methods seemed to be a more frequent experience than the hands-on methods, among which the use of everyday life was the least frequently reported; nevertheless, such a difference was negligible regarding the effect size. Comparing to the finding reported by some studies in last decade that textbook-based was a quite dominant teaching mode compared to hands-on in many American classrooms (Kober, 1993), this shows the gradual shift of teachers' instructional practice, but a fundamental change in response to the national and state science standards has not yet fully achieved.

Table 19.4 Means of instructional methods reported by the TIMSS 1999 USA booklet 8 sample.

Instructional experience	Description of questionnaire item	*n*	Mean (max = 4)
Textbook-based methods			
Notes from board	We copy notes from the board.	990	3.08
Quiz	We have a quiz or test.	986	3.06
Worksheet	We work from worksheets or textbooks on our own.	984	3.12
Hands-on methods			
Problem	The teacher shows us how to do science problems.	993	2.98
Project	We work on science projects.	983	2.82
Everyday life	We use things from everyday life in solving science problems.	985	2.50
Demonstrated experiment	The teacher gives a demonstration of an experiment.	982	2.99
Student experiment	We ourselves do an experiment or practical investigation in class.	978	2.82

Note.   Students were instructed to report how often each of the instructional methods happened in their science classrooms at a 4-point scale. After recoded, scores of 1, 2, 3, and 4 refer to 'Never', 'Once in a while', 'Pretty often', and 'Almost always', respectively.

Relation between Types of Knowledge and Instructional Experiences

Informed by the relevant research and science standards, we speculate that the text-book-based instructional methods are associated with students' learning of declarative knowledge, whereas the hands-on methods can be effective in improving students' procedural and/or schematic knowledge (for example, Haury and Rillero, 1994; Glasson, 1989; Sutton and Krueger, 2001). For instance, textbook-based lecture, an effective way of teaching declarative knowledge, is unlikely to be successful in teaching procedural knowledge. In contrast, using everyday life examples, instead of unconnected factual information, to solve science problems is recommended as a strategy to improve students' understanding across more than one content area (NRC, 1996). Therefore, this instructional method supports students' construction of schematic knowledge rather than rote-learning of declarative knowledge.

In order to test this predicted pattern, a set of structural equation modeling was executed while controlling for some important variables at student level (that is, highest education level of parents, index of home educational resources, out-of-school study science time, index of confidence in science ability, self-reported indication of usually doing well in science and positive attitude toward science). Decisions for selecting those controlling variables were made because they are commonly identified in the literature and available in the TIMSS 1999 dataset. We acknowledge that the complex sampling, including clustering of students within classrooms and sampling weights, should be taken into account in order to model instructional data and achievement scores (for example, Keeves et al., 2000). However, it is inappropriate with the data sample for each booklet to run the hierarchical linear modeling, for example, about half of the classrooms had only one or two students in Booklet 8, Kreft, 1996; Pollack, 1998.[5] To address this methodological dilemma, we chose to report the results from the structural equation modeling and focus our interpretations on the differences between the use of knowledge framework and total scores instead of statistical significance (see Table 19.5).

Overall, most of the coefficients of instructional experiences on students' performance on the test items were not significant, indicating that after controlling all the relevant background variables students' instructional experiences did not contribute much to their learning outcomes. Across both models, teaching methods as copying notes from board, using everyday life problems and examples, and having students' engaged in experiments seems to be associated with the students' performance whereas other instructional methods are negatively related with students' test scores. The magnitude of the standardized coefficients and the patterns of instruction-learning relationship are not consistent with the findings from a recent study by Papanastasiou and her colleagues who examine the relation between TIMSS 1999 instruction and students' achievement indicated by Rasch (2004) scores. We suspect this may be due to the use of data from different countries and different indicators for student performance.

In terms of differences between the two models, the relationship exhibits slightly different patterns for the instructional methods, such as the direction of the relationship for methods as quiz and everyday life or the strength of the relationship for methods as student experiment. Within the knowledge-factor model, the coefficiens vary

Table 19.5 Comparing structural coefficients when relating instructional variables to different science achievement models.

Instructional variable	One factor model		Knowledge-factor model					
	Science achievement		Declarative		Procedural		Schematic	
	Estimates	S.E.	Estimates	S.E.	Estimates	S.E.	Estimates	S.E.
Notes from board	0.022 (0.042)	0.030	0.036 (0.047)	0.047	0.054 (0.097)	0.040	−0.009 (−0.012)	0.049
Quiz	−0.029 (−0.044)	0.040	−0.080 (−0.084)	0.062	0.015 (0.022)	0.053	−0.021 (−0.022)	0.069
Worksheet	−0.055 (−0.095)	0.031	−0.042 (−0.050)	0.050	−0.075 (−0.124)	0.044	−0.111 (−0.131)	0.053
Everyday life	0.005 (0.009)	0.034	0.019 (0.024)	0.055	−0.100 (−0.018)	0.045	0.004 (0.005)	0.057
Problem	0.017 (−0.031)	0.033	−0.001 (−0.001)	0.052	−0.023 (−0.039)	0.047	−0.055 (−0.067)	0.057
Project	−0.049 (−0.096)	0.031	−0.082 (−0.110)	0.050	−0.006 (−0.011)	0.043	−0.093 (−0.125)	0.053
Demonstrated experiment	−0.038 (−0.073)	0.033	−0.043 (−0.056)	0.050	−0.048 (−0.088)	0.044	−0.063 (−0.082)	0.057
Student experiment	0.093 (0.181)	0.035	0.091 (0.121)	0.050	0.131 (0.245)	0.048	0.164 (0.219)	0.054

Notes.   The table reports both unstandardized structural coefficients and standardized structural coefficients (the estimates in the parentheses). The ratio of an unstandardized coefficient to S.E. typically presents the statistical strength of the relation.
In the structural equation modeling, one latent variable as overall science achievement was assumed and included in the tested model when using one factor model whereas three latent variables were assumed and included to represent the different types of knowledge when using the knowledge-factor model.

across the three types of knowledge, but some of the instruction–learning relationships are contradictory to our prediction that the textbook-based instructions are positively related to students' learning of declarative knowledge, whereas hands-on instructions are effective for learning procedural and schematic knowledge. To make sense of this contradiction, we reason that the instruction–learning relation is more accurately represented if considering the quality of teaching in terms of how teachers implement the instructional methods, rather than the implementation frequency or the amount of time teachers use those methods. For example, some researchers claim that frequent use of hands-on instructions does not automatically lead to the increase of students' learning if instructional activities are not focused or well-connected to scientific concepts (for example, Kober, 1993). Therefore, one real challenge for studies such as TIMSS 1999 would be to include items within the limited scope of student and teacher questionnaires to capture some important aspects of teaching quality. Equally challenging, secondary studies must strategically select and work with the data from those studies, such as being aware of the constraint of the data while interpreting the results, interpreting data combining several items to create appropriate indicators.

To conclude, the relations between instruction and student performance differ between the one factor and knowledge-factor models and across the three types of knowledge. Generally speaking, the knowledge-factor model shows a more detailed picture of the instruction-learning relation compared to one factor model. Of course, given problems associated with the self-report technique (for example, inaccuracy of reported information, discrepancy between students' and teachers' perceptions about instruction) and the limitation of our methodology discussed earlier, these conclusions need to be cautiously examined before generalization.

Conclusion

The main purpose of this chapter was to outline and apply a knowledge framework to examine the science achievement tested by the TIMSS 1999 Booklet 8 science items and its relation with students' instructional experiences. This framework conceptualizes science achievement as four types of knowledge: declarative knowledge or 'knowing that', procedural knowledge or 'knowing how', schematic knowledge or 'knowing why' and strategic knowledge or 'knowing when, where, and how knowledge applies'. We argue that test items with different combinations of item characteristics as affordances and constraints can elicit or restrict students' use of their science knowledge.

For this purpose, we presented three sets of analyses. First, we applied the framework in a logical analysis to code and classify the TIMSS 1999 items. We observed that declarative knowledge was heavily loaded in the test items, whereas none of the items tested strategic knowledge. Second, the results from factor analyses indicated that the underlying pattern for the item scores converged with the proposed item-knowledge links from the logical analysis. Those items that pre-classified as a certain knowledge type were clustered together. The knowledge-factor model was also found to fit better the item scores than any of the alternative models. And third, we examined the associations between students' science knowledge and the self-reported instructional experiences. The structural coefficients revealed that a modeling of the relationship with the knowledge factors provided more specific, meaningful interpretations of students' learning than with one factor model. In addition, the coefficients varied across the three types of knowledge, which was another source of evidence bearing on the differences between the knowledge types. Even though the pattern failed to fully confirm our predictions, most coefficients supported the distinctions between the knowledge types that we proposed.

Overall, we conclude that the science achievement measured by the TIMSS 1999 test can be distinguished as different types of knowledge. The study reported suggests that it is necessary to implement a framework for science achievement, which can be either the one we proposed here or any theoretical frameworks developed on the basis of cognitive research and related fields, to carefully examine what is measured in science tests. Such a framework provides not only accurate interpretations of students' achievement scores but meaningful insights and interesting questions on how instructional experiences are connected to students' learning. Furthermore, the framework offers valuable information for testing development and practice as well. This chapter, we believe, is a beginning to develop and evaluate those frameworks. It should be followed with more studies to provide systematic, vigorous evidence using different methods with additional sets of science items.

Notes

1. We recognize the simplification of our framework, for example, treating types of knowledge as if they were truly separable or presenting a static typology of knowledge.
2. In order to make it manageable to test any single student, TIMSS 1999 used an incomplete block design to sample test items and students. The TIMSS 1999 science items were organized into 21 clusters, which were then placed into eight test booklets so that a student only responded in one randomly assigned test booklet. Released items and data files can be found at http://isc.bc.edu/timss1999i/study.html.
3. The inter-coder agreement was calculated based on the codes for all the TIMSS 1999 science items.
4. Performance expectations implemented in TIMSS 1999 included five categories: (1) understanding simple information; (2) understanding complex information; (3) theorizing, analyzing and solving problems; (4) using tools, routine procedures; and (5) investigating the natural world. The framework was revised from a very similar one originally developed in TIMSS, whereas a substantive modified framework based on cognitive demands as (1) factual knowledge, (2) conceptual understanding and (3) reasoning and analysis was proposed in TIMSS 2003. This chapter applied the TIMSS 1999 framework to classify the test items.
5. Analysis data within each booklet allows us to analyze test items and validate the posited knowledge codes in CFA.

References

AIR (2000) *Review of science items from TIMSS 1999: A report from the science committee to the technical review panel*. Washington, DC: American Institutes for Research.

Alexander, P.A. and Judy, J.E. (1988) The interaction of domain-specific and strategic knowledge in academic performance. *Review of Educational Research*, 58, 375–404.

Anderson, J.R. (1983) *The architecture of cognition*. Cambridge, MA: Harvard University Press.

Ayala, C.C., Ayala, M.A. and Shavelson, R.J. (2000) On the cognitive interpretation of performance assessment scores. Paper presented at the AREA annual meeting, New Orleans, LA, April.

Baker, E.L., Linn, R.L., Herman, J.L. and Koretz, D. (2002) *Standards for educational account systems*. CRESST policy brief 5. Los Angeles: University of California, CRESST.

Baxter, G.P. and Glaser, R. (1998) Investigating the cognitive complexity of science assessments. *Educational Measurement: Issues and Practice*, 7(3), 37–45.

Bennett, R. and Ward, W. (eds.) (1993) *Construct versus choice in cognitive measurement*. Hillsdale, NJ: Erlbaum.

Britton, E.D., Dossey, J., Eubanks, L., Gisselberg, K., Hawkins, S., Raizen, S.A. and Tamir, P. (1996) Comparing examinations across subjects and countries. In E.D. Britton and S.A. Raizen (eds.) *Examining the examinations: An international comparison of science and mathematics examination for college bound students* (pp. 23–54). Boston, MA: Kluwer.

Bybee, R.W. (1996) The contemporary reform of science education. In J. Rhoton and P. Bowers (eds.), *Issues in science education* (pp. 1–14). Arlington, VA: National Science Teachers Association.

Bybee, R.W. (1997) *Achieving scientific literacy: from purpose to practice*. Portsmouth, NH: Heinemann.

Chi, M.T.H., Feltovich, P. and Glaser, R. (1981) Categorization and representation of physics problems by experts and novices. *Cognitive Science*, 5, 121–52.

Chi, M.T.H., Glaser, R. and Farr, M. (eds.) (1988) *The nature of expertise*. Hillsdale, NJ: Erlbaum.

de Jong, T. and Ferguson-Hessler, M.G.M. (1996) Types and qualities of knowledge. *Educational Psychologist*, 31, 105–13.

de Jong, T. and van Joolingen, W.R. (1998) Scientific discovery learning with computer simulations of conceptual domains. *Review of Educational Research*, 68(2), 179–201.

De Kleer, J. and Brown, J.S. (1983) Assumptions and ambiguities in mechanistic mental models. In D. Gentner and A.L. Stevens (eds.), *Mental models* (pp. 155–90). Hillsdale, NJ: Erlbaum.

Dreyfus, H.L. and Dreyfus, S.E. (1986) *Mind over machine: The power of human intuition and expertise in the era of the computer*. New York: The Free Press.

Duit, R. and Treagust, D. (1998) Learning in science – from behaviorism towards social constructivism and beyond. In B. Fraser and K. Tobin (eds.), *International handbook of science education* (pp. 3–25). Dordrecht: Kluwer.

Dutton, J.E. and Jackson, S.E. (1987) The categorization of strategic issues by decision makers and its links to organizational action. *Academy of Management Review*, 12, 76–90.

Gentner, D. and Stevens, A.L. (eds.) (1983) *Mental models*. Hillsdale, NJ: Erlbaum.

Glaser, R., Raghavan, K. and Baxter, G.P. (1992) *Cognitive theory as the basis for design of innovative assessment: Design characteristics of science assessment* (CSE Technical report No. 349). Los Angeles: University of California, CRESST.

Glass, G.V. (1990) Using student test scores to evaluate teachers. In J. Millman and L. Darling-Hammond (eds.), *The new handbook of teacher evaluation* (pp. 229–40). Newbury Park, CA: SAGE.

Glasson, G.E. (1989) The effects of hands-on and teacher demonstration laboratory methods on science achievement in relation to reasoning ability and prior knowledge. *Journal of Research in Science Teaching*, 26, 121–31.

Haertel, E. (1986) The valid use of student achievement measures for teacher evaluation. *Educational Evaluation and Policy Analysis*, 8(1), 45–60.

Hamilton, L.S., McCaffrey, D.F., Stecher, B.M., Klein, S.P., Robyn, A. and Bugliari, D. (2003) Studying large-scale reforms of instructional practice: An example from mathematics and science. *Educational Evaluation and Policy Analysis*, 25(1), 1–29.

Hamilton, L.S., Nussbaum, E.M., Kupermintz, H., Kerthoven, J.I.M. and Snow, R.E. (1997) Enhancing the validity and usefulness of large-scale educational assessments: II. NELS: 88 science achievement. *American Educational Research Journal*, 32, 555–81.

Haury, D.L. and Rillero, P. (1994) *Perspectives of hands-on science teaching*. Columbus, OH: ERIC Clearinghouse for Science, Mathematics and Environmental Education.

Keeves, J.P., Johnson, T.G. and Afrassa, T.M. (2000) Errors: What are they and how significant are they? *International Education Journal*, 11(3), 164–80.

Kober, N. (1993) *What we know about science teaching and learning. EdTalk*. Washington, DC: Council for Educational Development and Research.

Kohn, A. (2000) *The case against standardized testing*. Greenwich, CT: Heinemann.

Kreft, Ita G.G. (1996) Are multilevel techniques necessary? An overview, including simulation studies. Unpublished report, California State University, Los Angeles. Retrieved 12 August 2003, from http://www.calstatela.edu/faculty/ikreft/quarterly/quarterly.html

Kupermintz, H. and Snow, R.E. (1997) Enhancing the validity and usefulness of large-scale educational assessments: II. NELS: 88 mathematics achievement to 12th grade. *American Educational Research Journal*, 34, 124–50.

Li, M. (2001) A framework for science achievement and its link to test items. Unpublished dissertation, Stanford University.

Li, M., Shavelson, R.J. and White, R. (in press) Towards a framework for achievement assessment design: The case of science education. Manuscript submitted.

Linn, R.L. (2000) Assessment and accountability. *Educational Researcher*, 29, 4–16.

Lomax, R.G., West, M.M., Harmon, M.C. and Viator, K.A. (1995) The impact of mandated standardized testing on minority students. *Journal of Negro Education*, 64, 171–85.

Martin, M.O., Mullis, I.V.S., Gonzalez, E.J., Gregory, K.D., Smith, T.A., Chrostowski, S.J., Garden, R.A. and O'Connor, K.M. (2000) *TIMSS 1999 international science report findings from IEA's Repeat of the Third International Mathematics and Science Study at the eighth grade.* Chestnut Hill, MA: Boston College.

Martinez, M.E. (1999) Cognition and the questions of test item format. *Educational Psychology*, 34, 207–18.

Mayer, D.P., Mullens, J.E. and Moore, M.T. (2000) *Monitoring school quality: An indicators report* (NCES 2001–030). Washington, DC: National Center for Education Statistics.

Messick, S. (1993) Trait equivalence as construct validity of score interpretation across multiple methods of measurement. In R. Bennett and W. Ward (eds.), *Construction versus choice in cognitive measurement* (pp. 61–74). Hillsdale, NJ: Erlbaum.

Monfils, L., Camilli, G., Firestone, W. and Mayrowetz, D. (2000) Multidimensional analysis of scales developed to measure standards based instruction in response to systemic reform. Paper presented at the annual meeting of AERA, April.

Muthén, L.K. and Muthén, B.O. (1999) *Mplus: The comprehensive modeling program for applied researchers: User's guide.* Los Angeles, CA: Muthén and Muthén.

National Center for Education Statistics (2001) *The nation's report card: Science highlights 2000.* Washington, DC: National Center for Education Statistics.

National Research Council (NRC) (1996) *National Science Education Standards.* Washington, DC: National Academy Press.

New Standards (1997) *Performance standards: English language arts, mathematics, science, applied learning.* Washington, DC: Author.

OECD/PISA (2001) *Measuring student knowledge and skills. The PISA 2000 assessment of reading, mathematical, and scientific literacy.* Paris: Organisation for Economic Cooperation and Development/Programme for International Student Assessment.

Papanastasiou, E.C., Zacharia, Z. and Zembylas, M. (2004) Examining when instructional activities work well in relation to science achievement. Paper presented at the 1st IEA International Research Conference, Lefkosia, Cyprus, May.

Pellegrino, J., Chudowsky, N. and Glaser, R. (2001) *Knowing what students know: The science and design of educational assessment.* Washington, DC: National Academy Press.

Pollack, B.N. (1998) Hierarchical linear modeling and the 'unit of analysis' problem: A solution for analyzing responses of intact group member. *Group Dynamics: Theory, Research, and Practice*, 2, 299–312.

Quellmalz, E. (2002) Using cognitive analysis to study the validities of science inquiry assessments. Paper presented at the Annual AERA Meeting, New Orleans, LA.

Randel, J.M. and Pugh, H.L. (1996) Differences in expert and novice situation awareness in naturalistic decision making. *International Journal of Human Computer Studies*, 45, 579–97.

Rasmussen, J. (1986) *Information processing and human-machine interaction: An approach to cognitive engineering*. New York: North-Holland.

Ruiz-Primo, M.A. (2002) On a seamless assessment system. Paper presented to the AAAS annual meeting, Boston, MA, February.

Ruiz-Primo, M.A., Shavelson, R.J., Hamilton, L. and Klein, S. (2002) On the evaluation of systemic science education reform: Searching for instructional sensitivity. *Journal of Research in Science Teaching*, 39, 369–93.

Sadler, P.M. (1998) Psychometric models of student conceptions in science: Reconciling qualitative studies and distractor-driven assessment instruments. *Journal of Research in Science Teaching*, 35(3), 265–96.

Shavelson, R.J. and Ruiz-Primo, M.A. (1999) On the assessment of science achievement (English version). *Unterrichts wissenschaft*, 27(2), 102–27.

Shavelson, R.J., Carey, N.B. and Webb, N.M. (1990) Indicators of science achievement: Options for a powerful policy instrument. *Phi Delta Kappan* (May), 692–7.

Shavelson, R.J., Li, M., Ruiz-Primo, M.A. and Ayala, C.C. (2002) Evaluating new approaches to assessing learning. Presented at Joint Northumbria/EARLI Assessment Conference, University of Northumbria at New Castle, Longhirst Campus, August.

Smerdon, B.A., Burkam D.T. and Lee, V.E. (1999) Access to constructivist and didactic teaching: Who gets it? Where is it practiced? *Teachers College Record*, 101(1), 5–34.

Snow, R.E. (1989) Toward assessment of cognitive and conative structures in learning. *Educational Researcher*, 18(9), 8–14.

Sugrue, B. (1993) *Specifications for the design of problem-solving assessments in science. Project 2.1. Designs for assessing individual and group problem-solving.* ERIC ED: 372 081.

Sugrue, B. (1995) A theory-based framework for assessing domain-specific problem-solving ability. *Educational Measurement: Issues and Practices*, 14(3), 29–36.

Sutton, J. and Krueger, A. (eds.) (2001) *EDThoughts: What we know about science teaching and learning*. Aurora, CO: Mid-continent Research for Education and Learning.

Tytler, R. (2002) Teaching for understanding in science: Constructivist/conceptual change teaching approaches. *Australian Science Teachers Journal*, 48(4), 30–5.

Von Secker, C. (2002) Effects of inquiry-based teacher practices on science excellence and equity. *Journal of Educational Research*, 95, 151–60.

Wenglinsky, H. (2002) How schools matter: The link between teacher classroom practices and student academic performance. *Education Policy Analysis Archives*, 10(12). Retrieved 12 August 2003, from http://epaa.asu.edu/epaa/v10n12/

White, R. (1999) The nature and structure of knowledge: Implications for assessment. Unpublished manuscript.
Zeitz, C. (1997) Some concrete advantages of abstraction: How experts' representations facilitate reasoning. In P.J. Feltovich, K.M. Ford and R.R. Hoffman (eds.), *Experts in context* (pp. 43–65). Menlo Park, CA: AAAI Press.

20

Background to Japanese Students' Achievement in Science and Mathematics

Yasushi Ogura

Introduction

The TIMSS study and a number of secondary analyses of TIMSS data have investigated the detail factors that influence achievement at school level, but few factors related to out-of-school education have been examined.

Many upper-grade elementary school students (fifth and sixth graders) and more lower-secondary school students (seventh to ninth graders) study at cram school (*juku*) for a number of hours in the evening, a few days a week or even during the day on the weekends or during holidays. The influence of this type of out-of-school education on student achievement in Japan is assumed to be significant, but it has never been investigated. This chapter studies the effect of attending *juku* on achievement in Japan using data from TIMSS 1995. The next section presents the context of Japanese education, especially the role of *juku* teaching, followed by a section on the methods of addressing the research question. The results of our analyses are then presented, ending with discussion and conclusion.

Japanese Education and the Place of *Juku*

Most *juku* teach a specialized curriculum so that students can achieve higher scores in the paper-and-pencil achievement tests of school subjects and also to prepare them for entrance examinations for upper-level schools at the end of Grades 6, 9 or 12. A different type of *juku* emphasizes remedial instruction for students who have difficulty in studying the regular curriculum at school. Hiring private tutors and correspondence education are also widespread for examination preparation and/or remedial instruction. Parents sometimes play the role of tutor to their child.

The entrance examination at the end of Grade 6 is important for a few students, namely those who want to enter non-public lower-secondary schools, which are out of the municipal government service. The majority, about 93 percent, of elementary school pupils proceed to public lower-secondary schools without any examination. However, the majority of students in public lower-secondary schools have to take an entrance examination to go on to one of the upper-secondary schools at the end of Grade 9. In addition, almost all students who want to enter university have to take entrance examinations again at the end of Grade 12. The examination results determine the selection of applicants. This series of examinations makes student learning in Japan competitive, and this competition is closely related to the flourishing *juku* industry.

It is anticipated that *juku* and private tutoring, which are out-of-school education, considerably influence the level of student achievement in science and mathematics as measured by the TIMSS. What influence on achievement they have, however, is not known, as it is not possible to observe how a specific student's achievement has developed if that student has taken (or not taken) extra lessons out-of-school, and to compare the two cases.

The relation between attending *juku* and achievement is complex. Students who do not need extra help from others to reach the expected level of achievement and students who want to spend their time for other purposes may not attend *juku*. Students who live in remote areas, where there are no *juku*, have difficulty in attending *juku*. Attending *juku* is not always effective in improving student achievement because achievement may depend on other factors such as student motivation and the quality of instruction at *juku*. Some parents even use *juku* for childcare purposes. Moreover, attending *juku* is expensive. According to national statistics on educational expenses for children (MEXT, 2003), parents who use *juku* for their children at lower-secondary school level paid a total of about 215,000 yen per child (approximately US$2000) on average in the fiscal year 2002. Socio-economic factors certainly affect the use of out-of-school education.

Considering this complex situation, we should at least recognize that there are, for a considerable number of students in Japan, science and mathematics out-of-school curricula, and that this type of curriculum as a background factor may have a significant influence on the achievement results in the TIMSS.

The proportion of students who take out-of-school instruction in science and mathematics, including *juku* and private tutoring from Grades 5–9, is shown in Table 20.1, with separate sub-proportions of students who take advanced instruction, remedial instruction and 'not apparent' type of instruction based on the 2002 national assessment result (NIER, 2002). Table 20.1 shows the trend of a gradual increase across the grades of the proportion of students who take extra lessons in science and mathematics, reaching more than half of students for science and about two thirds for mathematics at Grade 9. Most lower-secondary school students have out-of-school lessons to improve their achievement level. From Grade 8, which is the target of the TIMSS (the upper grade of population 2) to Grade 9, there is about a 20 and 15 percent increase in the proportions of students taking extra lessons in science and mathematics, respectively. Apparently, the situation of out-of-school education in Japan differs considerably across grade levels.

Table 20.1 Percentages of students who take or do not take out-of-school science and mathematics lessons from grade 5 to 9, based on the results of national assessment in 2002.

	Taking extra lesson			Not taking extra lesson	No answer
	Advanced instruction	Remedial instruction	Not apparent distinction		
Science					
Grade 5	9.0	4.4	4.9	71.2	10.5
Grade 6	8.4	4.4	5.0	71.7	10.5
Grade 7	8.9	10.3	7.7	61.1	12.0
Grade 8	11.1	13.3	8.6	54.7	12.3
Grade 9	19.0	19.1	13.4	40.2	8.3
Mathematics					
Grade 5	21.0	8.3	6.8	55.6	8.3
Grade 6	22.3	9.3	6.9	53.2	8.3
Grade 7	22.9	16.4	8.1	42.9	9.8
Grade 8	25.0	20.4	8.8	35.9	9.8
Grade 9	32.1	24.1	11.1	26.1	6.6

Source. NIER (2002). Based on a randomly selected nationally representative sample (total N about 450,000).

Table 20.1 also shows that over 20 percent of Grade 5 and 6 elementary school students take advanced out-of-school mathematics instruction. Since there is no examination to go on to public lower-secondary schools to which more than 93 percent of cohort students belong, it is plausible that most of these students are applicants of the entrance examination for non-public lower-secondary schools at the end of Grade 6. It is assumed that, through extra study for the examination, their achievement level will strengthen.

Wolf (2002) found in his analysis of the TIMSS database that, in almost every country participating in the TIMSS, extra-school instruction is used more for remedy than enrichment. Table 20.1 illustrates that a different situation applies in Japan, as there are more students taking extra lessons in mathematics for enrichment rather than remedy in Japan.

The national assessment in 2002 (NIER, 2002) also showed that the average test score of students receiving remedial instruction is lower than that of students not receiving extra lessons, and that the average test score of students receiving advanced instruction is the highest. Students choose the type of extra instruction appropriate for their needs.

It is apparent that out-of-school education plays a significant role in improving students' achievement levels in Japan. The majority of students at lower-secondary school level intensively study school subjects using out-of-school education to prepare for entrance examinations. Other students who do not use out-of-school education also have to study school subjects for the entrance examination at the end of Grade 9.

This system, in which entrance examinations push students to study hard and in which out-of-school education provides special curricula for students to succeed, may work effectively to maintain the achievement level of Japanese students. However, we should bear in mind that, as Deci (1975, 1992) warns, extrinsic motivation can undermine intrinsic motivation. If students do not learn to study autonomously, they may stop studying after finishing examinations. Suppose that there were no examinations to enter schools, would Japanese students continue to seriously study school subjects?

In the national assessment in 2002 (NIER, 2002), students were asked: 'Do you think the study of <each subject> is important regardless of entrance examinations?' The results are shown in Table 20.2.

The results clearly show that the importance of studying science and mathematics not for entrance examinations decreases in the minds of Japanese students along with the grade level. About half of the students at lower-secondary school do not value science. However, a sense of the value of achievement may be directly connected to student achievement behavior, according to Eccles' expectancy-value model of achievement motivation (Eccles, 1983; Wigfield and Eccles, 1992). From the results shown in Table 20.2, it seems reasonable to hypothesize that Japanese students, with the increase in grade level, become less intrinsically motivated to study science and mathematics.

The TIMSS 1995 Japanese population 2 data can be used to analyze the situation of and relationship among variables, such as taking out-of-school lessons, motivation to study science and mathematics, student grade level, science and mathematics achievement, and type of school of student. In this chapter, the following research questions will be investigated by using these variables: (1) Who takes extra lessons? (2) What is student motivation like? (3) How are situations of investigating variables different between seventh and eighth graders? (4) What relationships are there among investigating variables?

Method

This section presents methods of addressing the research questions. The dataset used in the research, the classification of students, the variables relevant to the research questions, and the analysis methods are then discussed.

Table 20.2 Percentages of students (grades 5–9) who answered affirmatively ('strongly agree' or 'agree') to the question, 'do you think the study of < each subject > is important regardless of entrance examinations?'.

	Japanese	Social	Math	Science	English
Grade 5	76.3	71.9	79.8	63.6	–
Grade 6	76.6	71.2	80.7	58.9	–
Grade 7	73.4	59.7	73.5	51.4	79.6
Grade 8	72.9	58.7	66.5	49.6	79.7
Grade 9	76.2	69.6	61.0	49.0	81.5

Source. NIER (2002).

Data

The TIMSS 1995 Japanese population 2 data include 5,130 Grade 7 and 5,141 Grade 8 student data in 151 lower-secondary schools. Schools were randomly selected by two-stage stratified sampling. Students in each school were selected as a cluster by randomly selecting one class at each grade. The Grade 7 and 8 student data allow us to investigate the influence of selected variables at both grade levels, as well as differences between the two grade levels.

Student Classification

The Japanese dataset has a stratification variable. By using the variable with the sampling information from the TIMSS 1995 national report (NIER, 1996), participating schools were classified into four groups: (1) public schools located in 'towns/villages' with less than 50,000 people; (2) public schools located in 'cities' with less than 1 million people; (3) public schools located in 'big cities' with more than 1 million people; and (4) 'non-public' schools including private schools and national government schools. Public schools are those controlled by municipal governments and account for up to 93 percent of students in all lower-secondary schools. This classification variable is useful for investigating the influence of out-of-school education, because the difference in urbanization status among the three public school groups is connected with the use of *juku*. There are many *juku* in big cities and only few in towns and villages. There are also more choices of upper-secondary school in bigger cities for lower-secondary school students who are to take entrance examinations, and this makes the atmosphere in these cities more competitive. Most students in non-public schools, especially in private schools, will not take an entrance examination to go on to the upper-secondary level, and this makes the atmosphere different from that in public schools. Considering these differences in situation, the sample of students is analyzed not only as a whole but also separately per type of schools. According to the stratification variable, four strata as types of schools are distinguished by following labels: 'Public_Town/Villages', 'Public_Cities', 'Public_Big Cities' and 'Non-Public'.

Students' Extra Lesson Taking

A question in the TIMSS student questionnaire related to out-of-school education asks students about the time spent on taking extra lessons in science and mathematics in the TIMSS student questionnaire (Japanese version): 'During the week, how much time before or after regular classes do you spend taking extra science lessons or taking science lessons at *juku*?' ('Science' is replaced by 'mathematics' in another question.) The students answered on a five-point scale: (1) none; (2) less than 1 hour; (3) 1 hour or more and less than 3 hours; (4) 3 hours or more and less than 5 hours; (5) more than 5 hours.

Responses to this question may refer not only to the time spent on lessons at *juku*, but also to lessons with private tutors or supplemental instruction in schools. The reason for taking extra lessons, whether remedial or advanced, is not asked. The interpretation of the responses should be that students refer to time spent on extra lessons in science and mathematics not included in the school curriculum. The students who answered 'none' do not take extra lessons, but they may still be studying independently out of school using other materials. This research focuses on the difference between students who take extra lessons and those who do not.

Student Motivation

There are a number of questions in the TIMSS student questionnaire asking students 'What do you think about science?' Students indicated their level of agreement using a four-point Likert-type scale for the following statements: 'I enjoy studying science' [BSBSENJY], 'Science is boring' [BSBSBORE], 'Science is an easy subject' [BSBSEASY], 'Science is important for the life of everyone' [BSBSLIFE], and 'I would like a job that involved using science' [BSBSWORK].

Another series of questions starts with the phrase, 'I need to do well in science' and is connected by the four phrases: 'to get the job I want' [BSBSJOB], 'to please my parents' [BSBSPRNT], 'to get into the high school or college/university I prefer' [BSBSSCHL] and 'to please myself' [BSBSSELF]. Students indicate their level of agreement using a four-point Likert-type scale. There are similar questions for mathematics, and the related variables all begin with 'BSBM' (as opposed to 'BSBS').

These nine questions for science and mathematics reveal student motivation in these subjects. Principal component analysis was undertaken for the Japanese Population 2 data, resulting in two components explaining 48 percent and 45 percent of the total variance of the nine questions for science and mathematics, respectively. As for science, [BSBSLIFE] was not loaded highly for either component after the Varimax rotation and was therefore excluded from the analysis. As a result, two scales were created. The first scale is the 'intrinsic value' of science, and is the sum of responses (maximum value is 16) to the four questions: [BSBSENJY], [BSBSBORE] (inversed), [BSBSEASY] and [BSBSWORK]. The second scale is the 'extrinsic value' of science, the sum of responses (maximum value is 16) to the four questions: [BSBSJOB], [BSBSPRNT], [BSBSSCHL] and [BSBSSELF]. Cronbach's α is the reliability coefficient of each scale was 0.70 for intrinsic value, and 0.60 for extrinsic value. As for mathematics, principal component analysis produced the same results as for science, and two scales were created. Cronbach's α as the reliability coefficient of each scale in mathematics was 0.69 for intrinsic value, and 0.57 for extrinsic value.

Intrinsic value is the sense of value felt in undertaking the task itself. Extrinsic value is the sense of value without there necessarily being intrinsic value for the student. According to Deci's (1992) self-determination theory, extrinsically motivated behavior can be self-determined or non-self-determined. While non-self-determined behavior may undermine 'intrinsic motivation', self-determined behavior may not. Because the TIMSS data do not make this distinction, it should be noted that even high extrinsic value points do not necessarily directly suggest low 'intrinsic motivation'.

Statistical Analyses

Plausible values and the Jackknife Repeated Replication (JRR) method were used to calculate the mean values and the standard errors of science and mathematics achievement scores and other statistics using the SAS macro programs provided by IEA (2001). In the multiple comparison of mean values, the Dunn-Bonferroni procedure (Winer *et al.*, 1991) was used to adjust critical value α to compensate for the increase in the probability of error.

Results

This section presents the results of analysis of research questions: (1) Who takes extra lessons? (2) What is student motivation like? (3) How are situations of investigating variables different between seventh and eighth graders? (4) What relationships are there among the investigating variables?

Statistics such as achievement means or percentages presented in this section are estimated values with regard to the TIMSS population 2 calculated from the Japanese TIMSS sample ($N = 5,130$ at Grade 7, $N = 5,141$ at Grade 8), using the sampling weights given to individual students in the sample to represent a certain number of students in the population.

Who Takes Extra Lessons?

Table 20.3 from the column of 'Population N' shows the percentages of students in each stratum: 24 percent in 'Public_Towns/Villages', 54 percent in 'Public_Cities', 15 percent in 'Public_Big Cities' and 7 percent in 'Non-public'. The percentages of eighth grade students taking extra lessons in each stratum are also shown. The students in the 'Public_Big Cities' stratum take extra lessons the most, with 51 percent for science and 74 percent for mathematics. This suggests that out-of-school education plays an especially significant role in big cities for students studying science and mathematics. For students in 'Non-public' schools, where the majority will not take entrance examinations at the end of the ninth grade, the percentages of students who take extra lessons are not as large as in the strata of public schools in either subject.

Table 20.3 also shows the mean values and the standard errors of achievement scores by students taking extra lessons and students not taking extra lessons in each stratum. The mean value of the science scores of students who are not taking extra lessons was significantly higher than that of students taking extra lessons in the 'Public_Towns/Villages' stratum, suggesting that in towns and villages, extra lessons are mainly aimed at remediation of weaker students. The mean value of mathematics scores of students taking extra lessons was significantly higher than that of students not taking extra lessons in the 'Public_Big Cities' stratum, suggesting that these students take those lessons to prepare for the competitive entrance exams for the upper-secondary school level.

Figures 20.1 and 20.2 show the score distributions of students who take and who do not take extra lessons in each stratum in science and mathematics, respectively.

Table 20.3 Percentages of eighth-grade students who take extra lessons in science or mathematics, the mean values and the standard errors of achievement scores compared with students who do not take extra lessons in each stratum, and through all strata shown with the numbers of sample in the TIMSS 1995 Japanese population 2 data and the estimated numbers of population in each category.

Stratum	Science/achievement			Mathematics/achievement			Sample N	Population N
	Taking extra lessons		No extra	Taking extra lessons		No extra		
	pct% (s.e.)	mean (s.e.)	mean (s.e.)	pct% (s.e.)	mean (s.e.)	mean (s.e.)		
Public_Towns/Villages	36.0 (2.5)	556.2 (4.2)	572.7 (3.5)	54.5 (2.9)	580.6 (3.8)	597.7 (7.1)	1,394	398,640
Public_Cities	44.6 (2.0)	563.0 (2.5)	571.1 (2.6)	68.8 (1.9)	601.8 (2.7)	600.5 (3.5)	2,672	886,101
Public_Big Cities	51.5 (3.3)	565.9 (6.1)	557.3 (6.7)	73.8 (2.6)	612.2 (5.0)	581.3 (7.6)*	785	244,519
Non-public	21.1 (4.4)	621.2 (16.7)	640.6 (3.2)	42.1 (7.8)	687.4 (8.6)	697.4 (10.5)	290	112,681
All	41.9 (1.4)	564.1 (2.0)	576.2 (2.1)*	64.2 (1.5)	603.0 (2.2)	608.3 (3.4)	5,141	1,641,941

Notes. '*' in the 'no extra' column means that the difference of achievements between students who take extra lessons and students who do not take extra lessons within a group is statistically significant. (Critical value ($\alpha = 3.261$, $p < 0.05$) was adjusted for multiple comparison.) The numbers of sample N are slightly different from the numbers actually used for the estimations in this table and in Tables 20.4 and 20.5 because there are missing values in the variables used.

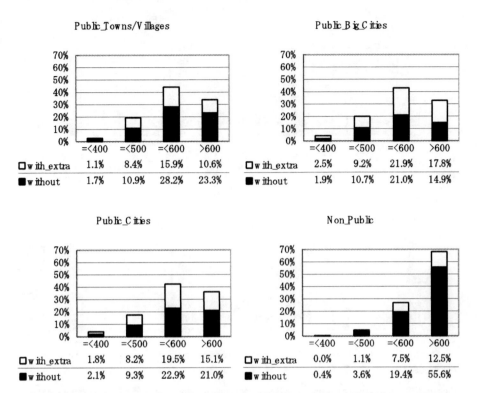

Figure 20.1 Science score distribution of students who take and who do not take extra lessons in each stratum.
Notes. ' = < 400' means less than 400 score points in the acievement test; ' = < 500' means a test score greater than 400 and equal to or less tan 500 points, and so on.

From these figures, it is apparent that the score distributions among the three strata of public schools are almost the same. However, the proportion of students who take extra lessons is larger in more urbanized areas than in less urbanized areas in both subjects. Comparing the data on public schools with those of non-public schools does not directly lead to the conclusion that taking extra lessons does not improve students' achievement, as schools belonging to the 'non-public' stratum are located in more urbanized areas, and most of the students who have moved to non-public schools were high achievers in public elementary schools in those areas. This 'pull-out effect' should be considered. Most high achievers at the elementary school level in less urbanized areas will remain in public schools at the lower-secondary school level.

What Is Student Motivation Like?

Table 20.4 shows the mean values and the standard errors of intrinsic value in science and mathematics for all students and students who take and who do not take extra lessons in each stratum and through all strata. The mean values of intrinsic value in science

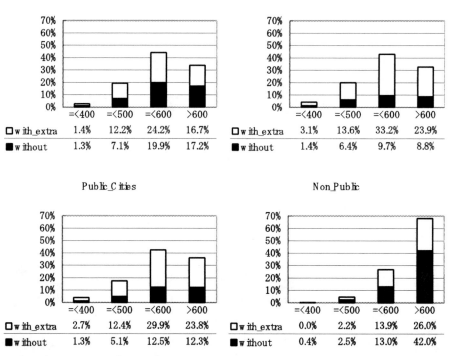

Figure 20.2. Mathematics score distribution of students who take and who do not take extra lessons in each stratum.
Notes. See Figure 20.1.

and mathematics of students who take extra lessons tend to be higher than those of students who do not take extra lessons. Students in the 'Public_Big Cities' stratum tend to show lower intrinsic value in science and mathematics than students in other public strata. Students in the 'Non-public' stratum tend to show higher intrinsic value in science and mathematics than did students in other strata.

Table 20.5 shows the mean values and the standard errors of extrinsic value in science and mathematics of all students and students who take and who do not take extra lessons in each stratum and through all strata. The mean values of extrinsic value in science and mathematics of students who take extra lessons tend to be higher than those of students who do not take extra lessons. Students in the 'Public_Big Cities' stratum tend to show lower extrinsic value in science and mathematics than students in other public school strata.

How Different Are the Situations of Investigating Variables Between Seventh and Eighth Graders?

Table 20.6 shows the mean values and the standard errors of science and mathematics achievement scores of seventh and eighth grade students. The mean scores of

Table 20.4 Mean values and standard errors of intrinsic value in science or mathematics of eighth-grade students who take extra lessons compared with students who do not take extra lessons in each stratum and through all strata in the TIMSS 1995 Japanese population 2 sample.

Stratum	Science/intrinsic value			Mathematics/intrinsic value			Sample N	Population N
	All	Taking extra lessons	No extra	All	Taking extra lessons	No extra		
Public_Towns/Villages	9.4 (0.1)	9.6 (0.1)	9.2 (0.1)	9.1 (0.1)	9.3 (0.1)	8.8 (0.1)*	1,394	398,640
Public_Cities	9.2 (0.1)	9.3 (0.1)	9.1 (0.1)	9.0 (0.1)	9.1 (0.1)	8.6 (0.1)	2,672	886,101
Public_Big Cities	8.6 (0.2)	8.8 (0.2)	8.4 (0.2)	8.9 (0.1)	9.1 (0.2)	8.5 (0.3)	785	244,519
Non-public	9.5 (0.3)	9.7 (0.4)	9.5 (0.3)	9.7 (0.2)	10.1 (0.4)	9.4 (0.2)	290	112,681
All	9.2 (0.1)	9.3 (0.1)	9.1 (0.1)	9.0 (0.1)	9.2 (0.1)	8.7 (0.1)*	5,141	1,641,941

Notes. N is shown in Table 20.3. '*' in the 'no extra' column means that the difference between students who take extra lessons and students who do not take extra lessons within a stratum or all strata is statistically significant. (Critical value ($\alpha = 3.494$, $p < 0.05$) was adjusted for multiple comparison.)

Table 20.5 Mean values and the standard error of extrinsic value of science or mathematics of eighth-grade students who take extra lessons compared with students who do not take extra lessons in each stratum and through all strata in the TIMSS 1995 Japanese population 2 sample.

Stratum	Science/extrinsic value			Mathematics/extrinsic value			Sample N	Population N
	All	Taking extra lessons	No extra	All	Taking extra lessons	No extra		
Public_Towns/Villages	10.8 (0.1)	11.0 (0.1)	10.6 (0.1)	11.2 (0.1)	11.4 (0.1)	10.9 (0.1)	1,394	398,640
Public_Cities	10.7 (0.0)	10.8 (0.1)	10.6 (0.1)	11.0 (0.0)	11.1 (0.1)	10.9 (0.1)	2,672	886,101
Public_Big Cities	10.3 (0.1)	10.5 (0.1)	10.1 (0.1)	10.8 (0.1)	10.9 (0.1)	10.7 (0.2)	785	244,519
Non-public	10.3 (0.1)	11.0 (0.3)	10.2 (0.1)	10.7 (0.1)	11.0 (0.3)	10.5 (0.3)	290	112,681
All	10.6 (0.0)	10.8 (0.1)	10.5 (0.0)*	11.0 (0.0)	11.1 (0.0)	10.8 (0.1)*	5,141	1,641,941

Notes. See Table 20.4.

Table 20.6 Mean values and the standard errors of achievement scores in science or mathematics of seventh and eighth-grade students shown by the numbers of sample in the TIMSS 1995 Japanese population 2 data and the estimated population in each category.

Achievement and grade level	Science			Mathematics			Sample N	Population N
	All	Taking extra lessons	No extra	All	Taking extra lessons	No extra		
Seventh graders	531.0 (1.9)	520.3 (2.2)	538.4 (2.7)*	571.1 (1.9)	565.6 (2.3)	580.1 (3.2)*	5,130	1,562,418
Eighth graders	571.0 (1.6)^	564.1 (2.0)^	576.2 (2.1)*^	604.8 (1.9)^	603.0 (2.2)^	608.3 (3.4)^	5,141	1,641,941

Notes. '*' in the 'no extra' column means that the difference between students who take extra lessons and students who do not within a grade is statistically significant. The mark '^' in the eighth graders row means that the difference between seventh graders and eighth graders within a group is statistically significant. (Critical value ($\alpha = 2.935$, $p < 0.05$) was adjusted for multiple comparison.) The numbers of sample N shown in Table 20.6 are slightly different from the numbers actually used for the estimations in Tables 20.7 and 20.8 because there are missing values in the variables used.

Table 20.7 Mean values and the standard errors of intrinsic value of science or mathematics of seventh and eighth-grade students in the TIMSS 1995 Japanese population 2 data.

Intrinsic value and grade level	Science			Mathematics			Sample N	Population N
	All	Taking extra lessons	No extra	All	Taking extra lessons	No extra		
Seventh graders	9.5 (0.1)	9.6 (0.1)	9.4 (0.1)	9.3 (0.1)	9.4 (0.1)	9.1 (0.1)*	5,130	1,562,418
Eighth graders	9.2 (0.1)^	9.3 (0.1)	9.1 (0.1)	9.0 (0.1)^	9.2 (0.1)	8.7 (0.1)*^	5,141	1,641,941

Notes. See Table 20.6.

Table 20.8 Mean values and the standard errors of extrinsic value of science or mathematics of seventh and eighth-grade students in the TIMSS 1995 Japanese population 2 data.

Extrinsic value and grade level	Science			Mathematics			Sample N	Population N
	All	Taking extra lessons	No extra	All	Taking extra lessons	No extra		
Seventh graders	10.6 (0.0)	10.8 (0.1)	10.5 (0.1)*	11.2 (0.0)	11.3 (0.0)	10.9 (0.1)*	5,130	1,562,418
Eighth graders	10.6 (0.0)	10.8 (0.1)	10.5 (0.0)*	11.0 (0.0)	11.1 (0.0)^	10.8 (0.1)*	5,141	1,641,941

Notes. See Table 20.6.

eighth grade students are significantly higher than those of seventh grade students, regardless of whether they take extra lessons or not in both subjects.

In contrast to Table 20.6, Table 20.7 shows that the mean values of intrinsic value in science and mathematics of eighth grade students tend to be lower than those of seventh grade students, regardless of whether they take extra lessons or not in both subjects. Intrinsic value goes down with the grade level.

As Table 20.8 shows, the mean values of extrinsic value in science and mathematics are more stable as grade levels change than those of intrinsic value.

What Relationships are There among Investigating Variables?
Tables 20.9 and Table 20.10 for science and mathematics, respectively, show the results of multiple regression analysis of the achievement scores as dependent variables, and intrinsic value, extrinsic value and taking extra lessons as independent variables in each classification of stratum and grade level.

Intrinsic value shows statistically significant differences in science and mathematics achievement scores in most of the classifications but extrinsic value do not. Taking extra lessons shows a statistically significant decrease in the science achievement scores for 'All' seventh graders and an increase in the mathematics scores for students in the 'Public_Big Cities' stratum. Among the independent variables, intrinsic value has a stronger relationship with student achievement scores in science and mathematics than the other two variables. This, however, is not obvious in the 'Non-public' stratum at the eighth grade.

Discussion and Conclusion

This study clearly shows how out-of-school education plays an important role in developing the achievement of Japanese students. The majority of eighth graders take extra lessons in science and mathematics. Conversely, there is a considerable number of high achievers who do not take extra lessons. Therefore, taking extra lessons is not a necessary condition for high achievement, but a major supplementary condition.

The urbanization status of places where students live is important in understanding the relationship between student achievement and the use of out-of-school education. In more urbanized areas, many high achievers in elementary school move from public education to non-public education through entrance examinations at the end of the sixth grade, which is explained as the 'pull-out effect'. The remainder of students in public lower-secondary schools in these areas has to take entrance examinations to go on to upper-secondary schools in a highly competitive atmosphere. This education system, with a structure that strongly depends on entrance examinations, may push the majority of students to take extra lessons. As a result, the educational standard of students in public schools is almost the same among different urbanized areas, regardless of the 'pull-out' effect.

The fact that extra lessons are taken for remediation or enrichment complicates the analysis of the effect of out-of-school education on student achievement. As Figures

Table 20.9 Results of multiple regression analysis of seventh and eighth-grade science achievement in each stratum and through all strata in the TIMSS 1995 Japanese population 2 data.

	N	Mult-RSQ	SS_Res	SS_Reg	SS_Total	Intercept (s.e.)	Intrinsic value b (s.e.)	Extrinsic value b (s.e.)	Taking extra lesson b (s.e.)
Seventh grade									
Public_Towns/Villages	1,340	0.033	2.44E+09	8.19E+07	2.52E+09	466.7 (16.5)	7.3 (1.8)*	−0.1 (0.8)	−7.1 (2.7)
Public_Cities	2,655	0.036	5.70E+09	2.11E+08	5.91E+09	460.4 (11.2)	7.5 (0.8)*	0.4 (0.6)	−5.3 (1.9)
Public_Big Cities	752	0.037	1.45E+09	5.50E+07	1.50E+09	465.4 (17.3)	6.5 (1.5)*	1.2 (1.3)	−5.9 (2.4)
Non-public	300	0.096	6.49E+08	6.91E+07	7.18E+08	524.3 (32.9)	11.4 (2.5)*	−1.9 (1.7)	−8.8 (4.3)
All	5,047	0.044	1.08E+10	5.02E+08	1.13E+10	467.8 (7.9)	8.3 (0.6)*	0.0 (0.5)	−8.5 (1.7)*
Eighth grade									
Public_Towns/Villages	1,380	0.038	2.83E+09	1.10E+08	2.94E+09	502.8 (17.3)	7.3 (1.3)*	0.7 (1.5)	−9.3 (3.1)
Public_Cities	2,634	0.053	6.36E+09	3.58E+08	6.72E+09	475.7 (12.7)	9.5 (1.0)*	0.9 (1.1)	−2.7 (2.0)
Public_Big Cities	768	0.052	1.86E+09	1.02E+08	1.96E+09	462.0 (15.0)	7.9 (1.7)*	2.3 (1.3)	4.3 (2.7)
Non-public	286	0.043	6.94E+08	3.11E+07	7.25E+08	618.2 (22.8)	7.0 (2.0)	−3.4 (2.0)	−8.6 (4.6)
All	5,068	0.048	1.23E+10	6.15E+08	1.29E+10	493.1 (9.1)	9.0 (0.8)*	0.3 (0.7)	−4.9 (1.5)

Notes. N, Mult-RSQ, SS_Res, SS_Reg, SS_Total, Constant, s.e., and b represent the number of cases, squared multiple correlation, residual, regression, and total sum of squares, constant of regression, standard error, and regression coefficient, respectively. The program used for this calculation is 'JackRegP.sas' which was introduced by Gonzalez and Miles (2001). The dependent variables are five plausible values of science. 'Intrinsic value' and 'Extrinsic value' ranging from 4 to 16, while 'Taking extra lessons' ranges from 1 to 5 as the original response is assumed to reflect variability. Responses with a missing value were removed. Regression coefficient b means the average difference in achievement score accompanied by a 1.0 point increase in the independent variable, and '*' means that the difference is statistically significant. (The critical value ($\alpha = 3.857$, $p < 0.05$) was adjusted for multiple comparison.)

Table 20.10 Results of multiple regression analysis of seventh and eighth-grade mathematics achievement in each stratum and through all strata in the TIMSS 1995 Japanese population 2 data.

	N	Mult-RSQ	SS_Res	SS_Reg	SS_Total	Intercept (s.e.)	Intrinsic value b (s.e.)	Extrinsic value b (s.e.)	Taking extra lesson b (s.e.)
Seventh grade									
Public_Towns/Villages	1,346	0.041	2.93E+09	1.26E+08	3.06E+09	488.8 (13.7)	9.4 (1.4)*	−1.7 (1.3)	−1.1 (2.0)
Public_Cities	2,661	0.045	6.85E+09	3.21E+08	7.17E+09	494.9 (14.1)	9.8 (0.9)*	−1.8 (0.9)	0.3 (1.9)
Public_Big Cities	748	0.054	1.84E+09	1.04E+08	1.94E+09	493.5 (24.8)	10.2 (1.1)*	−2.2 (1.7)	3.0 (2.7)
Non-public	299	0.089	6.99E+08	6.88E+07	7.68E+08	563.8 (39.7)	11.5 (3.6)	0.9 (1.2)	−9.1 (2.6)
All	5,054	0.052	1.34E+10	7.44E+08	1.42E+10	504.8 (11.3)	11.0 (0.7)*	−2.7 (0.7)	−2.4 (1.6)
Eighth grade									
Public_Towns/Villages	1,381	0.049	3.68E+09	1.89E+08	3.87E+09	521.5 (22.9)	11.4 (1.8)*	−2.2 (1.5)	−5.2 (3.6)
Public_Cities	2,639	0.052	8.24E+09	4.55E+08	8.69E+09	513.1 (16.1)	10.7 (0.8)*	−1.1 (1.3)	2.7 (1.8)
Public_Big Cities	771	0.049	2.26E+09	1.16E+08	2.38E+09	515.6 (25.9)	7.2 (1.7)*	−1.1 (2.0)	14.3 (2.8)*
Non-public	287	0.010	8.75E+08	8.69E+06	8.83E+08	662.7 (12.1)	3.7 (2.1)	0.1 (1.4)	−2.8 (5.9)
All	5,078	0.048	1.60E+10	8.01E+08	1.68E+10	530.2 (11.5)	10.9 (0.7)*	−2.2 (0.9)	0.7 (1.7)

Notes. See Table 20.9.

20.1 and 20.2 show, at all achievement levels, there are students who take extra lessons to improve their achievement level as much as possible. High achievers may need more of the enrichment and less of the remedial type of extra lessons, while low achievers need the more remedial type. However, the issue is not clear cut.

Although the achievement level is almost the same over the public school strata, the level of intrinsic value varies. It is relatively low in the 'Public_Big Cities' stratum, which can be partly explained by the pull-out effect. High achievers tend to have high intrinsic value, as the multiple regression analysis (Tables 20.9 and 20.10) suggests. A number of high achievers in public elementary schools in big cities move to non-public schools, and this may be one reason for the depression of intrinsic value in these areas. In addition, taking extra lessons may strengthen students' achievement, but may not improve their intrinsic motivation. Moreover, the competitive atmosphere in big cities undermines students' intrinsic motivation. Thus, education in more urbanized areas should pay more attention to elevating students' intrinsic motivation.

A reason that intrinsic value in students who take extra lessons tends to be higher than that in students who do not may be that students who do not take extra lessons include students who do not want to study at all.

Extrinsic value is relatively low in the 'Public_Big Cities' stratum as is the case for intrinsic value, but the reason may be different. High achievers do not tend to have high or low extrinsic value, as the multiple regression analysis (Tables 20.9 and 20.10) suggests. Students in the 'Non-public' stratum also show relatively low extrinsic value. It is hypothesized that students in more urbanized areas tend to have less chance to think about the value of studying itself, and just follow the given tasks; this depresses extrinsic value in these areas as a whole. Students in city areas are surrounded by a lot of stimulus and tend not to think about the value of their study tasks or their behavior. This implies that students in city areas need to make more intentional a connection between a study task and its value.

The reason that extrinsic value in students who take extra lessons tends to be higher than that in students who do not may be that these students tend to be extrinsically motivated for entrance examinations.

Student-grade level is significantly related not only to achievement but also to intrinsic value. Intrinsic value goes down with the grade level, while achievement scores rise. The decrease in intrinsic motivation has the same trend shown in Table 20.2. Japanese students become less intrinsically motivated to study science and mathematics as their grade increases. Students who are not motivated to study science and mathematics may not choose these subjects in upper-secondary schools or study them autonomously. Japanese science and mathematics educators should take these results seriously when considering what and how students at this stage should be taught.

In conclusion, for a good understanding of the achievement and motivation of Japanese students in science and mathematics, it is necessary to understand the entrance examination system, whether a school is public or not, whether students take extra lessons or not, how urbanized the area where the student lives is, and what grade the student belongs to. Without considering these factors, the results of international comparison are only superficial. International studies like the

TIMSS can be strengthened and will have more effect if they are combined with country-specific, relevant, contextual information to make more in-depth analysis possible.

References

Deci, E.L. (1975) *Intrinsic motivation*. New York: Plenum Press.

Deci, E.L. (1992) The relation of interest to the motivation of behavior: A self-determination theory perspective. In K.A Renninnger, S. Hidi, and A. Krapp (eds.), *The role of interest in learning and development* (pp. 43–70). Mahwah, NJ: Lawrence Erlbaum Associates.

Eccles, J. (1983) Expectancies, values and academic behaviors. In J.T. Spence (ed.), *Achievement and achievement motives* (pp. 75–146). San Francisco: Freeman.

International Association for the Evaluation of Educational Achievement (IEA), E.J. Gonzales, and T.A. Smith (eds.) (1997) *User guide for the TIMSS international database – primary and middle school years*. Boston, MA: Boston College.

International Association for the Evaluation of Educational Achievement (IEA), E.J. Gonzales, and J.A. Miles (eds.) (2001) *TIMSS 1999 user guide for the international database*. Boston, MA: Boston College.

Ministry of Education, Culture, Sports, Science and Technology (MEXT) (2003) *Household expenditure on education per student (in Japanese)*. Tokyo: MEXT.

National Institute for Educational Research (NIER) (1996) *Sho-tyu gakusei-no sansu, sugaku, rika-no seiseki (in Japanese)*. Toyokan.

National Institute for Educational Policy Research (NIER) (2002) *Heisei 13 nendo shou-tyu gakko kyoiku katei jissi jyokyo tyosa hokokusho (in Japanese)*. NIER.

Wigfield, A., and Eccles, J.S. (1992) The development of achievement task values: A theoretical analysis. *Developmental Review*, 12, 265–310.

Winer, B.J., Brown, D.R. and Michels, K.M. (1991) *Statistical principles in experimental design*. New York: McGraw Hill.

Wolf, R.M. (2002) Extra-school instruction in mathematics and science. In D.F. Robitaille and A.E. Beaton (eds.), *Secondary analysis of the TIMSS data* (pp. 331–41). Boston, MA: Kluwer Academic Publishers.

Part 5

Teaching Aspects/Classroom Processes, School Management Variables and Achievement

21

Linking TIMSS to Research on Learning and Instruction: A Re-analysis of the German TIMSS and TIMSS Video Data

Mareike Kunter and Jürgen Baumert

Introduction

In the literature on instructional research, the early models of instructional effectiveness have been challenged by several theoretical developments (see Collins *et al.*, 2001; Shuell, 1996, 2001). Following these suggestions, we examine the effects of two different instructional aspects – strategies of classroom management and a constructivist teaching approach – on students' learning and interest development. Drawing on a multi-perspective framework we combined general teaching features on the class level with students' individual experiences of their learning processes.

In the first part of this chapter, we give a short overview on current developments in instructional research. We then present results from a study based on data from the German sample of the TIMSS 1995 assessment. On account of the various national extensions to the international study design described below, this sample constitutes an excellent database for studying the following research questions: (1) What are the effects of classroom management strategies and constructivist teaching elements on students' achievement? (2) Do these instructional aspects influence students' personal experience of challenge as a possible prerequisite for students' interest development? (3) Do different perspectives on the teaching process – that is, the individual student's interpretation vs. objective features of the learning environment – show differential validities in predicting achievement and interest?

Diverse Views on Instructional Quality

In research on instructional quality characteristics of effective classroom management such as clarity of structure, discipline and feedback mechanisms have long been considered crucial to students' achievement gains (Brophy, 1999; Grouws and Cebulla, 2000; Walberg and Paik, 2000). By now, there is ample empirical evidence showing that students in well-organized classes in which time is used effectively show better learning gains than students in less structured environments (Anderson et al., 1989; Walberg and Paik, 2000).

Following recent theoretical developments, however, the focus has shifted from the description of structural conditions to a closer examination of the instructional features of the learning environment (Greeno, 1998; Grouws and Cebulla, 2000; Shuell, 1996). Recently, instructional research has addressed questions such as how the learning content is presented, what types of tasks are used and what kind of learning processes take place. Many of these analyses are based on constructivist learning theories which conceptualize learning as an active, constructive and cumulative process, in which students are engaged in high-level cognitive activities, developing new concepts and understandings based on their former knowledge or preconceptions (Collins et al., 2001; Greeno et al., 1996). Due to its broad usage, the term 'constructivist teaching' has lost much of its original meaning and is now applied to many different forms of active, hands-on instruction. However, the key notion of constructivist learning theories is to conceptualize learning as an activity of knowledge construction, that is, 'an active process in which learners are active sense makers who seek to build coherent and organized knowledge' (Mayer, 2004: 14). In truly constructivist learning environments, students themselves infer the meaning of new constructs by linking and contrasting them to established knowledge, with teachers acting as mediators who provide these personal learning opportunities. Constructivist environments thus differ from conventional teaching approaches, where the teacher is assumed to possess all the information that has to be learned and to transfer this information directly to the learners.[1] Especially in the field of mathematics and science education, it has been shown that learning situations that use tasks and patterns of discourse that explicitly incorporate students' preconceptions and former experiences, and thus trigger personal and meaningful engagement with the core ideas, offer unique opportunities for conceptual growth (Cognition and Technology Group at Vanderbildt, 1992; Duit and Confrey, 1996). Most of these studies examine specifically designed learning contexts and thus report experimental or quasi-experimental data. Whereas this methodological approach provides sound evidence on the effects of certain instructional elements, the question of whether elements of constructivist teaching can also foster student learning gains within a regular classroom situation remains to be addressed.

Coming from different theoretical backgrounds, the two facets of instruction – the classroom management and the provision of constructivist learning situations – have usually been studied independently from each other. However, it seems plausible to assume that both form the basis for high quality instruction (Brophy, 1999; Shuell, 2001). Thus, our first question is whether effective classroom management and constructivist approaches alike can be considered necessary conditions for a positive student development.

Multiple Criteria for Instructional Quality

Most research on instructional quality concentrates on students' domain-specific achievement gains as the most salient educational outcome. There is, however, growing consensus that certain motivational characteristics are important prerequisites for learning, particularly when higher order learning gains are involved (for example, Pintrich, 1999; Pintrich et al., 1993). One of these motivational characteristics is an intrinsic motivational tendency, that is, the disposition to pursue an activity for its own sake rather than because of the expected external consequences. In the context of instructional research, the concept of domain-specific interest has often been used to describe intrinsic tendencies in students' learning (Hoffmann et al., 1998). Within this approach, interest is conceptualized as a specific relationship between a person and a topic, object or activity characterized by positive emotional experiences and feelings of personal relevance (value commitment). Various studies have shown that students with higher interest in a domain use more deeper-level-processing strategies, or show more engagement and persistence (Alexander and Murphy, 1998; Hoffmann et al., 1998; Schiefele, 1998). In addition to the benefits of interest for the learning process, it has also been argued that intrinsic motivational tendencies may constitute desirable educational outcomes in themselves (Krapp, 2002).

How might teaching support this motivational development? In order to identify conditions of the learning environment that support the development of interest or other intrinsic motivational tendencies, research usually draws on the theory of self-determination developed by Deci and Ryan (2000; Ryan and Deci, 2000). The theory states that the experience of self-determination which strives when a person's basic needs – that is, the need for social relatedness, autonomy and competence – are met, fosters intrinsic motivational tendencies (Deci and Ryan, 2000). Several studies have applied this theoretical framework to the school setting and have shown that students develop intrinsic motivation such as domain-specific interest or motivational involvement when they feel challenged and personally involved in their learning environments (Miserandino, 1996; Skinner and Belmont, 1993; Turner et al., 1998).

Linking the two features of instruction mentioned above – classroom management and constructivist approaches – to the research on self-determination, one may expect different effects of both aspects on students' intrinsic motivational tendencies. With respect to classroom management, one may argue that this type of typically teacher-led structuring of the learning environment may undermine students' feelings of autonomy and thus be detrimental to students' motivational development (Weinert and Helmke, 1995). Conversely, it has been found that students report higher levels of self-determination in well-structured learning settings (Skinner and Belmont, 1993), perhaps because these provide them with sensible action alternatives and thus may support a sense of competence and success. This sense of competence, or challenge, may also be provided by a constructivist learning environment. For instance, a study by Turner et al. (1998) into students' sense of involvement immediately after mathematics lessons showed that students experience high levels of involvement, especially in lessons which focused on conceptual understanding, in which students were held accountable for their own understanding and in which the teacher stressed the intrinsic aspects of the learning process.

It thus can be expected that constructivist teaching settings are well-suited to offer the experience of self-determination as a precondition for the development of intrinsic motivational tendencies. As to effective classroom management techniques, the expectations are not quite as clear. Whether students' intrinsic motivational tendencies are fostered by this aspect of teaching may depend on whether students experience the external regulation of the learning environment as stimulating rather than as oppressing. Consequently, our second research question is whether effective classroom management and constructivist teaching approaches enhance students' sense of challenge, which in turn fosters interest development.

Multiple Perspectives on the Learning Environment

Recently, it has been questioned whether features of the classroom environment exert universal effects on all students in these classrooms (Snow *et al.*, 1996). Accordingly, many researchers stress the role of students' individual perceptions and interpretations of the classroom environment as mediating factors between 'objective' teaching input and educational outcomes (Church *et al.*, 2001; Krapp, 2002; Shuell, 1996). This may be particularly true for motivational outcomes: here, the functional significance of the learning environment – that is, the personal meaning the environment has to the learner – rather than the environment per se may represent the causal agent of development (Church *et al.*, 2001; Grolnick and Ryan, 1986). From a methodological point of view, this means that features of the classroom environment may be conceptualized, both on the class and on the individual person level. Whereas the classroom variables traditionally employed in instructional research, such as observer ratings or class means of questionnaire-based ratings, reflect the shared environment similar to all students in one class, the individual evaluations or interpretations of the learning environment may differ for various students within one classroom (for example, Nuthall and Alton-Lee, 1995). Only recently, with the emergence of multilevel analysis techniques, has it been possible to combine both observation levels and consider these different perspectives on the learning environment simultaneously. Consequently, a third question we address is whether these different perspectives show differential validities in predicting the two educational outcomes. In particular, we expect students' interpretation of the classroom as a challenging environment to be a better predictor for their interest development than the instructional features found in their classrooms in general.

Research Questions

In summary, the purpose of this study is to examine the effect of classroom management strategies and constructivist approaches on students' learning and interest development in the context of regular secondary school mathematics classrooms. Drawing on evidence from experimental studies, we are particularly interested whether elements of constructivist teaching that occur in regular classrooms also have positive effects on students' learning. With regard to classroom management, we expect to replicate the evidence of positive effects on students' learning gains, however, we have no directed hypotheses on the effects on interest development. As studies on interest development stress the importance of students' individual experiences within their learning context, we analyze whether both classroom management and constructivist approaches enhance students' individual perceptions of challenge, which, in turn, influence their interest development.

Research Approach and Method

Database

The study re-analyses data from the German sample of TIMSS (TIMSS, Baumert *et al.*, 1997; Beaton *et al.*, 1996) and the associated TIMSS Videotape Classroom Study (Stigler *et al.*, 1999). In Germany, the international design of the TIMSS study was extended to a longitudinal design (Baumert *et al.*, 1997). More specifically, additional achievement tests and questionnaires on motivation and perceptions of teaching were administered in Grade 7 and again one year later to the same students in Grade 8. Between these two measurement points, a sub-sample of classes took part in a video study, with one mathematics lesson being videotaped for each participating class.

Participants

The sample used in this study is drawn from the nationally representative TIMSS middle school sample, from which 100 random classes were chosen for the TIMSS video sample. Longitudinal *and* video data are only available for 80 of these classes, however. The participants in this study are 1900 students (48.5 percent girls) from all three major school tracks in Germany. Compared to the original TIMSS sample, this sample is slightly biased towards higher track schools. Students' mean age at the second measurement point (Grade 8) was 14.8 years. Overall, 34 percent of the students were enrolled in lower track schools (*Hauptschulen*), 28 percent in the intermediate track (*Realschulen*) and 39 percent in the academic track (*Gymnasien*).

Measures

For this study, measures from different data sources were combined and used on two different levels. Students' mathematics achievement, interest and sense of challenge were used on the individual student level and elements of effective classroom management and constructivist approaches were used on the class level.

Individual level: mathematics achievement

Students' mathematics achievement was assessed with the TIMSS achievement test. Conceptually, the TIMSS test covers various content areas and different performance categories, such as conducting routine and complex procedures, applying knowledge, or solving mathematical problems (Beaton *et al.*, 1996). As part of the longitudinal design in Germany, students completed tests with overlapping item sets in Grade 7 and again one year later in the international assessment, in Grade 8. At both times, items from the international item pool were combined with a set of items taken from the First and the Second International Mathematics Study (Husén, 1967; Robitaille and Garden, 1989) and an earlier German study by the Max Planck Institute for Human Development. Individual achievement scores were estimated based on item response theory (see Martin and Kelly, 1997, for details). To describe students' achievement on one dimension, an equating procedure was used in which achievement at Grade 8 served as a scale anchor.

Individual level: mathematics-specific interest
Mathematics-related interest was assessed by a four-item scale and administered to the TIMSS sample as part of the national extension in both Grades 7 and 8. Students answer questions regarding their feelings about mathematics and how important they think the subject is. Examples are 'How much do you look forward to mathematics lessons?' or 'How important is it for you to know a lot in mathematics?' (five-point answer scale). Reliability coefficients (Cronbach's α) were 0.81 in Grade 7 and 0.78 in Grade 8.

Individual level: sense of challenge
Students' sense of challenge was assessed in Grade 8 with a 14-item scale from a questionnaire tapping students' perceptions of their mathematics classroom and mathematics teacher. In it, students rate the appropriateness of pacing in lessons, the degree of challenge they experience, or the extent to which whether they feel motivated and inspired during lessons (for example, 'Our mathematics teacher works through lessons quickly, so that you always have to pay attention, but aren't completely overwhelmed' or 'The exercises often include tasks that really make you aware of whether you've understood something'; four-point answer scale, alpha = 0.67). Whereas the interest scale tapped students' values and feelings towards the subject of mathematics, this variable describes the general cognitive stimulation that students experience during class.[2]

Class level: combined methods
For the variables on class level, two different methodological approaches were employed. Features of classroom management were assessed using the class means of students' ratings on effective time use, whereas the constructivist approaches were assessed by video observations. We decided on this combination of methods based on a study by Clausen (2002), which demonstrated that class level variables from different data sources seem to possess specific validities depending on the type of construct one aims to measure.

Class level: effective classroom management
The scale consists of nine items describing the occurrence of disciplinary problems or other disturbances and the degree of continuity in the lesson structure (for example, 'At the beginning of mathematics class, it takes quite a while until students calm down

and start working'; four-point answer scale). In order to receive a measure of the overall level of effective time use in each mathematics class, students' ratings were averaged per class. The intraclass correlation coefficient (ICC_1), which describes the proportion of variance between classes, was 0.14 and the Spearman–Brown adjusted intraclass correlation coefficient (ICC_2), which estimates the reliability of the class-arranged student ratings, was 0.80.

Class level: constructivist teaching approaches

As part of the TIMSS Videotape Classroom Study, one mathematics lesson in each of the participating classes was videotaped between the two measurement points. All recordings took place during Grade 8. The recording procedures followed standardized guidelines (see Stigler *et al.*, 1999, for details). The teachers' participation was voluntary and they were asked to teach the lesson as 'normally' as possible. A questionnaire administered to the teachers after the recording revealed that most teachers felt that they had succeeded in doing so (Stigler *et al.*, 1999).

These observations were used for a re-analysis with a newly developed high-inference rating instrument based on a set of scales developed by Widodo *et al.* (2002). The rating instrument analyses the occurrence of constructivist learning situations. Using several subcategories, observers rate whether the teacher explores students' prior knowledge and understanding, provides tasks that make use of this prior knowledge and involve cognitive conflict or demand conceptual restructuring, or shapes the classroom discourse in this way. High inference ratings on these categories were made for the whole lesson on a four-point Likert scale, ranging from 1 = 'does not apply at all' to 4 = 'applies most of the time'. For the present study, the sub-categories are averaged into one score. In order to achieve a high degree of reliability and validity, a rating procedure based on consensus judgment was applied. Each of the 80 lessons was rated independently by two raters, who had been trained in the use of the rating instrument beforehand with similar material. In case of discrepancy between the ratings, raters discussed their answers and determined a consensual rating, referring to the video material if necessary. The mean inter-rater reliability (intra-class coefficients) prior to this consensus rating was $ICC = 0.73$, indicating satisfactory reliability (see Kunter, 2005, for details on the rating procedure).

Class level: school track

To take possible track differences into account, the type of secondary school was included as dummy codes in the analyses.

Statistical Analyses

Missing Data

Due to the longitudinal design and the combination of diverse data sources, complete data sets were not available for all students. Video ratings were available for all classes, but, on the individual level, the proportion of missing data ranged from 5 to 28 percent. Analyses revealed that a small portion of the missing data was caused by the drop out of low-achieving students, but that most of the missing data seemed to occur at random. In order to avoid a reduction of the sample size, missing values were imputed and complete datasets were produced for all students using a multiple imputation method (Rubin, 1996; Schafer, 1999). The software used for the analyses (WesVar, HLM, see below) handles these multiple datasets simultaneously and produces combined parameters.

Multilevel Analyses
The data used in this study are hierarchical in structure: students are nested within classes and each class comes from one school. Moreover, variables are used on two different levels. In order to test group differences in the individual-level variables, parameters were estimated using the WesVar software, which produces appropriate estimates and standard errors by taking into account the hierarchical structure of the data (Martin and Kelly, 1997; Westat, 2000). For the regression models, a multilevel regression technique (HLM; Raudenbush et $al.$, 2001) is used to consider the individual-level and class-level variables simultaneously. To facilitate the interpretation of the HLM models, all predictor variables were standardized to z-scores (m = 0, SD = 1).

Results

Descriptives
Table 21.1 reports the descriptive statistics for all measures; first for the whole sample and then separately for the different school tracks. From Grade 7 to 8, students show an average achievement gain of approximately one third of a standard deviation. For mathematics-related interest, the similar means for at both measurement points seem to suggest that students' interest did not change over the course of the year. In fact, the correlation between the two measurement points is $r = 0.55$ ($p < 0.05$), pointing to a limited degree of stability. Thus, there is room for individual change in interest, even though the sample parameters as a whole did not change. Students from different tracks differ in their achievement and interest, with students in the academic track showing higher levels of achievement, but lower levels of interest than students in the lower tracks.

Concerning the teaching aspects, one can see that constructivist approaches are rather rare in German mathematics classrooms, considering that a score of 1 on the video ratings means 'does not apply at all'. There are, however, some constructivist elements to be found and their frequency varies across school types, with higher ratings in the academic than in the vocational track. The same pattern can be found for classroom management. The correlation between the two variables is $r = 0.30$ ($p < 0.05$), indicating that both instructional aspects may occur jointly. Instructional quality thus seems to be highest in the academic track and lowest in the vocational track.

Table 21.1 Descriptive statistics for all variables, for the whole sample and by school track including results of the F-tests for significant mean differences between tracks.

	Whole sample		Vocational track		Intermediate track		Academic track		Test for track differences
	M	(SD)	M	(SD)	M	(SD)	M	(SD)	F
Individual level									
Math achievement Grade 7	-0.20	(1.08)	-0.96_a	(0.94)	-0.22_b	(0.84)	0.47_c	(0.90)	89.99*
Math achievement Grade 8	0.43	(1.20)	-0.47_a	(0.89)	-0.37_b	(0.89)	1.26_c	(1.05)	125.50*
Math interest Grade 7	3.23	(0.84)	3.35_a	(0.85)	3.20_a	(0.85)	3.14_b	(0.81)	3.88*
Math interest Grade 8	3.18	(0.88)	3.33_a	(0.89)	3.16_a	(0.91)	3.06_b	(0.84)	8.04*
Sense of challenge	2.53	(0.51)	2.59_a	(0.49)	2.51_a	(0.53)	2.50_a	(0.52)	0.28
Class level									
Classroom management	2.50	(0.24)	2.41_a	(0.17)	$2.47_{a,b}$	(0.23)	2.61_b	(0.28)	6.19*
Constructivist approaches	1.53	(0.38)	1.35_a	(0.30)	$1.52_{a,b}$	(0.32)	1.72_b	(0.43)	7.60*

Notes. $*p < 0.05$.
Track differences: In each row, track means which differ significantly ($p < 0.05$) from each other are marked with a different subscript (post hoc tests in WesVar).

For further analyses, these differences have to be kept in mind and type of track will always be included as a control variable.

It is important to note, however, that the differences in teaching variables do not concur with differences in the individual feelings of challenge, which do not vary systematically across tracks.

In addition to the track differences discussed above, it is important to qualify the systematic variation between classrooms as units of the learning environment. Without taking the type of track into account, students from different classes vary greatly in their mathematics achievement – intra-class correlations (ICC) = 0.44 (Grade 7) / 0.52 (Grade 8) – and to a lesser degree in their interest (ICC = 0.08 / 0.11) and sense of challenge (ICC = 0.15). Once the effect of track membership is taken into account, 8 to 14 percent of the variation found in the outcome variables and in the perception of challenge can be explained by class membership (ICC mathematics = 0.12 / 0.14; ICC interest = 0.08 / 0.09; ICC challenge = 0.14; all $p < 0.05$).

Effects of Instructional Features on Achievement and Interest

In order to determine the effect of the instructional features on achievement and interest development, we carried out multilevel regression analyses. Separate models were calculated for effects on achievement and interest, respectively. In these analyses, we predicted students' achievement (or interest) in Grade 8 using achievement (interest) in Grade 7 and the perception of challenge as individual-level predictors and the questionnaire-based ratings of effective classroom management and the observer-based ratings for constructivist teaching as class-level predictors. In subsequent analyses, the type of school was included to control for the observed track differences.

Table 21.2 shows the results of the analyses for mathematics achievement. With achievement in Grade 7 being included as a predictor for achievement in Grade 8, the coefficients of all other predictors can be interpreted as the effect on achievement gains.

The models only partially confirm the expected positive effects of the instructional features on students' mathematics achievement. As to effective classroom management, there is a small positive effect on achievement gains. Students in classes featuring more effective time use show slightly more learning gains than students from other classes. However, the subsequent analyses, in which the type of school track was controlled, show that this effect is largely due to differences between classes from different tracks. When comparing students from classes within a track, the type of teaching approach does not seem to make a difference to achievement gains. No significant effect of constructivist approaches on achievement gains could be detected. At the individual level, the perceived sense of challenge had no effect on achievement in Grade 8.

The analyses predicting interest development produce a different pattern, as can be seen in Table 21.3. Neither variable on the class level affects interest development. Rather, the individually perceived sense of challenge proved to be the best predictor of interest development. This means that students who experience their mathematics lessons as challenging develop a relatively high level of interest, regardless of the

Table 21.2 Results of HLM models predicting mathematics achievement at grade 8.

Criterion	Math achievement Grade 8			
Predictors	Without school type		With school type	
	b	(SE)	b	(SE)
Individual variables				
Achievement Grade 7	0.46	(0.03)*	0.45	(0.03)*
Sense of challenge	0.00	(0.02)	0.01	(0.02)
Class-level variables				
Low track[a]	–	–	–0.20	(0.04)*
High track[a]	–	–	0.22	(0.04)*
Classroom management	0.11	(0.05)*	0.02	(0.04)
Constructivist teaching approaches	0.09	(0.05)	–0.03	(0.03)
R^2 total	0.45	–	0.55	–
R^2 individual level	0.21	–	0.22	–
R^2 class level	0.67	–	0.87	–

Notes. b = HLM regression weight; SE = standard error of b; R^2 = proportion of explained variance. *$p < 0.05$. [a]dummy-coded, reference category: intermediate track.

Table 21.3 Results of HLM models predicting mathematics-related interest at grade 8.

Criterion	Interest Grade 8			
Predictors	Without school type		With school type	
	b	(SE)	b	(SE)
Individual variables				
Interest Grade 7	0.49	(0.02)*	0.49	(0.02)*
Sense of challenge	0.22	(0.02)*	0.21	(0.02)*
Class-level variables				
Low track[a]	–	–	0.05	(0.04)
High track[a]	–	–	–0.06	(0.04)
Classroom management	0.02	(0.03)	0.05	(0.03)
Constructivist teaching approaches	0.00	(0.03)	0.00	(0.03)
R^2 total	0.35	–	0.35	–
R^2 individual level	0.34	–	0.31	–
R^2 class level	0.64	–	0.68	–

Notes. b = HLM regression weight; SE = standard error of b; R^2 = proportion of explained variance. *$p < 0.05$. [a]dummy-coded, reference category: intermediate track.

Table 21.4 Results of HLM models predicting the individual perception of challenge with constructivist teaching elements.

Criterion	Sense of challenge			
Predictors	Without school type		With school type	
	b	(SE)	b	(SE)
Class-level variables				
Classroom management	0.12	(0.05)*	0.17	(0.05)*
Constructivist teaching approaches	0.05	(0.04)	0.11	(0.04)*
Low track[a]	–	–	0.12	(0.07)*
High track[a]	–	–	–0.05	(0.05)
R^2 total	0.02	–	0.05	–
R^2 individual level	0.00	–	0.00	–
R^2 class level	0.12	–	0.32	–

Notes. b = HLM regression weight; SE = standard error of b; R^2 = proportion of explained variance. *$p < 0.05$. [a]dummy-coded, reference category: intermediate track.

overall style of teaching in their mathematics classroom or the type of secondary school they attend.

One theoretical assumption was that the sense of challenge students experience is fostered by constructivist learning approaches. In addition, the question whether effective classroom management would support or undermine students' sense of challenge was raised. This was tested in another set of multi-level analyses in which students' sense of challenge was predicted by the ratings of constructivist approaches and by the effective time use ratings. As can be seen in Table 21.4, students' sense of challenge is, in fact, higher in classes with more effective classroom management. This supports the assumption that students feel involved and challenged given a certain degree of structure within their learning environment, rather than experiencing the type of structuring as oppressive. In general, no effect of constructivist elements on students' sense of challenge could be detected. However, if one takes the type of track into account, one sees that both classroom management and constructivist learning approaches are positively related to the level of experienced challenge. It should be pointed out that these effects are rather small and that only a small proportion of variance is explained. Because only class variables were considered in these models, the amount of variance explained between classes is of particular interest, with 12 percent of this variance being explained by instructional measures alone. Nevertheless, these results suggest that both features of instructional quality may have an indirect effect on interest, which is mediated by students' experience of challenge.

Conclusion

The aim of this study was to describe effects of two distinct features of instructional quality on students' learning and interest development in a large, representative sample. In addition, students' individual perception of challenge was considered in order to examine the theoretical assumption that both features may foster students' interest indirectly via this sense of challenge.

The results confirm the theoretical assumptions of self-determination theory (Deci and Ryan, 2000; Ryan and Deci, 2000), which states that students' experience of autonomy and competence is one underlying cause for interest development. They also point to the possibility that this sense of challenge may be enhanced in learning environments in which a clear structure ensures an effective use of time and in which learning situations enabling students to actively construct their own knowledge are provided. The results thus supplement the existing body of research on instructional quality, which has thus far focused mainly on structural elements of classroom organization, indicating that a content-oriented perspective to the learning environment may be promising. They cannot, however, confirm the results from studies on the beneficial effects of constructivist teaching approaches in general, as no direct effects of constructivist teaching elements could be found on interest development nor achievement.

One critical point that should be discussed in the light of these results concerns the generally small effects of the class-level variables. This finding may be discussed from a theoretical as well as a methodological point of view. From a theoretical standpoint, it is important to note that we did not expect either constructivist teaching or effective classroom management alone to emerge as the prime route to desired student outcomes. The literature on instructional quality makes it quite clear that, rather than one single approach, the adaptive and responsive orchestration of various instructional features is the key to successful teaching (Brophy, 1999). Undoubtedly, the most successful teachers are those who combine various methods flexibly and appropriately. The question of how to capture this flexibility seems to be one of the most imminent challenges facing instructional research.

From a methodological point of view, in particular, the validity of the video ratings could be questioned. As the original TIMSS Video Study aimed at describing general features of instruction within one culture, it was not designed to provide data on differential teaching effects within one country (Stigler et al., 1999). Using a single recording as an estimate of the typical instruction found in this class seems problematic. We cannot rule out the possibility that teachers prepared specifically for this occasion and showed teaching behavior that differed from their regular instruction. However, because teachers were not aware of the research question – especially in the case of the present re-analysis – a systematic bias in this direction does not seem plausible. Moreover, research on teaching behavior shows that teachers develop distinct styles of teaching that seem to be rather stable over time. For instance, a recent German video study observed classroom instruction in 13 physics classes over the course of six lessons, revealing rather stable patterns of teaching for each participating teacher in terms of classroom organization and structure of discourse (Seidel et al., 2002). Similar evidence was found in an observational study by Mayer (1999), who rated instructional practices in 17 high school mathematics classes for several weeks. In Mayer's study, only 24 percent of the total variability in teaching practices was explained by within-class differences, indicating that teachers' instructional style varies comparatively little over time. Undeniably, increasing the number of recorded lessons would very much improve the reliability and validity of the observational measure. Still, bearing in mind that such a procedure would almost inevitably entail a reduction of sample size and would thus reduce the sample variability, we conclude that the advantage of being able to use the large TIMSS sample balances out the limitations of the one-lesson design.

One strong point of the study seems to be the combination of objective class variables with students' individual perceptions of the classroom. The results showed that the two types of student outcomes were affected by different aspects of the classroom environment. Whereas mathematics achievement seemed to be influenced mainly by class-level predictors, these predictors did not influence students' interest development. Here, their individual perception of challenge was the best predictor. Even within a given learning environment, students may thus differ in their experiences, which may then influence their cognitive and motivational development in a specific manner. Thus, the use of different data sources to tap general instructional features and students' individual perceptions as two distinct aspects of the classroom context and the subsequent combination of these different aspects within one multilevel model, constitutes an improvement on earlier models of instructional quality and studies of classroom climate.

Notes

1. Note that constructivist learning should not be equated with the use of elaborative learning strategies, in which new learning material is consciously linked with known material in order to secure thorough encoding in memory. In the constructivist framework, knowledge is *created*, not encoded through prior knowledge.
2. Theoretically, but also empirically, all three variables derived from the student questionnaires (interest, challenge and classroom management, see below) represent distinct constructs. In a principal component analysis, no significant item overlap was detected. Of the items in each dimension, no single target item loaded higher than 0.20 on either of the other two dimensions.

References

Alexander, P.A. and Murphy, P.K. (1998) Profiling the differences in students' knowledge, interest and strategic processing. *Journal of Educational Psychology*, 90(3), 435–47.

Anderson, L.W., Ryan, D. and Shapiro, B. (eds.) (1989) *The IEA classroom environment study*. New York: Pergamon Press.

Baumert, J., Lehmann, R., Lehrke, M., Schmitz, B., Clausen, M., Hosenfeld, I., Köller, O. and Neubrand, J. (1997) *TIMSS – Mathematisch-naturwissenschaftlicher Unterricht im internationalen Vergleich: Deskriptive Befunde [TIMSS – mathematics and science teaching in international comparison: Descriptive results]*. Opladen: Leske + Budrich.

Beaton, A.E., Mullis, I.V.S., Martin, M.O., Gonzalez, E.J., Kelly, D.L. and Smith, T.A. (1996) *Mathematics achievement in the middle school years: IEA's Third International Mathematics and Science Study* (TIMSS). Chestnut Hill, MA: Boston College.

Brophy, J. (1999) *Teaching* (Vol. 1). Genf, Schweiz: International Academy of Education/International Bureau of Education.

Church, M.A., Elliot, A.J. and Gable, S.L. (2001) Perceptions of classroom environment, achievement goals and achievement outcomes. *Journal of Educational Psychology*, 93(1), 43–54.

Clausen, M. (2002) *Unterrichtsqualität: Eine Frage der Perspektive? [Instructional quality: A question of perspectives?]* Münster: Waxmann.

Cognition and Technology Group at Vanderbildt (1992) The Jasper series as an example of anchored instruction: Theory, program, description and assessment data. *Educational Psychologist*, 27, 291–315.

Collins, A.M., Greeno, J.G. and Resnick, L.B. (2001) Educational learning theory. In N. Smelser and P.B. Baltes (eds.), *International encyclopedia of the social and behavioral sciences* (pp. 4276–9). Oxford: Elsevier.

Deci, E.L. and Ryan, R.M. (2000) The 'what' and 'why' of goal pursuits: Human needs and the self-determination of behavior. *Psychological Inquiry*, 11(4), 227–68.

Duit, R. and Confrey, J. (1996) Reorganizing the curriculum and teaching to improve learning in science and mathematics. In D.F. Treagust, R. Duit and B.J. Fraser (eds.), *Improve teaching and learning in science and mathematics* (pp. 79–93). New York: Teachers College Press.

Greeno, J.G., Collins, A.M. and Resnick, L.B. (1996) Cognition and learning. In D.C. Berliner and R.C. Calfee (eds.), *Handbook of educational psychology* (pp. 15–46). New York: Macmillan.

Greeno, J.G. and the Middle School Mathematics Through Applications Project Group. (1998) The situativity of knowing, learning and research. *American Psychologist*, 53(1), 5–26.

Grolnick, W.S. and Ryan, R.M. (1986) Origins and pawns in the classroom: Self-report and projective assessments of individual differences in children's perceptions. *Journal of Personality and Social Psychology*, 50(3), 550–8.

Grouws, D.A. and Cebulla, K.J. (2000) *Improving student achievement in mathematics* (Vol. 4). Genf, Schweiz: International Academy of Education/International Bureau of Education.

Hoffmann, L., Krapp, A., Renninger, K.A. and Baumert, J. (1998) *Interest and learning*. Kiel: Institut für die Pädagogik der Naturwissenschaften an der Universität Kiel.

Husén, T. (ed.) (1967) *International study of achievement in mathematics: A comparison of twelve countries*. New York: Wiley.

Krapp, A. (2002) An educational-psychological theory of interest and its relation to SDT. In E.L. Deci and R.M. Ryan (eds.), *Handbook of self-determination research* (pp. 405–27). Rochester, NY: University of Rochester Press.

Kunter, M. (2005) Multiple Ziele im Mathematikunterricht [Multiple goals in mathe-
    matics classes]. Unpublished doctoral dissertation, Free University, Berlin.
Martin, M.O. and Kelly, D.L. (eds.) (1997) *Third International Mathematics and
    Science Study. Technical report: Vol II. Implementation and analysis. Primary and
    middle school years.* Chestnut Hill, MA: Boston College.
Mayer, D.P. (1999) Measuring instructional practice: Can policy makers trust survey
    data? *Educational Evaluation and Policy Analysis*, 21(1), 29–45.
Mayer, R.E. (2004) Should there be a three-strikes rule against pure discovery
    learning? *American Psychologist*, 59(1), 14–19.
Miserandino, M. (1996) Children who do well in school: Individual differences in
    perceived competence and autonomy in above-average children. *Journal of
    Educational Psychology*, 88(2), 203–14.
Nuthall, G. and Alton-Lee, A. (1995) Assessing classroom learning: How students use
    their knowledge and experience to answer classroom achievement test questions in
    science and social studies. *American Educational Research Journal*, 32(1),
    185–223.
Pintrich, P.R. (1999) The role of motivation in promoting and sustaining
    self-regulated learning. *International Journal of Educational Research*, 31,
    459–70.
Pintrich, P.R., Marx, R.W. and Boyle, R.A. (1993) Beyond cold conceptual change:
    The role of motivational beliefs and classroom contextual factors in the process of
    conceptual change. *Review of Educational Research*, 63(2), 167–99.
Raudenbush, S., Bryk, A. and Congdon, R. (2001) *Hierarchical linear and nonlinear
    modeling* (HLM) (Version 5.05). Lincolnwood, IL: Scientific Software
    International.
Robitaille, D.F. and Garden, R.A. (eds.) (1989) *The IEA Study of Mathematics II:
    Contexts and outcomes of school mathematics*. Oxford: Pergamon Press.
Rubin, D.B. (1996) Multiple imputation after 18+ years. *Journal of the American
    Statistical Association*, 91(343), 473–89.
Ryan, R.M. and Deci, E.L. (2000) Intrinsic and extrinsic motivations: Classic defini-
    tions and new directions. *Contemporary Educational Psychology*, 25(1), 54–67.
Schafer, J.L. (1999) *NORM for Windows 95/98/NT. Multiple imputation of incomplete
    data under a normal model*. University Park, PA: Penn State Department of
    Statistics.
Schiefele, U. (1998) Individual interest and learning: What we know and what we
    don't know. In L. Hoffmann, A. Krapp, K.A. Renninger and J. Baumert (eds.),
    Interest and learning (pp. 91–104). Kiel: Institut für die Pädagogik der
    Naturwissenschaften an der Universität Kiel.
Seidel, T., Prenzel, M., Duit, R., Euler, M., Geiser, H., Hoffmann, L., Lehrke, M.,
    Müller, C.T. and Rimmele, R. (2002) 'Jetzt bitte alle nach vorne schauen!'
    Lehr-Lernskripts im Physikunterricht und damit verbundene Bedingungen für
    individuelle Lernprozesse ['Can everybody look to the front of the classroom
    please?' Patterns of instruction in elementary physics classrooms and its implica-
    tion for students' learning]. *Unterrichtswissenschaft*, 30(1), 52–77.

Shuell, T.J. (1996) Teaching and learning in a classroom context. In D.C. Berliner and R.C. Calfee (eds.), *Handbook of educational psychology* (pp. 726–64). New York: Simon and Schuster Macmillan.

Shuell, T.J. (2001) Learning theories and educational paradigms. In N. Smelser and P.B. Baltes (eds.), *International encyclopedia of the social and behavioral sciences* (pp. 8613–20). Oxford: Elsevier.

Skinner, E.A. and Belmont, M.J. (1993) Motivation in the classroom: Reciprocal effects of teacher behavior and student engagement across the school year. *Journal of Educational Psychology*, 85(4), 571–81.

Snow, R.E., Corno, L. and Jackson, D., III. (1996) Individual differences in affective and conative functions. In D.C. Berliner and R.C. Calfee (eds.), *Handbook of educational psychology* (pp. 243–310). New York: Macmillan Library Reference.

Stigler, J.W., Gonzales, P., Kawanaka, T., Knoll, S. and Serrano, A. (1999) *The TIMSS videotape classroom study: Methods and findings from an exploratory research project on eighth-grade mathematics instruction in Germany, Japan and the United States.* Los Angeles: US Department of Education, Office of Educational Research and Improvement.

Turner, J.C., Meyer, D.K., Cox, K.E., Logan, C., DiCintio, M. and Thomas, C.T. (1998) Creating contexts for involvement in mathematics. *Journal of Educational Psychology*, 90(4), 730–45.

Walberg, H.J. and Paik, S.J. (2000) *Effective educational practices* (Vol. 3). Genf, Schweiz: International Academy of Education/International Bureau of Education.

Weinert, F.E. and Helmke, A. (1995) Learning from wise Mother Nature or Big Brother Instructor: The wrong choice as seen from an educational perspective. *Educational Psychologist*, 30(3), 135–42.

Westat. (2000) WESVAR 4.0. Rockville: Westat.

Widodo, A., Duit, R. and Müller, C.T. (2002) *Constructivist views of teaching and learning in practice: Teachers' views and classroom behaviour.* Paper presented at the Annual Meeting of the American Educational Research Association, New Orleans, LA, April.

22

Where to Look for Student Sorting and Class-size Effects: Identification and Quest for Causes Based on International Evidence

Ludger Wößmann

Introduction

This chapter tries to address two research questions, as well as to make two methodological points. The two interrelated research questions are:

1. Do differences in class size cause differences in student performance – that is, do students learn more in smaller classes? This is the much-researched question of *class-size effects*.
2. Are students sorted into smaller or larger classes based on their performance – that is, do forces such as parental residential choice, tracked school systems, and remedial or special-enrichment classes within schools, lead to a non-random placement of students into differently sized classes between and within schools? This is one aspect of the much less researched question of *student sorting*.

Note that these two questions are interrelated. The first considers the effect of class size on student performance, while the second basically investigates the reverse effect of student performance on class size. This interrelatedness is depicted in the rather simplified Figure 22.1.

We can usefully divide the different kinds of sorting into two broad categories: sorting taking place *between schools*, such as residential choice or tracking by schools, and sorting taking place *within schools*, such as teachers or heads of schools assigning students to different classes. Thus, while there is one known causal effect of

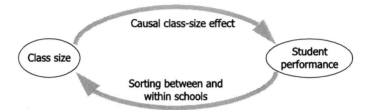

Figure 22.1 Interrelation between class size and student performance.

class size on student performance, there may be a multitude of effects driving the sort-ing pattern. It might, therefore, be helpful to think of the 'sorting effects' in terms of any sort of correlation between class size and student performance that is not caused by the direct effect of class size on performance.

In dealing with these two interesting questions, I also want to make two method-ological points:

1. Given the interrelatedness of the two research questions, one needs to come up with a proper *strategy of identification* in order to disentangle and thus be able to answer the two questions.
2. *International evidence* can inform the discussion of these questions, as it provides a cross-country pattern of results that can be used to examine potential causes, or underlying factors, that determine whether there are class-size effects and whether student sorting exists. Due to the limited number of country observations and the multitude of potential underlying factors, some of which might be unobservable, such cross-country evidence can never be definitive. Still, the international evi-dence can be used as a helpful starting point when asking 'where to look' for the effects, thereby increasing our understanding of how school systems work.

This chapter is structured as follows. The next section introduces the specific features of the international TIMSS database that will be used in the following identification strategies. The following two sections deal with class-size effects and sorting effects, respectively. These two sections present largely non-technical summaries of the much more detailed research papers, Wößmann and West (2002) and West and Wößmann (2003), respectively, supplemented by some preliminary new evidence on what might lie behind the cross-country pattern of class-size effects. Both sections start by giving the in-tuition of the respective identification strategy and then briefly present the cross-country pattern of results. The sections end by using this cross-country pattern to propose causes of these different effects. The final section draws some very brief policy conclusions.

The International Database

The international database used in the analyses draws on the data collected as part of the large-scale international student achievement test, the Third International

Mathematics and Science Study (TIMSS).[1] I mainly use the nationally representative samples of middle-school students tested in the TIMSS 1995 where the target population was defined as all students enrolled in the two adjacent grades that contained the largest proportion of 13-year-old students at the time of testing (Grades 7 and 8 in most countries).[2] In addition to students' test scores in mathematics and science, I use the extensive background questionnaires that were administered to students, teachers and school principals in order to gather information on the students themselves and on their institutional environments. Schools were sampled randomly within each country, and, as a general rule, one class per grade was selected at random within each sampled school. Schools serving only students with special needs were excluded from the target population, implying that any estimates should be unaffected by cross-country differences in the treatment of these students. Within sampled classes, however, all students were generally required to participate.

The identification strategies outlined below require a restriction of the samples to those schools in which both a seventh- and an eighth-grade class were actually tested. Furthermore, only those schools were included which had data on the actual class size and data on the grade-average class size in both the seventh and the eighth grades. Consequently, the analyses were ultimately conducted on the 18 school systems for which data from at least 50 schools in both mathematics and science remained after applying these criteria. Despite its reduced size, the remaining sample of countries includes systems from four different continents with a wide range of distinctive institutional configurations.

The database includes student-level data on student performance in mathematics and science, which were measured separately in TIMSS using international achievement scores with an international mean of 500 and an international standard deviation of 100. It also has data on family-background information for individual students from the student background questionnaires; data on the actual class size of each mathematics and science class from the teacher background questionnaires; and data on the school-level average class size in Grades 7 and 8 from the school-principal background questionnaires.

Class-size Effects

Identification

As is obvious from the introductory discussion, the causal effect of class size on student performance cannot be estimated simply by regressing student performance on class size. Coefficients estimated in such an ordinary least squares (OLS) regression will be biased estimates of the causal class-size effect because there may be sorting effects in place. Therefore, an identification strategy is needed to extract the causal class-size effect from the total correlation between class size and performance. The central aim of such a strategy is to identify exogenous variation in class size, which might be evoked by so-called 'natural experiments'. Exogenous means that the variation in class size is not caused by variables in our model – most importantly, it is not caused by student performance. Instead, the variation comes about for a reason that is unrelated – exogenous – to how students perform in the first place. A comparison of student performance in classes that differ in size for exogenous reasons can then be used to test whether the class-size differences cause differences in student performance.

The intuition of the identification strategy for class-size effects, derived more formally in Wößmann and West (2002), is as follows. In a nutshell, causal class-size effects are identified by relating differences in the relative performance of students in seventh and eighth grades within individual schools to that part of the between-grade difference in class size in the school that reflects between-grade differences in average class size. This approach effectively excludes both between-school and within-school sources of student sorting. Between-school sorting is eliminated by controlling for school fixed effects, which exclude any systematic between-school variation from the analysis. Within-school sorting is filtered out by instrumenting actual class size by the average class size in the relevant grade at each respective school.[3] Arguably, this remaining class-size variation between classes at different grades of a school is caused by random fluctuations in cohort size between the two grades in the catchment area of each school. Being exogenous to student performance, this random variation in class size can be used to identify the causal effect of class size on student performance.

This identification strategy is designed to take advantage of certain unique features of the TIMSS database. The use of school-level fixed effects is made possible by the fact that the study sample includes more than one class from each school. Using each school's average class size in each grade as an instrument imposes the additional requirement of data on achievement, actual class size and grade-average class size for different grades taking part in the same achievement test. Among large-scale international studies of student achievement, TIMSS is the only database with this particular set of characteristics.

Cross-country Pattern of Results

The results of an implementation of this identification strategy for the 18 countries in the sample are reported in Table 22.1, where mathematics and science performance are pooled together to give the basic overall pattern. In the table, statistically significant positive coefficients are presented in italics, statistically significant negative coefficients in boldface, and highly imprecise estimates in italic boldface. The first column, reporting a least-squares (LS) regression of test scores on class size, shows that, in most school systems, student performance is statistically significantly related to *larger* class sizes. That is, if we were to interpret such a LS coefficient as an estimate of the causal class-size effect, which traditionally has often been done, we would come to the counterintuitive conclusion that, in most countries, students fare better in larger classes. The second column, reporting a specification where school fixed-effects (FE) are controlled for, excludes the effects of between-school sorting, and is of limited interest at this stage.

The third column, reporting the combined school-fixed-effects instrumental-variables specification (SF-IV), gives a pattern of results that is totally different from the initial least-squares one. Based on this identification method, we can conclude that

Table 22.1 Coefficient of class size in the three base regressions.

	LS		FE		FE-IV	
	Coeff.	Std. error	Coeff.	Std. error	Coeff.	Std. error
Australia	4.927*	(0.730)	3.098*	(0.559)	−1.741	(4.867)
Belgium (Fl)	2.422‡	(1.372)	0.633	(1.036)	3.924	(2.611)
Belgium (Fr)	0.143	(1.186)	−1.058†	(0.860)	0.156	(0.859)
Canada	0.297	(0.249)	0.247	(0.227)	0.052	(0.616)
Czech Republic	2.222†	(0.983)	−1.545	(1.407)	0.512	(1.274)
France	2.532*	(0.665)	0.659	(0.530)	−1.380	(1.028)
Greece	0.468†	(0.210)	−0.013	(0.121)	−2.011†	(0.943)
Hong Kong	5.587*	(1.028)	2.982*	(0.543)	−8.043	(8.178)
Iceland	−0.306	(0.369)	−0.169	(0.501)	−2.197†	(0.877)
Japan	3.264*	(0.762)	−0.356	(0.294)	−0.101	(0.394)
Korea	−0.057	(0.075)	−0.137*	(0.046)	−0.796	(0.570)
Portugal	1.082*	(0.296)	0.931*	(0.298)	0.447	(0.556)
Romania	2.854*	(0.654)	0.294	(0.587)	1.391	(1.731)
Scotland	0.333	(0.435)	0.073	(0.285)	−22.741	(34.651)
Singapore	5.499*	(0.511)	3.079*	(0.420)	0.492	(0.489)
Slovenia	0.963	(0.671)	0.113	(0.572)	0.769	(1.051)
Spain	0.450*	(0.174)	0.036	(0.144)	−0.533	(0.427)
USA	−0.140	(0.125)	−0.000	(0.092)	3.345	(5.515)

Notes. Dependent variable: TIMSS test score.
LS = Least-squares regression (controlling for grade and subject); FE = school fixed-effects specification (controlling for grade, subject and school fixed effects); FE-IV = Fixed-effects instrumental-variables specification (controlling for grade, subject, school fixed effects and 12 family-background variables, with actual class size instrumented by grade-average class size); clustering-robust standard errors in parentheses. For details on the different specifications, see West and Wößmann (2003).
Significance levels: *1 percent; †5 percent; ‡10 percent.

there are statistically significant (and sizable) causal class-size effects in two countries, Greece and Iceland, where students perform better if they are taught in smaller classes. In all other countries, though, there is no evidence of statistically significant causal class-size effects. As shown in Wößmann and West (2002), Wald tests can reject the possibility of even very small beneficial class-size effects in six of these countries (Flemish Belgium, Canada, Japan, Portugal, Singapore and Slovenia), and they can reject large effects in a total of 11 countries (the previous countries plus French Belgium, Czech Republic, Korea, Romania and Spain). For no single country does the counterintuitive least-squares finding survive the scrutiny of a well-founded identification strategy. Note that in four countries, including the USA, the FE-IV estimates are so imprecise that they do not allow for a meaningful assessment of the existence of class-size effects in these countries. Note also that the substantial difference between the LS and the FE-IV pattern of results already implies that there is substantial sorting going on in most of the countries, with students being sorted into differently sized classes based on their performance.

Quest for Causes

The cross-country pattern of class-size effects evident in Table 22.1 raises the question: Why are there class-size effects in some countries, but not in others? Relating this pattern to cross-country differences in the school systems that produced this pattern may give a hint on where to look for potential answers to this question – and thereby potentially also for answers to the question of causes of class-size effects more generally. This is done very roughly in Table 22.2, which reports several characteristics for the groups of countries with class-size effects, without large class-size effects and without any class-size effects. Apparently, neither average class size nor GDP per capita can differentiate countries with class-size effects from countries without such effects, suggesting that diminishing returns are not a major impact factor behind the cross-country pattern. However, average student performance seems to be substantially lower in countries with class-size effects.

Data on overall educational expenditure per student suggest that the countries with class-size effects spend less per student than countries without class-size effects. Given that the former on average have smaller class sizes than the latter, this suggests that the former spend less per teacher. This is reflected in data on average teacher salaries, which are substantially lower in countries with class-size effects. Arguably, a low average salary level for teachers probably means that a country draws its teaching population from a relatively low level of the overall capability distribution of the total labor force in the country. This is because higher wages in the labor market tend to attract more capable people, so that paying teachers more on average should tend to attract people with higher capabilities into the teaching profession. This argument is corroborated by the only direct measure of teacher quality available, namely teachers' education. While in the countries without class-size effects, about 60 percent of the teachers received at least a BA with additional teacher training, this share is only 13 percent in Greece, and one third of teachers in Iceland do not even have a secondary degree.

This cross-country pattern of class-size effects and country characteristics suggests that noteworthy class-size effects are observed only in countries with a relatively low-quality teaching force. Thus, capable teachers seem to be able to teach well in large classes, thereby promoting student learning equally well regardless of class size (at least within the range of variation that occurs naturally between grades). Less capable teachers, however, while perhaps doing reasonably well when faced with smaller classes, do not seem to be up to the job of teaching large classes. This explanation could jointly explain why there exist class-size effects in some countries but not in others and why average student performance in the countries with class-size effects is relatively low: both phenomena may be a consequence of relatively low-paid teachers. A piece of evidence from the TIMSS teacher background questionnaires is supportive of this explanation: While about 45 percent of teachers in the countries with class-size effects reported that their teaching was limited 'a great deal' by a high student/teacher ratio, this share was only 20–25 percent in the countries without class-size effects – quite telling, given that actual class sizes in the former countries are, on average, smaller than in the latter countries.

Another possibility to test this proposition would be to include a proxy for teacher quality, as well as an interaction term between this teacher-quality proxy and class

Table 22.2 Country characteristics and the existence of class-size effects.

	Class-size effect	Mean class size	GDP per capita	Mean test score	Expend. per student		Teacher salary		
						Rel. to GDPpC		Rel. to GDPpC	Per teach. hour
CSE[a]	-2.1	24.2	17,336	467	2,374	15	14,946	1.0	36
No Large CSE[b]	0.1	31.4	13,031	514	3,478	26	27,496	1.8	48
No CSE[c]	0.9	28.1	18,558	537	5,667	28	29,038	1.8	44

	Teachers' education					Teachers report limiting student–teacher ratio
	Training, no sec.	Secondary	BA	BA+ train.	MA (+train.)	
CSE[a]	16	5	45	32	1	45
No Large CSE[b]	2	19	15	35	27	25
No CSE[c]	2	32	7	55	5	20

Notes. [a]Mean of the two countries with statistically significant class-size effects. [b]Mean of the five countries where large-scale class-size effects are ruled out. [c]Mean of the six countries where any noteworthy class-size effects are ruled out. 'Rel. to GDPpC' = relative to GDP per capita (ratio first estimated in each country, then averaged across countries); 'Per teach. hour' = per teaching hour. 'Train.' = training. For details, see Wößmann and West (2002).

size, in the estimation of education production functions. If the existence of class-size effects were indeed related to teacher quality, the interaction term should be significantly positive. A fundamental problem for within-country estimations of such a specification is the paucity of reliable proxies for teacher quality. Rivkin *et al*. (2004) demonstrate that teacher quality, as measured by the average student achievement gains they elicit, varies widely, but that this variation is only weakly related to differences in teachers' education and experience, which are generally the only quality information available. However, if one accepts – in line with the argument raised above – that average teacher salary in a country might be a reasonable proxy for average teacher quality, then this specification can be implemented using cross-country data.

In order to have a reasonable sample size, given that the teacher-quality proxy is measured only at the country level and that OECD data on average teacher salaries is used which is only available for a limited number of countries, the results of TIMSS 1995 and its replication study, TIMSS 1999, were pooled together, yielding data for 41 countries. Preliminary evidence on an international country-level regression of mathematics test scores on class size, teacher salary, and an interaction between the two, controlling among others for GDP per capita (as a proxy for the overall quality of the workforce) and for seven regional dummies, suggests that the interaction term between class size and teacher salary turns out to be statistically significantly positive.[4] That is, the (negative) class-size effect is significantly smaller (in absolute terms), the higher are teacher salaries.

Table 22.3 converts the estimated coefficients of this regression into relative performance levels of students faced with different combinations of class size and teacher salary. When reading the table horizontally, we can see that student performance

Table 22.3 Student performance by combination of class size and teacher salary.

Class size	Teacher salary[a]						
	10	15	20	25	30	35	40
20	102	102	103	103	104	104	105
22	91	94	96	98	100	102	104
24	81	85	89	92	96	100	104
26	71	76	82	87	92	98	103
28	61	68	75	82	89	96	103
30	51	59	68	76	85	93	102
32	40	51	61	71	81	91	101
34	30	42	54	65	77	89	101
36	20	33	47	60	73	87	100
38	10	25	40	55	70	85	99
40	0	16	33	49	66	82	99

Notes. TIMSS and TIMSS 1999 mathematics performance, relative to students in classes of 40 students with a teacher salary of 10,000 international dollars (measured in percent of an international standard deviation), derived from a country-level regression of math test scores on class size, teacher salary, and an interaction term between the two, controlling for a TIMSS 1999 dummy, average student age, GDP per capita, and seven regional dummies. Reported class sizes and teacher salaries range roughly from the 10th to the 90th percentile of the increases with teacher-salary levels at all class-size levels, although particularly when class sizes are large. When reading the table vertically, we can see that student performance is larger in smaller classes (that is, there is a class-size effect), but this effect basically peters out at high levels of teacher salary. One should certainly be cautious not to over stretch this simple OLS pattern of international country-level evidence, but it is still suggestive of the conclusion that there may be something to the proposition that there are class-size effects at low levels of teacher salaries, but not at high salary levels.

Thus, the international evidence seems to suggest that class-size effects might be related to teacher quality, in that class size does matter with relatively low-paid teachers, but it does not with relatively high-paid teachers. In fact, such an explanation might also be able to reconcile Krueger's (1999) finding of substantial class-size effects in Tennessee's Project STAR experiment with Hoxby's (2000) refutation, based on quasi-experimental evidence, of even small class-size effects in Connecticut.

Student Sorting

Identification

The pattern of results in Table 22.1, showing large differences in the estimated coefficients on class size across the three specifications, suggests substantial endogeneity in the relationship between class size and student achievement in most of the school systems analyzed. The placement of students of different achievement levels in differently sized classes, which I will refer to as 'student sorting', can stem from any number of causes, both intended and unintended. Actors involved in the sorting of students include parents (for example, through residential choice), teachers and school administrators (through the placement of students into classrooms), and the general rules of the school system (for example, through ability tracking between schools). The net outcome of these different forces may be either compensatory sorting, in that lower-performing students end up in smaller classes, regressive sorting, in that higher-performing students end up in smaller classes, or no sorting by performance at all.

To capture the aggregate effect of the sorting process, one can define sorting effects simply as the observed relationship between student performance and class size, exclusive of any causal effect of class size on performance. The most obvious sorting effects occur when students are placed into specific schools and classrooms explicitly according to their prior academic performance. However, school systems may also sort students according to a variety of other characteristics, such as race, sex, disruptiveness, or socio-economic status (SES). To the extent that these characteristics are correlated with academic achievement, these alternative forms of sorting may play a substantial indirect role in generating the overall patterns of sorting by performance. In this sense, one may think of sorting effects as the 'omitted-variables bias' in OLS estimates of the causal class-size effect.

Consequently, sorting effects can be estimated by comparing the coefficients from the three different specifications reported in Table 22.1, yielding estimates of sorting effects both between and within schools. While the standard LS coefficient estimate reflects a combination of between- and within-school sorting and causal class-size effects, the FE coefficient estimate, which controls for school fixed effects, reflects only within-school sorting and causal class-size effects. Thus, the extent of between-school sorting can be identified as the difference between the FE estimate and the LS estimate. Likewise, the FE-IV coefficient estimate reflects the causal effect of class size on student performance. Thus, subtracting the FE-IV coefficient from the FE coefficient yields an estimate of within-school sorting in a school system.[5]

Cross-country Pattern of Results

Table 22.4 presents the estimates of between- and within-school sorting effects for the 18 sampled countries based on this identification strategy. Positive estimates indicate compensatory sorting, with low-performing students placed in relatively small classes. The size of the estimates specifies how many fewer test-score points students placed in a class that is one student smaller tend to score. Conversely, negative estimates indicate regressive sorting, with smaller classes targeted at more advanced students.

The statistical significance of the sorting effects are estimated using Hausman tests, which, intuitively, test whether the bias affecting a parameter in a given specification relative to another specification – here, the bias in the class-size coefficient resulting

Table 22.4 Estimates of between- and within-school sorting effects.

	Between-school sorting		Within-school sorting	
Australia	*1.829**	(0.469)	***4.839***	(4.835)
Belgium (Fl)	*1.789*[†]	(0.899)	−3.291	(2.397)
Belgium (Fr)	1.201	(0.817)	**−1.214***	(0.040)
Canada	0.051	(0.103)	0.195	(0.572)
Czech Republic	*3.767**	(1.007)	**−2.056***	(0.598)
France	*1.873**	(0.401)	*2.039*[†]	(0.880)
Greece	*0.481**	(0.172)	*1.998*[†]	(0.936)
Hong Kong	*2.604**	(0.873)	***11.025***	(8.160)
Iceland	−0.137	(0.340)	*2.029**	(0.719)
Japan	*3.620**	(0.703)	−0.255	(0.263)
Korea	0.080	(0.059)	0.659	(0.568)
Portugal	*0.151**	(0.032)	0.485	(0.469)
Romania	*2.560**	(0.290)	−1.097	(1.629)
Scotland	0.406	(0.328)	***22.667***	(34.650)
Singapore	*2.420**	(0.290)	*2.587**	(0.250)
Slovenia	*0.849*[†]	(0.351)	−0.656	(0.881)
Spain	*0.414**	(0.097)	0.569	(0.402)
USA	**−0.140**[‡]	(0.084)	***−3.345***	(5.514)

Notes. A positive estimate reflects compensatory sorting (weaker students placed in smaller classes); a negative estimate reflects the opposite, regressive sorting. Standard errors from Hausman specification tests in parentheses. For details on the calculation of the sorting effects, see West and Wößmann (2003). Significance levels: *1 percent; [†]5 percent; [‡]10 percent.

from sorting effects – is statistically significant. The estimated significance levels indicate that the identification method generates reasonably precise estimates of between-school sorting effects in all examined school systems, while four of the estimates of within-school sorting effects are extremely imprecise, which is a direct consequence of the lack of precision in these countries of the FE-IV estimates in Table 22.1.

The most notable feature of these results is the prevalence of statistically significant estimates, which indicates that students performing at different levels are indeed sorted into differently sized classes in the majority of countries. At the between-school level, positive estimates are predominant, suggesting that most school systems (12 out of 18) feature a compensatory pattern of between-school sorting. Notably, the single exception with statistically significant (albeit small) regressive between-school sorting is the USA. At the within-school level, fewer of the sorting estimates achieve statistical significance, and those that do are more evenly divided between positive and negative results (although the majority again suggests compensatory sorting). In sum, statistically significant sorting effects at either the between- or within-school level are evident in 15 of the 18 countries. These results confirm that students in each of these school systems are not assigned to differently sized classes randomly, but rather in a way that systematically reflects their performance. Variations in class size are indeed at least as much a consequence as a cause of differences in achievement.

Quest for Causes

What might determine the different sorting patterns between and within schools observed across countries? Which institutional characteristics of the school systems may be responsible for the divergent patterns? West and Wößmann (2003) develop some hypotheses on how class-size resources are allocated, intended as initial thoughts towards a theory of student sorting (Table 22.5). Among the diverse set of actors whose decisions may be reflected in the aggregate estimates of sorting effects, parents, teachers, administrators and central policy-makers are identified as key actors.

At the between-school level, the residential choices of parents, particularly in combination with decentralized educational finance, may tend to further regressive allocations, because the children of wealthy parents, who will tend to be high performers due to their families' backing, will also have access to superior resources if the level

Table 22.5 Hypotheses on main determinants of student sorting.

	Regressive	Compensatory
Between-school sorting	**Parental** residential choice (combined with local finance)	**Policy-makers** influenced by the political economy of tracking
Within-school sorting	**Teachers** govern student placement	**Administrators** govern student placement (especially when held accountable for efficiency and equity)

of resources available in local schools is positively correlated with the price of housing. Also at the between-school level, political pressures may force policy-makers in systems with ability tracking to compensate students placed in the tracks for lower performers with tangible additional resources, such as smaller classes.

At the within-school level, sorting may tend to be regressive where teachers control student placements, and compensatory where administrators control the placement of students and teachers into classrooms and where external exams make efficiency considerations more salient. If an individual teacher – for example the most senior – is allowed to govern the placement into classes, he may choose to place the best students into small classes and to teach these classes himself, leaving more junior teachers to deal with relatively large classes of relatively poor performers – which results in a regressive pattern. If, by contrast, administrators without a personal interest in teaching conditions in specific classrooms govern student and teacher placements, they may follow political pressures for compensatory placements. Furthermore, especially when external exams are in place to introduce accountability, administrators may follow efficiency considerations. While there is scant empirical evidence on efficiency in the assignment of students to small and large classes, Lazear's (2001) theory of educational production suggests that classroom teaching is a public good with congestion effects, so that it is optimal to place disruptive students in relatively small classes. Assuming disruptive students are also more likely to be low achievers, an efficient school would be characterized by compensatory within-school sorting effects.

The international evidence, using the international variation in the presented estimates of sorting effects in combination with institutional background data on the school systems included in TIMSS, is consistent with each of these hypotheses. While hardly definitive, given the limited number of country observations and the relatively weak proxies for the concepts which the hypotheses suggest are important, regressions of the estimated sorting effects on various institutional measures suggest the following relationships (Table 22.6; see West and Wößmann, 2003, for the full regression results).

Between-school sorting is positively related to measures of tracking and the prevalence of performance-based admissions; that is, sorting is more compensatory in tracked systems. Between-school sorting is negatively related to measures of residential mobility (measured by the percentage of students transferring into schools during an academic year) and of the extent of local school finance; that is, sorting is more regressive where parents exert much residential choice and where educational finance is decentralized. Within-school sorting is negatively related to teacher responsibility for student placement; that is, sorting is more regressive where teachers place students. And within-school sorting is positively related to responsibility of department heads for student as well as teacher placement and to the prevalence of external exams; that is, sorting is more compensatory where administrators place students and teachers and where external exams are in place.

To give one descriptive example of this evidence, Figure 22.2 plots the estimates of between-school sorting against a proxy of tracking (the percentage of schools in the system using academic performance as their admission criterion). The two are clearly

Table 22.6 Determinants of between- and within-school sorting effects.

	Between-school sorting			Within-school sorting		
Tracking	1.180‡	–	–	–	–	–
	(0.600)	–	–	–	–	–
Performance-based admissions	–	0.184*	0.217†	–	–	–
	–	(0.018)	(0.050)	–	–	–
Transfer rate	–	–0.399*	–	–	–	–
	–	(0.089)	–	–	–	–
Local finance	–	–	–0.037†	–	–	–
	–	–	(0.009)	–	–	–
Teachers place students	–	–	–	–0.053°	–0.030	–
	–	–	–	(0.033)	(0.030)	–
Department heads place students	–	–	–	0.035†	–	–
	–	–	–	(0.018)	–	–
Department heads assign teachers	–	–	–	–	0.101†	–
	–	–	–	–	(0.041)	–
External exams	–	–	–	–	–	0.051†
	–	–	–	–	–	(0.022)
Constant	0.391	0.793*	–0.374	0.722	0.338	–0.62
	(0.501)	(0.210)	(0.225)	(0.793)	(0.748)	(0.422)
Observations	18	10	6	17	17	15
Adjusted R^2	0.145	0.919	0.783	0.269	0.326	0.248

Notes. Dependent variable: estimated sorting effects (see Table 22.4).
Least-squares regressions. Standard errors in parentheses. All observations
are weighted to account for estimated dependent variable. For details on data
and specifications, see West and Wößmann (2003).
Significance levels: *1 percent; †5 percent; ‡10 percent; °15 percent.

positively related, with the USA as a strong negative outlier. However, data on residential mobility and local finance suggests that, when interacting these two latter factors, one is basically left with an indicator variable for the USA, which is, of course, a country with high residential mobility and a system of education finance dominated by local property taxes. This peculiarity of the US school system may account for the USA being an outlier in Figure 22.2 and for the fact that the USA is the only country with statistically significant regressive between-school sorting. This same logic may also explain that the estimate of between-school sorting for Canada is quite precisely estimated to be statistically indistinguishable from zero, as Canada is the country whose institutional arrangements for financing education probably come closest to those of the USA.

Policy Conclusions

The international evidence on class-size effects suggests that an answer to the question of whether differences in class size cause differences in student performance depends on the school system under consideration. However, the evidence also sug-

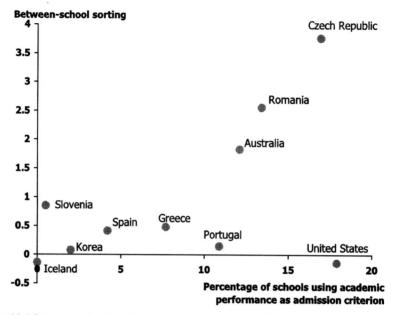

Figure 22.2 Between-school sorting and performance-based admissions.

gests that class-size reductions do not seem to be a promising policy strategy for the improvement of student performance in most countries, given that a large causal effect of class size on student performance can be ruled out in 11 out of 14 countries with meaningful estimates. Even more, given that the cross-country pattern of class-size effects suggests that they only exist where the level of teacher salaries is low, and given that teacher quality as proxied by teacher salaries seems to exert a strong positive effect on student performance, class-size reductions may not even be an attractive policy option in the two countries where class-size effects exist. The different countries seem to have chosen different points on the tradeoff between teacher quantity (class size) and teacher quality (salary): Greece and Iceland have relatively many but poorly paid teachers, while the countries without class-size effects have relatively few but well paid teachers. Given the relative cost of reducing class size and increasing teacher salaries, it may turn out in many countries that it may be better policy to devote the limited resources available for education to employing more capable teachers rather than to reducing class sizes – moving more to the quality side of the quantity–quality tradeoff in the hiring of teachers.

The international evidence also suggests that there is substantial within- and between-school sorting of students based on their performance, severely biasing conventional LS estimates of class-size effects. Furthermore, given the diversity of factors influencing student sorting, the net sorting pattern ultimately emerging in a school system may differ from what policy-makers might have initially intended. The interest that parents, teachers and administrators express in reducing class sizes as a means of improving achievement suggests that the distributional outcomes of the sorting process may be a useful indicator of the extent to which school systems reinforce or counterbalance existing academic and social inequities. Thus, policy-makers should be aware of the political economy of education, whereby the extent of fiscal decentralization, parental choices, tracking, and control over the assignment of teachers and students to classrooms all appear to exert important effects on the aggregate patterns of student sorting. However, given the limited empirical support for beneficial effects of smaller classes on measured academic achievement, the relationship between the specific sorting effects reported in this chapter and educational equity broadly conceived is not straightforward. To an extent, it depends on whether low-performing students in countries with compensatory sorting effects also receive a disproportionate share of other resources, or whether class size is instead used to compensate for other inequities in inputs. Given budget constraints, investments in teacher quantity of the type necessary to reduce class size for low-achieving students might themselves serve to lower the quality of these students' teachers, so that compensatory class-size policies may even serve to decrease educational opportunities for low-performers.

Acknowledgement

I would like to thank participants at the meeting of the Korea Task Force on K-12 Education at the Hoover Institution, Stamford University, for their comments and discussion.

Notes

1. See part I of this volume for a comprehensive introduction to TIMSS and Wößmann (2003) for additional background on the specific database used in this chapter.
2. In later parts of this article, I also use data from the TIMSS 1999 study conducted in 1999.
3. For a discussion of the validity of this instrument, see Wößmann and West (2002).
4. The same result is obtained for science test scores.
5. For a formal derivation of this identification strategy, see West and Wößmann (2003).

References

Hoxby, C.M. (2000) The effects of class size on student achievement: New evidence from population variation. *Quarterly Journal of Economics*, 115(4), 1239–85.

Krueger, A.B. (1999) Experimental estimates of education production functions. *Quarterly Journal of Economics*, 114(2), 497–532.

Lazear, E.P. (2001) Educational Production. *Quarterly Journal of Economics*, 116(3), 777–803.

Rivkin, S.G., Hanushek, E.A. and Kain, J.F. (2005) Teachers, schools, and academic achievement. *Econometrica*, 3(2), 417–58.

West, M.R. and Wößmann, L. (2003) Which school systems sort weaker students into smaller classes? International evidence. Kiel working paper 1145. Kiel: Institute for World Economics. Revised version forthcoming in the *European Journal of Political Economy*.

Wößmann, L. (2003) Schooling resources, educational institutions, and student performance: The international evidence. *Oxford Bulletin of Economics and Statistics*, 65(2), 117–70.

Wößmann, L. and West, M.R. (2002) Class-size effects in school systems around the world: Evidence from between-grade variation in TIMSS. Harvard University, Program on Education Policy and Governance Research Paper PEPG/02-02. (Revised version forthcoming in *European Economic Review*).

23

Grouping in Science and Mathematics Classrooms in Israel

Ruth Zuzovsky

Theoretical Framework and Educational or Academic Importance

To many educators ability grouping seems a sensible response to the academic diversity among students, in that it allows teachers to attune their instruction to students' capabilities. High achievers are challenged and stimulated and low achievers get more support. Thus all students seem to gain from ability grouping (Gamoran, 1986; Sorensen, 1970). Critics of ability grouping, however, contend that this practice has unintended harmful consequences. Ability grouping usually involves the assignment of poor teachers and inferior instruction to low ability groups (Finley, 1984; Oakes, 1985; Oakes *et al.*, 1992; Page, 1991; Talbert, 1990). Moreover, critics assert that, due either to the lower expectations of teachers who teach low-ability students (Rosenbaum, 1976), or to the stigmatizing effect ability grouping has on the self-esteem and aspiration of these students, the achievement gap between students of high and low ability increases over time (Findley and Bryant, 1971; Gamoran and Berends, 1987; Hallinan, 1992; Murphy and Hallinger, 1989; Rosenbaum, 1980).

When segregation by academic criteria is associated with social and economic characteristics, grouping contributes and perpetuates social inequality (Braddock, 1990; Gamoran *et al.*, 1995; Oakes, 1990; Persell, 1977; Rosenbaum, 1980).

Research on the effect ability grouping has on achievement has taken two broad forms. One type compares achievement gains of students in grouped units of learning to ungrouped heterogeneous classes and another type focuses on the differential effect of grouping on low and high ability, and low and high social class.

Reviews on the effect of grouping versus non-grouping have consistently shown that grouping has little or no effect on the overall student achievement in elementary and secondary schools (Slavin, 1987, 1990). Findings from the research on the differential effect of grouping are not conclusive (Hallinan, 1992).

The motivation to deal with the issue of grouping in Israel on the basis of TIMSS 1999 data stemmed from the fact that grouping was found to be quite widely used in Israel in both science and mathematics teaching, especially in the latter. Students from the 139 classes that participated in TIMSS 1999 were identified as being taught by more than one teacher per class. For these 139 classes, there were 261 science teachers and 326 mathematics teachers — in most cases teaching separate groups of students. This indicated the spread of a grouping phenomenon at the lower secondary school level in Israel, which is in contrast to a declared policy of social integration and of equal provision of educational opportunities for all students, which was introduced into the educational system in the late 1960s. As part of this policy, middle (lower secondary) schools were created hosting students who differed in terms of both socio-economic background and academic ability. A recommended heterogeneous class composition and suitable instructional methods to deal with this heterogeneity were implemented. It was supposed that this policy would raise the achievement level of all students, close the persistent achievement gaps between students from low and high socio-economic backgrounds and create conditions for social integration. However, these expectations were not realized. The gap between students from an affluent background and students from deprived social and cultural backgrounds continues to exist. Moreover, the pedagogical difficulties in managing the variation in student background and ability strengthened the tendency for segregation within this integration to re-occur.

Public and academic debate regarding the gains and losses of the integration policy has been steady in Israel (Adler, 1984, 1986; Chen *et al.*, 1978; Dar and Resh, 1986; Resh and Dar, 1990). Education-related ability grouping or segregation, which, to a significant extent, parallels (and reflects) social segregation phenomena in Israel, was found to have a small negative effect on academic achievement and a differential effect on the achievements of 'strong' students, who also come from an affluent background, versus 'weak' students, who tend to come from a poor background: the strong students were shown not to have been affected by enhanced segregation while the weak students were apparently harmed (Resh and Dar, 1990). Segregation in Israel thus appeared to widen the achievement gaps and reduced the weak students' chances to enter high-level tracks in their secondary studies.

The publication of TIMSS 1999 data provided an opportunity to revisit the old debate and to study the overall effect of grouping on mathematics and science achievement, and its differential effect on the achievements of students from different social backgrounds. This aim was operationalized through four questions:

1. To what extent do different modes of learning organization (whole class, groups formed within or between home classes) enhance the overall level of achievement in science and mathematics studies?
2. To what extent do the different modes of learning organization act to widen or close achievement gaps associated with social stratification?
3. Do certain school/class, or group characteristics, and especially those related to the organization of learning, act to attenuate the effect of student background variables on students' learning outcomes?

4. Are there any interaction effects between the organization of learning and other higher level variables such as instruction that can explain the variability in outcome measures?

Method

Initial Phase

Preliminary analyses (Zuzovsky, 2004) in the tradition of grouping versus no grouping studies enabled me to formulate some working hypotheses. Comparing the achievement of individual students studying in groups to those studying in whole class settings revealed significant differences in favor of students studying in groups in both school subjects. In a school level analysis the differences were found to be significant only in science. At this stage of the study it was assumed that grouping in both school subjects was ability-based, often associated with socio-economic stratification. As will be seen later, this assumption did not hold. In order to control the possible effect of students' socio-economic background on these differences, two-way analyses of variance were conducted separately for each school subject. The analyses were carried out both at the student level and at the school level. 'Grouping' and 'school SES' were the two independent variables, and the mean plausible[1] scores in mathematics and science were the dependent outcome variables. Results of the student level analyses in the two school subjects revealed significant main effects of both *grouping* and *socio-economic background* as well as a significant interaction effect between them. However, the effect of grouping was found to be much smaller in mathematics than in science and at the school-level analysis, *grouping* in mathematics had no significant main effect at all, nor was there a significant interaction effect with the school's *socio-economic index*.

Breakdown of achievements of subpopulations sorted according to the type of learning organization they were exposed to (whole class or groups) and according to the school's SES index (high, medium, or low) reveal that, while grouping enhances the overall level of attainment in both school subjects, it tends to favor more students from low and medium SES schools, especially in science.

Findings from these preliminary analyses supported the hypothesis that grouping enhances the overall level of attainment in both mathematics and science and at least in science it clearly acts in closing achievement gaps associated with social stratification.

Further Analyses

Acknowledging the hierarchical organizational structure of the educational system and the need to study simultaneously the effect of both student level variables and group/school level variables on the variability in school outcomes, I decided to apply hierarchical linear modeling (HLM) for this purpose. This modeling enables the simultaneous representation of outcome variation at each level of the educational hierarchy in one model and the identification of the important variables 'explaining' this variation at each level in terms of the magnitude of their regression coefficients (for a more detailed account on HLM, see Bryk and Raudenbush, 1992). Allowing the regression coefficients of student level variables on achievement (slopes) to vary randomly over groups or schools can detect variation in these regression coefficients that can be explained by the effect of higher level variables in a 'slopes as outcomes' model (Raudenbush and Bryk, 1986). This 'slopes as outcomes' model is equivalent to an interaction model where student level variables with random slopes and group/school level variables interact. If we specify, in the random models, additional possible interaction effects among the group/school variables themselves, this will cause the hierarchical linear modeling of school effects to become more interactive (Aitkin and Zuzovsky, 1994).

Analysis

In specifying the multilevel model, a decision had to be taken regarding the model hierarchy. Estimating the variance components of a three-level unconditional (null) model (students nested in groups, nested in school) revealed significant variance components only related to two levels in the hierarchy. Also supported by the belief that most meaningful learning interactions occur on the group level, we decided upon a two-level model — students nested within group or, alternatively, students nested within school (represented by the one home-class sampled). I then allowed the coefficients of all student level variables on achievement to vary among the second level units. The small variation found with almost all of them convinced me to specify a fixed two-level model, where the effects of lower level variables on the outcome variance remains constant over the higher-level units of analysis.

For each school subject, the null model (model (a)) enabled me to simultaneously estimate the 'within-units-of-analysis' and 'between-units-of-analysis' variance in mathematics and science achievement scores. I then developed several explanatory models specifying their explanatory variables. In model (b) only student level variables were introduced to the regression equation. In model (c) aggregates of these variables describing the student body composition were added. In model (d) schooling variables describing disciplinary climate, the implementation of ability grouping and indices describing instruction, were added to the previous sets of variables. Models (d) and (c) were considered alternatives and were compared with model (b) (individual background variables only). The last model, model (e) included all effective significant variables found at both levels in previous analyses.

By examining the change occurring in the estimates of the variance components of each model after introducing its sets of variables, it was possible to determine the explanatory power of each model and also to identify the strongest predictors of achievements according to the magnitude of the regression coefficients of the variables involved.

The significant, though small, variation that was found in one of the individual background variables (self-esteem as science and mathematics learner) enabled me, thereafter, to specify a random model and to trace some interaction effects between self-esteem as science and mathematics learners and some higher level variables.

The last step of the analysis was to specify, in the group-level random models, two theoretically important interactions between grouping and the two indices of instruction, and look for their effects. With these additional analyses I followed claims that attribute the effect of grouping to two types of mechanisms: 'institutional' – usually mediated through symbolic properties that affect self-esteem, expectations and motivations of students and teachers in the different learning organizations, and 'instructional' – related to instruction which is supposed to be carried out differently in the types of learning organizations (Gamoran, 1986).

Data Source

The data source consists of student responses and information collected in TIMSS 1999 Trend Study in Israel. Organization of learning in small groups occurred, in the case of mathematics in 91 schools, and in science in only 49 schools (out of 139 participating schools). Schools differed in the number of groups formed from one home-class and also in the number of students attending each group. Some groups hosted a very small number of students. Due to statistical constraints and some consideration regarding the minimal number of students required for a group-type instruction, I excluded from the HLM analysis small groups with less than seven students and was left with 203 mathematics groups, of which 46 were home-class units and 157 small groups the product of grouping policy. In science there were 199 groups, of which 92 were home-class units and 107 small groups. Each of these groups was taught by a different teacher. The student samples that served the analyses comprised 3,159 students in mathematics and 3,242 students studying science. In spite of the reduction in sample size (less than 5 per cent in the number of students and less than 10 per cent in the number of groups) the descriptive statistics of the student-level variables and the school-level variables did not change and the data set employed remained representative.

Most of the student variables derived from student questionnaires describing student characteristics were of a general nature, except for student self-esteem, which was specified separately for science and mathematics studies. Group-level variables were also derived from aggregated or averaged student responses. This enabled me to create a file with no missing data for the higher level variables of the model. Variables describing instruction and disciplined climate were based on student perceptions. Out of the student responses, two factors for each school subject – one describing teacher-led instruction and the other student-centered learning – were constructed. Another group level variable, an index of positive school-climate, was also based on student perception. (A full description of the variables employed in the analyses appears in Zuzovsky, 2004.)

Science and mathematics achievement scores, the mean of five estimations of plausible scores in each of these subject areas, were the dependent variable.

Results and Interpretations

We now present the findings from the second phase of the analysis in which hierarchical linear modeling was employed. This analysis was conducted separately, once decomposing the variance between and within groups of students taught by different teachers

($n = 203$ in mathematics and $n = 199$ in science) and another time when the variance was decomposed into between and within schools ($n = 134$ in both mathematics and science). I will first present findings from the analyses of the fixed models (where the effect of lower level variables on achievement is assumed to be constant over the higher level units of analysis) (for details see Zuzovsky, 2004 and Table 23.1 here).

Between-group/Between-school Variance in Achievements and the Percentage Thereof Explained by the Different Models

- The percentage of *between-groups* variance out of the total variance in outcomes is higher for mathematics (44 per cent) than for science achievement (36 per cent). However, the *between schools/classes* variance component is similar in the two school subjects (36 per cent of the total variance in mathematics and 38 per cent in science). It should be noted here that, while the between-schools variance in mathematics found in Israel is similar to the international average (Martin *et al.*, 2000), in science the between-schools variance component found in Israel is much higher than the international average (23 per cent) that was estimated by the same source.
- Several models were used to explain this variability. First, I estimated and controlled for the effect of student-level variables. These models were found to explain about half of the between-groups or -schools variation in achievement in both school subjects. The other specified models contained, in addition to the individual student-level variables, their aggregates or means at the group or school level, representing the student body composition variables. Another model introduced, in addition to the individual student-level variables, only the group or school indices representing a variety of school factors such as: disciplinary climate, SES, type of instruction and the type of learning organization employed. The full model contained, beyond individual student-level variables, all the significant variables

Table 23.1 Explanatory power of several school effectiveness models.

	Mathematics		Science	
	Groups model $n = 203$	Schools model $n = 134$	Groups model $n = 199$	Schools model $n = 134$
Between-groups/schools variance component (Model (a))	44%	36%	36%	38%
Model specifications	Percentage of between groups/schools variance explained by the different models			
Model (b): Individual student variables	50	55	49	52
Model (c): Model 1 + group composition variables	72	73	65	71
Model (d): Model 1 + group/school variables	58	63	57	63
Model (e): Model 1 + all significant variables	75	76	69	74

that were found in the previous analyses. Each of these models have their own explanatory power. Table 23.1 presents the amount of between-groups or schools outcome variance that was explained by each of the alternative models. Accordingly, the data in the table indicate that:

- The explanatory power of the combined full models is the highest. They explain, in almost all of the cases, about three-quarters of the between-group or -school variance in achievement.
- Group composition variables add to the already explained variance due to individual student level variables – about 16 per cent to 22 per cent more of the unexplained variance that is left.
- The net contribution of school or group variables that represent the conditions for learning set by the school and the type of instruction that takes place in groups or classes is minor, reaching no more than 8 per cent of the unexplained between-groups/schools variance in achievement left beyond the effect of the individual student variables.

Variables Explaining the Between-groups/schools Variance in Achievement

- Pupil-level variables in order of the magnitude of their effects are: ethnic origin; number of books at home; self-esteem as a pupil; educational academic aspiration; gender, and parental education.

Table 23.2 presents the regression coefficients of these variables in the fixed science and mathematics group and school-level models. Since most of the student level variables in these models were dummy variables, they were not centered. Their regression coefficient indicates the increase in achievement score points (on a scale of 0–1,000 and a standard deviation of 100), due to being a Hebrew-speaking student, having high self-esteem and having high academic aspirations. In the case of the number of books at home, the regression coefficient indicates the increase in achievement due to increase in the number of books at home.

The effects of student-level variables are similar in both school subjects. The interpretation of the regress coefficients in Table 23.2 are as follows:

- Jewish students achieve about one standard deviation higher than Arab students.
- Students in homes with more than 200 books achieve between one-third to half standard deviation more than students from homes with less than 10 books.
- Students with high self-esteem achieve about one-third standard deviation more than students with low self-esteem.
- Students with high academic aspirations achieve around a quarter of a standard deviation more than students with no academic aspirations at all.

Table 23.3 presents the regression coefficients of the student-body composition variables in all models studied; among them are: percentage of *parents with academic*

Table 23.2 Regression coefficients and significance of student-level variables in the group and school fixed models (model (b)).

Student variables	Mathematics		Science	
	Group model $n = 203$	School model $n = 134$	Group model $n = 199$	School model $n = 134$
Ethnic origin 0 = Arabic-speaking; 1 = Jewish speaking	**92.7*** (7.8)	**88.8*** (8.5)	**100.8*** (10.2)	**101.7*** (10.5)
Number of books at home 1 = less than 10; 5 = more than 200	**6.9*** (1.0)	**8.2*** (1.1)	**10.9*** (1.0)	**10.3*** (1.1)
Self-esteem as a pupil 0 = low; 1 = high	**34.0*** (2.5)	**33.9*** (2.6)	**36.1*** (2.8)	**36.4*** (0.8)
Educational academic aspiration 0 = low; 1 = high	**21.9*** (2.5)	**26.8*** (2.5)	**27.4*** (2.5)	**28.4*** (2.6)

Note. Standard errors in parentheses.

education and the groups or school average of the *number of books in the students' home*. Both variables usually associated with students' socio-economic background are the most significant predictors of achievement. Findings for Table 23.3 can be interpreted as follows:

- In groups or classes where all students have at least one parent with academic education, the group mean achievements are almost one standard deviation higher than in groups or classes where none of the students have parents with academic education.

Table 23.3 Regression coefficients and significance of group composition variables in the group and school fixed models (model (c)).

Group composition variables	Mathematics		Science	
	Group model $n = 203$	School model $n = 134$	Group model $n = 199$	School model $n = 134$
No. of books at home/mean	**22.4** (6.4)	**19.5*** (8.8)	**14.7*** (7.2)	16.5 (10.4)
At least one parent with academic education (%)	**81.0*** (15.5)	**100.6*** (21.4)	**92.5*** (16.3)	**121.7*** (23.1)
Students with high academic aspiration (%)	32.3 (15.3)	−4.3 (26.6)	3.2 (20.2)	−8.7 (29.8)
Students with high self-esteem as learners (%)	−0.4 (14.1)	−23.6 (24.9)	0.5 (17.7)	−20.6 (23.7)

Note. Standard errors in parentheses.

- In groups or classes where the mean of the book variable exceeds 5 (more than 200 books at home) the group mean achievements are about one standard deviation higher than those in groups where the mean of the book variable is 1 (homes with a few books). This group composition variable is somewhat less effective in science.
- School- or group-level variables describe the following characteristics:
  - Disciplinary climate – average student responses to a battery of questions about disciplined climate or lack of violence they experience in school.
  - Learning in groups – whole integrated class or segregated groups.
  - Types of instruction, only used in the group models – average on two factor scales describing frequency of student centered vs. teacher-led instruction, and 'school SES index', only used in the school models.

The regression coefficients of these variables in the fixed group and school models are presented in Table 23.4.

Here too the variables were not centered and their interpretation is as follows:

- In both mathematics and science, student-centered instruction (negatively associated with achievement) plays a significant role in explaining the between-group variability in outcomes. Wherever this type of instruction is frequent, students achieve less (about half of a standard deviation for each unit of the index).
- Disciplinary climate (positively associated with achievement) significantly explains the variance in outcomes only in the case of mathematics. The achievement gap between students studying in high disciplinary climate groups and students studying in low disciplinary climate groups reaches one standard deviation in favor of those in high *disciplinary climate groups*.

Table 23.4 Regression coefficients and significance of school or group variables in the group and school fixed models (model (d)).

Group/school variables	Mathematics		Science	
	Group model $n = 203$	School model $n = 134$	Group model $n = 199$	School model $n = 134$
Disciplinary climate (1 = low; 4 = high)	**20.4*** (9.4)	16.8 (12.2)	14.2 (11.6)	16.5 (13.2)
Grouping (0 = whole class; 1 = grouping)	−3.5 (6.9)	−3.6 (6.8)	**23.9*** (6.4)	**20.2** (7.5)
Pupil centered instruction (1 = seldom; 4 = often)	**−58.6*** (11.9)	–	**−41.6*** (10.8)	–
Teacher led instruction (1 = seldom; 4 = often)	−19.9 (11.8)	–	−13.2 (10.7)	–
School SES index (1 = high; 3 = low)	–	**−5.3*** (1.3)	–	**−6.3*** (1.4)

Note. Standard errors in parentheses.

- Organizing learning in groups is positively and significantly associated with achievement only in science. Students studying science in groups achieve between one-fifth to one-quarter of a standard deviation more than students studying in whole-class settings.
- As expected, the school SES index is also a significant predictor of achievement. Students studying in high-SES schools achieve about 0.15–0.18 of a standard deviation more than students in low-SES schools.

Findings from the Random Models

Since only self-esteem as a science or mathematics' learner was found to have significant, although small, slope variation, I specified a set of models similar to those used in the previous analyses, allowing the regressions coefficients of science and mathematics achievement on students' self-esteem as learners in these subjects to vary randomly across groups or schools (in addition to groups' school intercept).

Academic self-esteem as a science or mathematics learner can be regarded either as a realistic measure of one's own level of attainment or as a symbolic property associated with the group the individual student is affiliated to. It is positively associated with achievement (Shavelson and Bolus, 1981) in both school subjects.

In the present study, this association was found to be stronger in groups or classes with a high disciplinary climate. In science this relationship is also stronger in groups with high mean of teacher-led instruction (Zuzovsky, 2004, for more details). These findings will be dealt with later when the association between ethnic origin and achievement will be discussed.

Among the possible interaction terms between the group or school level variables, I decided to focus on the interaction between *grouping and instruction*. I followed Gamoran's response to Slavin's (1987: 342) 'best evidence synthesis' on grouping, where he comments:

> The study of grouping alone provides little information of value. What appear to be positive, negative or neutral effects of grouping may have little to do with grouping in and of itself, but may derive from how grouping is used to provide appropriate or inappropriate instruction.

I then ran another alternative random model for the specified variable adding two interaction terms between grouping and the *two indices of instruction*: student-centered and teacher-led.

The full random interactive models were found to have maximal explanatory power regarding the between-group achievement variance (Zuzovsky, 2004). They explain, in the case of mathematics, a further 25 per cent of the between-group yet unexplained variance, in addition to the already 48 per cent of between-group variance explained by the individual student-level variables (48 per cent + 25 per cent = 73 per cent). In science, this model explains 26 per cent more of the between-group variance in addition to the 40 per cent explained by the student variables model (40 per cent + 26 per cent = 66 per cent).

Only one significant and positive interaction effect in science was found between *grouping and instruction*. Students working in small groups in science achieve about half a standard deviation more for each degree of the student-centered type of instruction they receive than if they study in a whole-class setting. Such interaction effects do not occur in mathematics.

Group and School Variables and their Effect on the Association between Ethnic Origin and Achievement

The initially intense but fading strength of the association between ethnic origin and achievement, as a result of introducing higher-level variables to the alternative models of school effectiveness, points to the role school and group variables can play in reducing the high achievement gap found between Hebrew-speaking students and Arabic-speaking students.

These group/school-level variables are not associated with socio-economic factors, rather they represent the ways schools operate. Their effects are consistent in both Arabic and Hebrew-speaking populations. *Disciplinary climate and learning in groups* are positively associated with achievement while *student-centered instruction* is negatively associated with achievement.

Comparing the two populations in terms of these variables (see Table 23.5) shows that variables that are positively associated with achievement are lower in the Arabic-speaking groups, while those negatively associated with achievement are higher in these groups. Without ignoring the fact that there are significant differences in socio-economic characteristics between Arabic-speaking and Hebrew-speaking populations, which also can explain the achievement gap between the two populations, it is evident that schools can play a role in closing these gaps by providing adequate conditions for learning.

With less frequent student-centered instruction (negatively associated with achievement, and more frequent in the Arabic-speaking sector than in the Hebrew-

Table 23.5 Comparison of statistics of group-level variables in Arabic and Hebrew speaking groups.

	Arabic		Hebrew		t-value and significance
	(Math $n = 38$; Science $n = 28$)		(Math $n = 171$; Science $n = 171$)		
Disciplinary climate – science	1.7	(0.19)	3.3	(0.34)	−23.1***
Disciplinary climate – math	1.8	(1.8)	3.2	(0.38)	−33.9***
Studying in groups – science	7%	–	61%	–	417.5**
Studying in groups – math	33%	–	72%	–	395.2***
Pupil-centred instruction – science	2.2	(0.25)	1.6	(0.27)	9.8***
Pupil-centred instruction – math	2.1	(0.27)	1.6	(0.27)	10.7***
Teacher-led instruction – science	3.1	(2.0)	3.2	(0.27)	−0.18
Teacher-led instruction – math	3.3	(0.18)	3.5	(0.26)	−6.0***
Percent of parents with academic education	26%	(0.13)	48%	(0.22)	−7.6***

speaking sector), and with higher disciplinary climate positively associated with *mathematics* achievement and with more group-learning positively associated with *science* achievement, both low in the Arabic-speaking sector, the achievement gap between the two sectors in Israel can be narrowed.

Findings from the random model analyses shed additional light on this gap. Since the disciplinary climate was found to interact with academic self-esteem and since it is higher in the Hebrew-speaking than in the Arabic-speaking groups or classes, the association between academic self-concept and achievement in both subject areas is stronger in Hebrew-speaking classes. This occurs in spite of the fact that there are more students with high self-esteem in the Arab classes than in the Hebrew ones (55 per cent versus 35 per cent in science and 51 per cent versus 44 per cent in mathematics).

Disciplinary climate in both school subjects is also higher in a group organization of learning, which is more frequent in the Hebrew-speaking population, thus, the advantage of Hebrew-speaking students over Arabic-speaking students gets even larger. Table 23.6 presents these findings.

Conclusion

The main purpose of this study was to look for the effect of grouping on learning achievement in science and mathematics. Grouping is very common in mathematics classes and somewhat less so in science classes in Israel.

Four questions were posed at the beginning of this study. Briefly, to remind the reader, these were as follows: (1) To what extent does grouping enhance the overall level of achievement in science and mathematics studies? (2) To what extent does grouping act to widen or close achievement gaps associated with social stratification? (3) Do any of the group or school variables – especially those related to class organization – act to attenuate the effect of student variables or student learning outcomes? (4) Are there any interaction effects between grouping and other group or school level variables (instruction) that can explain the variability in outcome measures? At the start of this study, it was assumed that grouping was based on ability considerations.

Findings from the preliminary analyses showed that grouping enhances the overall level of attainment in both subject areas and at least in science it is clearly associated with higher achievement of medium and low SES students.

Findings from the HLM analyses highlight a series of variables at different levels that significantly affect the variance in student outcomes. The effects of student-level variables were found to be quite similar in both school subjects, with ethnic origin being the most influential. Among student-body-composition variables the percentage of parents with academic education is a strong predictor in both subjects. Ethnic origin at this level does not play a role as groups are homogeneous, consisting of either Hebrew or Arab-speaking students. However, since this variable is confounded with other group-level variables, the ethnic affiliations appear to be associated with variability in learning outcomes that have an effect on achievement. Among the group- or school-level variables, we found disciplinary climate to have a positive significant effect on the variance only in the case of mathematics, while pupil-centred instruction

Table 23.6 Differences in means of several group-level variables by language of speaking and learning organizations.

Group characteristics	Arab speaking		Hebrew speaking		t-value and significance	Whole class		Groups		t-value and significance
	M	SD	M	SD		M	SD	M	SD	
Teacher-led instruction math	3.3	(0.2)	3.5	(0.3)	**–6.0**	3.4	(0.2)	3.5	(0.3)	–1.1
Teacher-led instruction science	3.1	(0.2)	3.2	(0.8)	–1.8	3.2	(0.3)	3.2	(0.3)	**–2.2**
Discipline climate math	1.8	(0.2)	3.2	(0.4)	**–33.9**	2.6	(0.8)	3.1	(0.6)	**–3.8**
Discipline climate science	1.7	(0.2)	3.2	(0.3)	**–23.1**	2.7	(0.7)	3.3	(0.4)	**–6.6**
Students with high self-esteem – math	51%	(16)	44%	(17)	**2.1**	47%	(14)	45%	(18)	0.6
Students with high self-esteem – science	55%	(14)	35%	(15)	**6.7**	39%	(16)	37%	(16)	0.8

Notes. M = mean; all significant differences $p \leq$ are in bold.

and school SES have a significant negative effect in both subjects. Grouping was found to be associated with high achievement only in science.

In looking for some interaction effects between lower level variables and higher level ones, the only student level variables that seemed to vary significantly over the higher level units of analyses was self-esteem as a science or mathematics learner. This variable, which is not necessarily associated with socio-economic factors, nor with the type of learning organization, interacted in both school subjects with disciplinary climate and in science also with teacher-led instruction. Since disciplinary climate was found to be high in both school subjects, in group settings of learning it is through this interaction effect that grouping enhanced achievement and more so in science groups, where the disciplinary climate is higher (see Table 23.6).

Grouping was found to attenuate the effect of the ethnic origin variable on students' learning outcomes through the mediation of other group-level variables, typical of Arabic-speaking groups, some positively associated with learning outcomes; however, less frequent in Arabic groups and others negatively associated with outcomes, but more frequent in Arabic-speaking groups.

The last question (question 4) was meant to provide information about a possible link between grouping and instruction. This analysis provided an answer to my growing astonishment facing the repeated findings that grouping does not affect outcomes in mathematics, while in science it does. The significant interaction effect found between grouping and student-centred instruction in science can explain why students working in small groups in a student-centred type of instruction benefited more than when studying in whole class. In mathematics, such an interaction effect was not found.

Questions and Further Investigation

The accumulated evidence that grouping does not affect the variance in mathematics achievement led me to further investigate my assumption of the common nature of grouping in both subject areas which took for granted that grouping was ability based.

I conducted an additional set of analyses of variance in all schools where grouping was implemented using the mean plausible score in mathematics as my dependent variable (serving as a proxy for ability). I indeed found that, while in mathematics the groups were found to be statistically different from each other in their mathematics scores (ability measure), in science this was not the case. Science groups formed from one home-class did not differ in their mathematics (ability) mean. Grouping in mathematics can, thus, be viewed as basically ability grouping, a reaction to initial individual (ability or academic) differences that exist among students in the heterogeneous lower secondary schools.

The use of ability grouping in mathematics seems to be in line with the individual difference paradigm that accepts differential outcomes and is led by the belief that individual differences in ability account for the vast range of performance existing both within and between schools (Miller and Brookover, 1986; Persell, 1977). Grouping in mathematics is a technique that is responsive to the need to conduct meaningful instruction for every student, giving them the opportunity to learn to their potential; thus it does not advance students beyond what they can attain, nor does it close the gaps between high and low ability students.

Grouping in science, on the other hand, is not associated with initial ability differences among students, but rather results from other considerations. These considerations are related to the empirical nature of science and the collaborative working habits of scientists that consequently lead to a preferred mode of learning in science, that is, small-group learning which enables intensive hands-on learning, and fosters helping relations, intensive peer interaction and increased information processing (Springer *et al.*, 1999).

Organizing learning groups in science can thus be viewed as a school factor open to manipulation that can advance attainment. Findings from my study indeed support this view, as *grouping* was found to affect both the productivity and equality of science outcomes.

Acknowledgement

This chapter is based on an elaborated paper presented at the 1st IEA International Research Conference in Lefkosia, Cyprus, May 2004 and appears in C. Papanastasiou (Ed.), *Proceedings of the IRC – 2004 TIMSS* (Vol. 2, pp. 137–60). Full tables in that paper were summarized, both in table and text format of the present chapter, due to limited space. However, readers interested are invited to refer to the original available at: http://www.iea.nl

Note

1. Estimation of students' proficiency scores in science and mathematics based on IRT scaling (Yamamoto and Kulick, 2000).

References

Adler, H. (1984) School integration in the context of the development of Israel's educational system. In Y. Amir, S. Sharon and R. Ben Ari (eds.), *School desegregation* (pp. 21–46). Mahwah, NJ: Erlbaum.

Adler, H. (1986) Israeli education addressing dilemmas caused by pluralism: A sociological perspective. In D. Rothermund and J. Simon (eds.), *Education and integration of ethnic minorities* (pp. 64–87). London: Frances Pinter.

Aitkin, M. and Zuzovsky, R. (1994) Multilevel interaction models and their use in the analysis of large-scale school effectiveness studies. *School Effectiveness and School Improvement*, 5(1), 45–73.

Braddock, J.H. (1990) Tracking the middle grades. National patterns of grouping for instruction. *Phi Delta Kappan*, Feb., 445–9.

Bryk, A.S. and Raudenbush, S.W. (1992) *Hierarchical linear models*. Newbury Park, CA: SAGE.

Chen, M., Lewy, A. and Adler, C. (1978) *Processes and outcomes in the educational system*. School of Education, Tel Aviv University, Tel Aviv and The NCJW Research Institute for Innovation in Education, School of Education, The Hebrew University, Jerusalem (in Hebrew).

Dar, Y. (in collaboration with Resh, N.) (1981) *Homogeneity and heterogeneity in education*. The NCJW Research Institute for Innovation in Education, School of Education, The Hebrew University of Jerusalem, pub. No. 73. (Hebrew).

Dar, Y. and Resh, N. (1986) Classroom intellectual composition and academic achievement. *American Educational Research Journal*, 23, 357–74.

Findley, W.G. and Bryant, M.M. (1971) *Ability grouping: 1970*. Athens, GA: Center for Educational Improvement.

Finley, M.K. (1984) Teachers and tracking in a comprehensive high school. *Sociology of Education*, 57, 233–43.

Gamoran, A. (1986) Instructional and institutional effects of ability grouping. *Sociology of Education*, 57, 233–43.

Gamoran, A. and Berends, M. (1987) The effect of stratification in secondary schools: Synthesis of survey and ethnographic research. *Review of Educational Research*, 57, 415–35.

Gamoran, A., Nystrand, N., Berends, M. and Le Pore, P.C. (1995) An organizational analysis of the effects of ability grouping. *American Educational Research Journal*, 32(4), 687–715.

Hallinan, M.T. (1992) The organization of students for instruction in the middle school. *Sociology of Education*, 65, 114–27.

Martin, M.O., Mullis, I.V.S., Gregory, K.D., Hoyle, C. and Shen, C. (2000) *Effective schools in science and mathematics – IEA's TIMSS*. Chestnut Hill, MA: TIMSS International Study Center, Boston College.

Miller, S.K. and Brookover, W.B. (1986) School effectiveness vs. individual differences: Paradigmatic perspectives on the legitimation of economic and educational inequalities. Paper presented as part of a symposium at the AERA annual meeting, San Francisco.

Murphy, J. and Hallinger, P. (1989) Equity as access to learning: Curriculum and instructional treatment differences. *Journal of Curriculum Studies*, 19, 341–60.

Oakes, J. (1985) *Keeping track: How schools structure inequality*. New Haven, CT: Yale University Press.

Oakes, J. (1990) *Multiplying inequalities: The effect of race, social class and tracking on opportunities to learn mathematics and science*. Santa Monica: Rand.

Oakes, J., Gamoran, A. and Page, R.N. (1992) Curriculum differentiation: Opportunities, outcomes and meanings. In P.W. Jackson (ed.), *Handbook of research on curriculum* (pp. 570–608). Washington, DC: American Educational Research Association.

Page, R.N. (1991) *Lower track classroom: A curricular and cultural perspective*. New York: Teachers College Press.

Persell, C.H. (1977a) *Education and inequality: A theoretical and empirical synthesis*. New York: Free Press.

Persell, C.H. (1977b) *Education and inequality: The roots and results of stratification in America's schools*. New York: Free Press.

Raudenbush, S.W. and Bryk, A.S. (1986) A hierarchical linear model: A review. *Sociology of Education*, 59, 1–17.

Resh, N. and Dar, Y. (1990) *Segregation within integration. Educational separation in junior high schools. Factors and implications*, Publication No. 125, The NCJW Research Institute for Innovation in Education, School of Education, The Hebrew University, Jerusalem (Hebrew).

Rosenbaum, J.E. (1976) *Making inequality: The hidden curriculum of high school tracking*. New York: Wiley and Sons.

Rosenbaum, J.E. (1980) Social implications of educational grouping. *Review of Research in Education*, 8, 361–401.

Shavelson, R.J. and Bolus, R. (1981) Self concept, interplay of theory and methods. *Journal of Educational Psychology*, 74, 3–17.

Slavin, R.E. (1987) Ability grouping and achievement in elementary schools. A best evidence synthesis. *Review of Educational Research*, 57, 293–336.

Slavin, R.E. (1990) Achievement effects of ability grouping in secondary schools. A best evidence synthesis. *Review of Educational Research*, 60, 471–99.

Sorensen, A.B. (1970) Organizational differentiation of students and educational opportunity. *Sociology of Education*, 43, 355–76.

Springer, L., Stanne, M.E. and Donovan, S.S. (1999) Effects of small group learning on undergraduates in science, mathematics, engineering and technology: A meta-analysis. *Review of Educational Research*, 69(1), 21–52.

Talbert, J. (1990) Teacher tracking: Exacerbating inequalities in the high school. Paper presented at the AERA annual meeting, Boston.

Yamamoto, K. and Kulick, E. (2000) Scaling methodology and procedures for the TIMSS mathematics and science scales. In M.O. Martin, K.D. Gregory and S.E. Stemler (eds.), *Technical report*. Chestnut Hill, MA: ISC, Boston College.

Zuzovsky, R. (2004) Grouping and its effect on 8th graders' science and mathematics achievement. In C. Papanastasiou (ed.), *Proceedings of the IRC-2004 TIMSS* (Vol. 2, pp. 137–60). Cyprus: Cyprus University Press.

24

Factors Associated with Cross-national Variation in Mathematics and Science Achievement based on TIMSS 1999 Data

Ce Shen

Introduction

This study analyzes the data from the Third International Mathematics and Science Study (TIMSS) 1999 to test the effects of selected variables which may account for the cross-national variation of mathematics and science achievement for eighth-grade students from 38 school systems in various countries. Using multiple regression analysis, this study finds that economic development levels measured by gross national product (GNP) per capita and literacy levels are positively associated with achievement. The four variables reflecting school and classroom management problems – absenteeism, the rate of students repeating a grade, student body mobility, and frequency of teachers getting interrupted in class – demonstrate significant negative effects on achievement. Consistent with the findings from TIMSS 1995 data, students' self-perceived easiness of the two subjects, which can be used as a proxy of the rigor of academic programs, is negatively associated with achievement. The policy implications of these findings are discussed.

Both TIMSS 1995 and 1999 reveal substantial variations in student mathematics and science achievement cross-nationally, with a significant gap between the highest and lowest performing countries. Policy-makers and researchers of each participating country are encouraged to examine how their country's student achievement levels compare to those of other countries and what can be learned from other school systems to improve achievement. Using aggregate data at the country level, this study attempts to identify contributing factors that account for the cross-national variance in terms of mathematics and science achievement for all TIMSS 1999 participating countries ($N = 38$).

The purpose of this study was twofold: (1) to identify variables accounting for the cross-national variation of mathematics and science achievement of TIMSS 1999 data; and (2) to retest the findings and hypotheses from previous studies based on analyses of TIMSS 1995 data (Shen and Pedulla, 2000; Shen, 2001).

TIMSS studies have collected a wide range of invaluable data, including test results in various areas of mathematics and science, background information on students, teachers and school principals, and curriculum information from each school system. However, due to the diversity of educational, cultural, social and economic variables in the participating countries, no simple and straightforward factors are identified to explain the substantial cross-national variation of achievement in mathematics and science. This study intends to test the effects of selected country-specific variables on student achievement. These variables reflect the countries' economic development level, school and classroom environment, the literacy of students' families, and students' self-perceived easiness of the subjects of mathematics and science.

Data and Variable Measurements

Thirty-eight school systems participated in the TIMSS 1999 study. The dependent variables of this study are international achievement scores in mathematics and science for 8th grade students in each TIMSS 1999 participating country. The aggregated data used to represent the achievement level of a country may have both validity and reliability constraints. The average scores are aggregated over a considerable amount of subject areas. They did not take into account the variation between specific areas of study that fall under the broader categories of mathematics and science. In addition, the aggregated data for both the dependent and independent variables did not take into account individual variation among the sizable number of students and schools (about 3,000 to 5,000 students and about 150 schools for most participating school systems) within a participating country. As we all know, for many countries, within-country variation is tremendous.

The selection of independent variables was based on both theoretical and empirical considerations, including hypotheses based on previous research, data comparability and availability. In addition to TIMSS data, two measures of country-level data were initially included: economic development level as measured by real GNP per capita and public expenditure on education as percentage of GNP. The data of these two variables are listed in the Appendix.

The choice of predictor variables from the TIMSS database for this study was made after extensive exploratory analyses of the available data, including correlation analysis, factor analysis and regression analysis. The TIMSS database provides data for hundreds of variables such as school and teacher characteristics, and instructional practices including school and class size, school location, organization of classroom instruction, school resources, teachers' educational background, school and classroom environment, how teachers and principals spend their time at school, responsibilities

Table 24.1 Pearson correlation matrix for variables in the study (grade 8) (*N* = 38 school systems).

	1	2	3	4	5	6	7	8	9	10	11	12	13	Mean
1 Mathematics score	1.00													487
2 Science score	0.96**	1.00												488
3 Real GNP per capita	0.56**	0.52**	1.00											12,174
4 Expenditure on educ.	-0.14	-0.13	-0.05	1.00										4.69
5 School days a year	-0.14	-0.13	-0.10	-0.06	1.00									197
6 Absenteeism	-0.41**	-0.31	-0.06	0.32*	-0.05	1.00								4.17
7 Repeating the grade	-0.50**	-0.58**	-0.26	0.18	0.03	0.08	1.00							3.36
8 Student mobility	-0.62**	-0.64**	-0.08	0.20	0.01	0.52**	0.33*	1.00						3.15
9 Math class interrup.	-0.47**	-0.40**	0.08	0.17	-0.05	0.43**	0.03	0.54**	1.00					2.00
10 Science class interrup.	-0.56**	-0.50**	0.01	0.19	-0.03	0.39*	0.11	0.58**	0.98**	1.00				1.96
11 Number of books	0.63**	0.72**	0.36**	0.18	-0.28	0.18	-0.38*	-0.31	-0.17	-0.28	1.00			3.03
12 'Math is easy'	-0.72**	-0.74**	-0.32*	0.11	0.09	0.14	0.38**	0.61**	0.43**	0.52**	-0.64**	1.00		2.32
13 'Science is easy'	-0.75**	-0.74**	-0.46**	0.12	-0.01	0.09	0.54**	0.43**	0.29	0.39*	-0.67**	0.84**	1.00	2.51

Note. $*p < 0.05$, $**p < 0.01$.

of principals, how often teachers assign homework, and so on. Unfortunately, just like the study based on TIMSS 1995 data (Shen, 2001), the majority of these variables fail to demonstrate consistent effects on mathematics and science achievement for most school systems. Even fewer show significant effects on student achievement for cross-national analysis.

Some potentially relevant predictors were rejected because they were not internationally comparable (such as parents' education) or because of missing data. As a result, the number of predictors chosen for this study is quite limited.

To retest the findings and hypotheses yielded from a previous study based on TIMSS 1995 data, TIMSS 1999 data for each independent variable examined by Shen's (2001) study were included for the initial analysis. Several additional variables from school and student questionnaires, which were not used in Shen's study, are included and examined in the present study.

Variables from the TIMSS 1999 school questionnaire include the average number of instruction days per year, the average percentage of student absenteeism on a typical school day, the average percentage of students repeating a grade and a measure of student body mobility – the average percentage of students who started the school year but did not finish the school year at the same school. The variable of instruction days per year was dropped at a later point due to its non-significant effect. The data with standard deviation are listed in the Appendix.

Variables from the student questionnaires include the average number of books at students' homes, students' educational aspiration, measured by the highest educational level students expected to complete (recoded from seven categories to three due to inconsistency of school systems: 1 = Finish some secondary school, 2 = Finish secondary school, and 3 = Beyond secondary school), and a measure of classroom environment – the frequency of teachers getting interrupted by messages, visitors, and so on in class (on a four-point Likert scale, from 'never' to 'almost always'). Lithuania did not administer the question of class interruption. In order to keep this case, the international mean substitution is used for Lithuania. Also included is students' self-perceived easiness of mathematics and science, i.e., responses to the statement 'Mathematics (science) is an easy subject' (on a four-point Likert scale, from 'strongly disagree' to 'strongly agree'). This student self-perception variable is a general measure of students' attitude toward the two subjects (Kifer, 2002). The IEA believes that the curricula are key in the evaluation of educational achievement and many researchers place curricula in the central position in the framework of educational evaluation (Shavelson et al., 1987; Howie, in press). However, it may not be able to develop an index to measure the relative rigor and demanding levels of curricula at the school system level for cross-national comparison. The author believes that once the students' perceived easiness of the two subjects are aggregated at the country level it might be used as a proxy of the rigor of the curricula for the two subjects for various school systems.

Conceptually, these independent variables fall into three groups. The first group includes GNP per capita, public expenditure on education as percentage of GNP, and the length of the school year. These three measures reflect a country's economic development level and commitment to education. GNP per capita in 1997, converted at purchasing power parity (ppp), was included as an indicator of economic development level in this study. A country's economic development level affects the educational processes and outcomes by providing educational material, information, and human resources needed for educational attainment. As shown in the Appendix, the range of GNP per capita ranges from 1,500 international dollars for Moldova to 29,010 international dollars for the USA. Compared with the TIMSS 1995 sample, the TIMSS 1999 sample shows a greater range of differences in terms of economic development level.

The public expenditure as percentage of GNP and the number of instruction days of a school year can be regarded as indicators of a government's and/or a society's commitment to education. When governments spend a high percentage of GNP on education, they would expect to produce high-achieving students. Similarly, when students spend more days going to school, we also assume that the country is more likely to obtain a high achievement level.

The second conceptual group includes indicators reflecting how conducive the school and classroom environments are to learning and teaching. Previous research, including studies on TIMSS data, have confirmed the important effect of school and classroom condition and environment on students' achievement (for example, Howie, 2003). This study includes the average percentage of absenteeism, the average percentage of students repeating a grade, student body mobility and the frequency of teachers getting interrupted in class as indicators of school and classroom condition. Frequent absenteeism from school and skipping classes are a reflection of students' sloppy habits, poor attitude toward schooling, and ineffective classroom management, which are major contributors to low academic standards and unsatisfactory character formation (Wray, 1999: 14). The measure of this variable is somewhat different from the one used in Shen's (2001) study due to the modification of the TIMSS 1999 student questionnaire.[1]

Although a high percentage of students repeating a grade is associated with the educational policy of a school system, region and/or a specific school, it is also a reflection of ineffective school management and of a student's poor attitude toward schooling and academic efforts. The empirical research on the effect of student body mobility on achievement tends to indicate that an overall decline in achievement is associated with mobility (Johnson and Lindblad, 1991; Schuler, 1990; Wood, Halfon, Scarlata, Newacheck and Nessim, 1993). High mobility may disrupt student learning and classroom management, most notably when students enter classrooms that are at a different point in the curriculum from where they left off at previous schools, and especially in decentralized school systems. The TIMSS 1999 student questionnaire asked students how frequently their teachers get interrupted by messages, visitors, and so on in class. Frequent interruptions disrupt student learning and teacher instruction, reflecting poor school and classroom management.

The third conceptual group includes variables reflecting students' educational resources at home, students' educational aspiration, and self-perceived easiness of the two subjects. Following a number of previous studies, the number of books at a student's home is included in this study as an indicator of educational resources at home. As an aggregate measure, the average number of books at student homes is also a measure of a society's literacy level. The IEA Reading Literacy Study (Elley, 1994:

226) revealed that countries with many books in homes were mostly high performing countries because homes with plentiful source books apparently provide more advantages for children's literacy development. TIMSS 1995 data (Shen, 2001) reveal that the average number of books at students' homes demonstrates a positive association with student achievement in the country. This is a categorical variable: 1 = less than 10 books; 2 = 11 to 25 books; 3 = 26 to 50 books; 4 = 51 to 200 books; and 5 = more than 200 books.

Students' educational aspiration, measured by the level of schooling they expect to complete, was included in the initial analysis. The recoded data of educational aspiration with its standard deviation are listed in the Appendix. Although Shen (2001) found a positive relationship between this measure and student achievement for TIMSS 1995 data, the effect was not stable. Even though it can be a valid measure of students' educational aspiration for within-country data analysis, it is not internationally comparable. For example, TIMSS 1999 data show that 78 percent of American eighth-grade students expect to finish university compared to 38 percent of Japanese students (Mullis *et al.*, 2000: 124). We cannot infer from this gap that Japanese students have lower educational aspiration than their American counterparts, because the average student educational goal in a country is affected by a number of country-specific factors such as labor structure, school admission policy at various levels, and so on. Therefore, this variable was eventually excluded for further analysis.

Students' perceived easiness of the two subjects is a measure of students' attitude and perceptions towards the subjects. As Kifer (2002: 252) asserts, studying attitudes and perceptions within international comparative studies could be used to answer such a question: Do those students and educational systems with the most positive attitudes also tend to have the highest test scores? With the intention to answer such questions, Shen and Pedulla (2000), using correlation analysis based on TIMSS 1995 data, examined the relationship between students' achievement in mathematics and science and their self-perceived easiness of the two subjects. They found that within-country data generally show a positive relationship between self-perceived easiness and achievement. However, when one examines this relationship between countries, the opposite relationship occurs, that is, countries with a high proportion of students perceiving the subjects as easy performed poorly on the TIMSS tests and vice versa. This pattern exists for both mathematics and science at Grades 3, 4, 7 and 8. The authors suggest that this pattern may reflect low academic expectations and standards in low performing countries and high academic expectations and standards in high performing countries.

Using multiple regression analysis based on TIMSS 1995 data, Shen (2001) found that, cross-nationally, students' self-perceived easiness of mathematics and science (a proxy of academic standards of mathematics and science) demonstrates a significant negative effect on achievement at the country level even when other variables were controlled, including economic development level as measured by GNP per capita, students' school attendance, the length of a school year, students' educational aspiration and the average number of parents living with the student.

It is noted that the two previous studies based on TIMSS 1995 data analysis have some inherent limitations. First, the analyses are not based on a random sample of countries, as the 41 countries were those that voluntarily participated. Second, the sample of countries is quite small considering the total number of countries is over 200. Therefore, the findings need to be retested with different countries and in different time points. The analysis of the TIMSS 1999 data and sample will provide an opportunity to retest the hypotheses and findings from the analysis of the TIMSS 1995 data and sample.

Methods and Significance Testing

Because the unit of analysis is country or school system, the means of all the variables used in the analyses are computed at the country level. First, correlation analysis was conducted to test the relationship among the dependent and independent variables involved in this study. Ordinary least square (OLS) regression analysis was then used to estimate the relative importance of predictors in accounting for cross-national variance of achievement in mathematics and science. Regression models were constructed following the conventional rule to keep the ratio of cases to predictors to the suggested limit (Tabachnick and Fidell, 1996); in one regression equation, no more than four predictors were included.

Given that none of our samples are simple random samples, the use of tests of significance and t-statistics are presented for heuristic purposes. Bollen et $al.$ (1993) assert that statistical significance tests can be justified with samples such as this study in terms of a hypothetical 'superpopulation', where the observed sample is treated as one possible sample that could be drawn from that hypothetical population.

The significance levels based on statistics were used to indicate the relative strength of a variable in accounting for the cross-national variation of mathematics and science achievement. The ratio of the regression coefficient to its standard error was used as a measure of the level of significance. Following some previous sociological studies (London and Williams, 1990), coefficients are considered significant if the coefficient is at least 1.5 times the size of its standard error, which is equivalent to the conventional 0.10 level. According to Pedhazur (1982), in an analysis of this type, where the units of analysis are large aggregates and where the number of cases is relatively small, this measure is the most reliable guide to interpreting the significance level of coefficients.

Results

Table 24.1 presents the Pearson's zero order correlation coefficients among the variables involved in the study for 38 TIMSS 1999 participating countries. On the right hand side of the table, the mean values of all the dependent and predictor variables based on the 38 school systems are listed.

A few points from Table 24.1 are noted. Similar to the correlation found from TIMSS 1995 data (Shen, 2001), the public expenditure on education as a percentage of GNP has a negligent association with achievement. In addition, unlike the finding from TIMSS 1995, the number of school days a year has no significant association with achievement. To obtain a parsimonious solution for the multiple regression analyses, public expenditure on education and the length of school year are excluded from the regression analyses.

As shown in the table, the correlation of the two achievement scores is high (0.96), indicating that countries that did well in mathematics usually also did well in science. The correlation between achievement scores and GNP level are 0.56 and 0.52 for mathematics and science respectively, higher than those found from TIMSS 1995 data ($r = 0.24$ for mathematics and $r = 0.18$ for science respectively). There are at least two explanations for this discrepancy. First, Kuwait was in TIMSS 1995 but not in TIMSS 1999. It is a country of high GNP per capita, but with poor achievement in both mathematics and science for TIMSS 1995 tests. The presence of Kuwait drags down the correlation between GNP per capita and student achievement scores for the TIMSS 1995 sample. Second, as mentioned earlier, about 10 high-income countries did not participate in TIMSS 1999 and a number of less developed, middle- and low-income countries participated. The TIMSS 1999 sample makes the correlation between GNP per capita and student achievement more manifest. As shown in Table 24.1, out of 38 countries/school systems in TIMSS 1999, 19 countries have a GNP per capita of less than 10,000 international dollars circa 1997, compared to only 10 countries of this income level for the TIMSS 1995 sample (Shen, 2001).

As a measure of school management and students' attitude toward schooling, average percentage of absenteeism has a negative correlation with achievement. Similarly, the other two measures of school management problems – the percentage of students repeating a grade and student mobility – demonstrate a significant negative correlation with mathematics achievement. A measure reflecting classroom environment – the frequency of teachers getting interrupted in class – also shows a significant negative correlation with student achievement at the country level.

As was expected and consistent with the finding from TIMSS 1995 data analysis (Shen, 2001), the measure of students' educational resources at home – the average number of books at home – shows a strong positive correlation with student achievement at the country level. Consistent with findings from Shen and Pedulla (2000), the average self-perceived easiness of the two subjects demonstrates a significant negative association with achievement in both mathematics and science. It suggests that countries whose students are likely to rate the subjects as easy are usually poor performing countries in TIMSS tests and vice versa. To help readers acquire a visual sense of the correlation between the mathematics score and students' self-perceived easiness of mathematics, a scatterplot is presented in Figure 24.1.

The scatterplot of the science achievement scores and students' self-perceived easiness of science at the country level demonstrates a similar pattern as shown in Figure 24.1. In order to save space, however, it is not presented here.

Multiple regression analyses are performed to further test the results from correlation analyses. The regression analyses included all the independent variables discussed above, with the exception of public expenditure on education as a percentage of GNP and the length of the school year, which have no significant correlation with

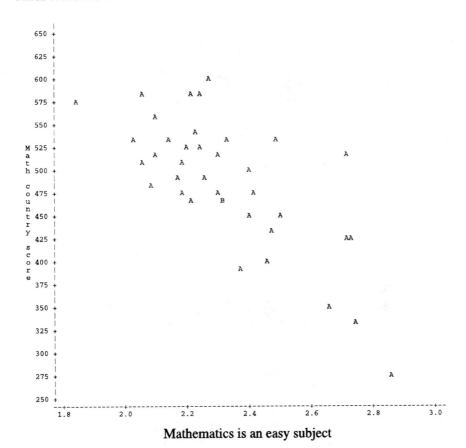

Mathematics is an easy subject

Note: A = 1 country; B = 2 countries.

Figure 24.1 Mathematics achievement versus 'mathematics is an easy subject'.

student achievement. Therefore, the following two tables report the regression analysis results based on only seven predictors for each subject.

Table 24.2 presents standardized coefficients for the seven predictors of mathematics achievement in TIMSS 1999 based on OLS regression analyses. Again, these seven predictor variables are ordered according to the three conceptual groups: (1) GNP per capita, the economic development indicator; (2) variables reflecting school and classroom environment; and (3) variables reflecting students' educational resources at home and their self-perceived easiness of mathematics or science. It is impossible and unnecessary to present the analysis results for all possible combinations of the seven predictors on each subject here. The table summarizes the results based on 11 models only (regression equations) with an attempt to allow all these predictors maximum opportunities to compete with each other.

As shown in Table 24.2, each predictor was included in six models (equations). The parameter estimate of each variable varies somewhat from model to model,

Table 24.2 Standardized OLS estimates of predictors of mathematics achievement (TIMSS 1999, $N = 38$ school systems).

Independent variables	Equation										
	2.1	2.2	2.3	2.4	2.5	2.6	2.7	2.8	2.9	2.10	2.11
Real GNP per capita	0.46**	–	–	–	0.41**	0.30**	0.33**	–	0.41**	0.41**	–
Absenteeism	–0.13	–0.11	–	–	–	–0.42**	–0.31**	–0.41**	–	–	–0.41**
Repeating the grade	–0.22**	–0.38**	–0.24**	–	–	–	–0.19**	–	–	–	–0.24**
Student mobility	–0.45**	–0.31*	–0.29**	–0.27**	–	–	–	–0.22*	–0.25**	–0.37**	–
Math class interrup.	–	–0.25**	–0.24**	–0.15	–0.34**	–	–	–	–0.35**	–	–0.19*
Number of books	–	–	0.40*	0.34**	0.23**	0.39**	–	0.63**	–	–	0.58**
'Math is easy'	–	–	–	–0.28*	–0.30**	–0.32**	–0.50**	–	–0.35**	–0.37**	–
Adj. R^2	0.66	0.49	0.63	0.61	0.72	0.79	0.74	0.67	0.74	0.70	0.70

Note.   * is at least 1.5 times its standard error, ** is at least 2 times its standard error.

depending on what other predictors are in the equation. The relative importance of a predictor in accounting for cross-national achievement variance can be claimed only when the predictor demonstrates a relatively stable effect on the dependent variable in most (if not all) models in which it was included. We also should keep in mind that a strong predictor in the model may prevent other predictors from achieving significance in the same model, especially when these predictors are correlated with each other. Three to four predictors are included for each model.

The adjusted R^2 ranges from 0.49 to 0.79, indicating that a substantial proportion of the cross-national variance of student mathematics achievement is accounted for by these regression models. For each regression model, efforts were made to check for multivariate outliers (cases with an unusual combination of values from two or more variables) by looking at the values of several relevant statistics and histograms. No significant influential outliers were identified. Consistent with the findings from the correlation analysis shown in Table 24.1, the seven predictors in Table 24.2 demonstrate significant effects on mathematics achievement. GNP per capita as a measure of a country's economic development level and average number of books at home as a measure of literacy level of a society demonstrate stable positive effects on mathematics achievement. As expected, the four measures of school and classroom management problems – absenteeism, grade repeating rate, student body mobility and class interruption – demonstrate negative effects on student achievement. Conversely, student perception of the easiness of mathematics (a proxy of the rigor of a curriculum) shows a stable, significant and negative effect on the dependent variable. This finding is consistent with the results based on TIMSS 1995 data analyses (Shen and Pedulla, 2000; Shen, 2001).

Table 24.3 presents standardized coefficients for the seven predictors of science achievement in TIMSS 1999 based on OLS regression analyses. The seven predictor variables are in the same order as in Table 24.2 for mathematics; the variables included in each of the eleven models are also the same as those used in Table 24.2. The results are quite similar to those reported in Table 24.2 for mathematics. The adjusted R^2 ranges from 0.55 to 0.78, reconfirming that a substantial proportion of the cross-national variation in science achievement is explained by these regression models. The only differences between the two tables are that the effects of GNP per capita and absenteeism are a little weaker than those in Table 24.2 for mathematics, while the effect of average number of books becomes a little stronger than that in Table 24.2 for mathematics. These differences are minor and consistent with the minor differences between the two subjects shown in the correlation matrix (Table 24.1).

Conclusion and Discussion

The author tried to make full use of the large TIMSS database and the predictor variables selected were based on extensive exploratory analyses of the available data. However, only a limited number of variables demonstrate a significant effect on student achievement for cross-national analysis. This fact reconfirms the argument that learning and teaching are cultural activities (Kawanaka et al., 1999) and pedagogical

Table 24.3 Standardized OLS estimates of predictors of science achievement (TIMSS 1999, $N = 38$ school systems).

Independent variables	Equation										
	3.1	3.2	3.3	3.4	3.5	3.6	3.7	3.8	3.9	3.10	3.11
Real GNP per capita	0.40**	–	–	–	0.29**	0.20**	0.23**	–	0.23**	0.30**	–
Absenteeism	-0.01	0.01	–	–	–	-0.37**	-0.23**	-0.29**	–	–	-0.33**
Repeating the grade	-0.31**	-0.44**	-0.28**	–	–	–	-0.24**	–	-0.24**	–	-0.30**
Student mobility	-0.50**	-0.35**	-0.30**	-0.34**	-0.31**	–	–	-0.28**	–	-0.43**	–
Science class interrup.	–	-0.25*	-0.16*	-0.08	–	–	–	–	-0.24**	–	-0.17*
Number of books	–	–	0.47**	0.39**	0.37**	0.55**	–	0.68**	–	–	0.61**
'Science is easy'	–	–	–	-0.30**	-0.23*	-0.25**	-0.48**	–	-0.48**	-0.41**	–
Adj. R^2	0.67	0.55	0.75	0.73	0.72	0.77	0.65	0.73	0.65	0.71	0.78

Note. * is at least 1.5 times its standard error, ** is at least 2 times its standard error.

methods and many school management issues are culturally embedded (Serpell and Hatano, 1997). It is impossible to define an educational model which fits all social and cultural environments. Nonetheless, a universal prerequisite for high achievement in mathematics and science is a school and social environment where education in mathematics and science is highly valued. With this consensus, dedicated educators and teachers will develop a high-standards curriculum and find the pedagogy appropriate for its own cultural and social contexts (Shen, 2001).

The findings from this study reveal consistency with findings from earlier studies in important aspects. Findings of the significant negative effect of self-perceived easiness of the two subjects on country-level student achievement are consistent with the findings from Shen and Pedulla (2000) and Shen (2001). The relationship between students' self-perceived easiness and actual achievement scores is of special interest and importance. The positive relationship at the individual level changes to negative at the country level. It is quite common that the magnitude of correlation coefficients somewhat change when the unit of analysis moves from an individual level to an aggregate level. However, it is not that common that the sign of the correlation changes when the unit of analysis moves from an individual level to an aggregate level. The findings from the study do not mean that we should encourage students to believe that mathematics and science are difficult and thereby decrease their self-perceived competence. That would be committing an ecological fallacy. The negative relationship is found at the country level, not at the individual level, therefore, the inference should be made at the country level too. By the same token, the negative correlations found for between-country analysis does not deny the existing motivation and self-efficacy theories. As mentioned earlier, these theories operate at the individual level, not at the country or culture level. The aggregate measures of students' self-perceptions have transcended individual characteristics and reflect a specific country's educational, cultural, and social contexts, which have gradually created individuals' attitudes, values and beliefs.

The negative relationship between the aggregate level of students' perceived easiness of the two subjects and the country level achievement scores is very likely to reflect the academic standards and the rigor of the curricula of each participating country. One possible explanation for poor student achievement in mathematics and science in the low performing countries is the low academic standards and unchallenging programs of those school systems. Countries with demanding curricula and high standards are more likely to produce students with high academic achievement levels in mathematics and science and vice versa.

Based on TIMSS 1995 data, Shen (2001) found that, as an indicator of students' attitude toward schooling, the index of school attendance shows a significant negative effect on student achievement at the country level. The present study finds the percentage of absenteeism to be similarly associated with achievement at the country level.

This chapter included three additional measures reflecting school and classroom environment: the percentage of students repeating a grade; student body mobility; and frequency of teachers getting interrupted in class. They all demonstrate consistent negative effects on student achievement for both mathematics and science. This has important policy implications for educational reform. A school and classroom environment conducive to teaching and learning is of paramount importance for improving student achievement. To create such an atmosphere requires effective school and classroom management and a serious attitude toward schooling and learning on behalf of students and their parents. Consistent with previous findings from TIMSS 1995 data analysis (Shen, 2001), public expenditure on education as a percentage of GNP does not demonstrate significant effects. The expenditure on education is affected by a number of factors and may not reliably reflect the commitment of the society on mathematics and science education.

Shen (2001) found that the average number of parents living with the students has a significant positive association with student achievement at the country level. Because nine TIMSS 1999 countries did not administer this question, this variable was not included in this analysis. However, when TIMSS 1995 data for this variable are substituted for those countries with missing data (not reported here), this variable still demonstrates significant positive association with achievement in the two subjects when other variables are controlled. This indicates that societies with a high rate of single parents are likely to be in an unfavorable position to compete with other countries in student academic achievement.

The evidence presented in this study supports the argument that education reform aimed at improving mathematics and science achievement requires establishing demanding intellectual standards and creating a school and classroom environment conducive to teaching and learning. It is not easy to realize these goals. Without a consensus on the importance of education in general and mathematics and science in particular from students, teachers, parents and the whole society, it would be very difficult, if not impossible, to successfully implement a high-standard curriculum. Only when policy-makers and society realize that high academic standards are vital to success in world competition can such consensus be reached. For many societies, it requires the change of people's attitudes and values on high standards, efforts, discipline and a good learning climate (Finn, 1991; Stevenson and Stigler, 1992; Wray, 1999; Lewis et al., 1995).

This study yields some different findings from those based on TIMSS 1995 data (Shen, 2001). The analysis based on the TIMSS 1999 data and sample finds that economic development level and the literacy level, measured by the average number of books at student homes, demonstrate a somewhat greater effect on student achievement in mathematics and science. As mentioned earlier, this discrepancy may be attributed to the fact that the TIMSS 1999 sample included more developing countries than the TIMSS 1995 sample. This makes the effects of economic-related factors more manifest.

TIMSS 1999 data also show that the length of the school year lost its significance when compared to the findings of Shen (2001) based on the TIMSS 1995 data and sample. The data for the length of school years were collected through school questionnaires, asking the number of 'instructional days a year'. The reliability of the data for some countries is questionable. For example, as shown in Table 24.1, the standard deviation for Hong Kong's instruction days a year is big (28.3 days). Other sources also show that the length of a school year in Hong Kong is about 200 days (Robitaille,

1997). Some schools have up to three sets of examinations at Grade 8 every academic year, each requiring 10–15 days (Martin *et al.*, 1999: 66). Apparently people who answered the question had inconsistent understandings on 'how many instructional days are in the school year?', which led to the large standard deviation of this measure. The number of instructional days for Singapore is reported from the Ministry of Education: 180 days a year for all schools. Another source also reports about 200 days a year for Singapore middle schools (Robitaille, 1997). It is very likely that,s in Hong Kong and Singapore, the days when teachers do not give new instructions, but are present in classroom and prepare for the exams, are not considered as 'instructional days'.

As mentioned earlier, the tremendous diversity of participating school systems in terms of educational, social and cultural contexts constitutes some of the major limitations of this kind of study. Based on the analysis of the Second International Mathematics Study (SIMS), Robitaille and Garden (1989) observed that such differences complicate the process of drawing valid conclusions and generalizing findings. The discrepancies found between this study and the study based on TIMSS 1995 data warrant the necessity of taking caution when generalizing the findings from any specific cross-national analysis. The findings and hypotheses from the present study and studies based on TIMSS 1995 data can be retested with data from IEA's TIMSS 2003 study.

Appendix

Appendix   Data for selected country-level variables initially involved in this study ($N = 38$ school systems).

Country/ school system	Real GNP per capita (International dollars)[a]	Public expenditure on education as % of GNP[b]	Instructional days a year (SD)	Educational aspiration (SD)
Singapore	28,460	2.30	180 (NA)	2.98 (0.15)
Korea, Rep. of	13,590	3.50	224 (3.8)	2.95 (0.23)
Chinese, Taipei	16,500[c]	4.69	221 (5.2)	2.97 (0.20)
Hong Kong, SAR	24,350	2.80	176 (28.3)	2.86 (0.38)
Japan	24,070	3.80[d]	223 (5.8)	2.74 (0.46)
Belgium (Flemish)	22,750	3.00	175 (NA)	2.78 (0.42)
Netherlands	21,110	4.90	190 (13.0)	2.63 (0.50)
Slovak Republic	7,910	4.60	192 (12.8)	2.61 (0.52)
Hungary	7,200	4.30	185 (3.6)	2.58 (0.51)
Canada	22,480	6.40	189 (7.2)	2.95 (0.25)
Slovenia	11,800	5.40	180 (9.3)	2.73 (0.52)
Russian Federation	4,370	3.90	196 (22.8)	2.89 (0.31)
Australia	20,210	5.20	194 (12.5)	2.70 (0.57)
Finland	20,150	7.00	186 (3.6)	2.38 (0.57)
Czech Republic	10,510	4.40	197 (8.1)	2.40 (0.64)
Malaysia	8,140	4.10	197 (11.8)	2.92 (0.34)
Bulgaria	4,010	3.00	170 (14.3)	2.75 (0.46)
Latvia (LSS)	3,940	6.10	176 (13.5)	2.89 (0.33)
United States	29,010	5.30	179 (3.3)	2.93 (0.28)

Appendix. (continued)

England	20,730	4.90	190 (5.1)	2.58 (0.68)
New Zealand	17,410	6.90	189 (5.7)	2.74 (0.52)
Lithuania	4,220	5.20	195 (NA)	2.88 (0.39)
Italy	20,290	4.60	210 (NA)	2.49 (0.64)
Cyprus	14,201	4.10	160 (NA)	2.71 (0.58)
Romania	4,310	3.30	159 (12.7)	2.60 (0.57)
Moldova	1,500	10.30	206 (12.9)	2.77 (0.54)
Thailand	6,690	3.60	201 (4.7)	2.62 (0.59)
Israel	18,150	6.90	200 (21.4)	2.85 (0.40)
Tunisia	5,300	6.60	204 (20.6)	2.89 (0.38)
Macedonia, Rep. of	3,210	4.90	176 (2.1)	2.63 (0.64)
Turkey	6,350	3.20	181 (2.5)	2.83 (0.48)
Jordan	3,450	6.40	191 (9.6)	2.86 (0.44)
Iran, Islamic Rep. of	5,817	3.30	205 (22.2)	2.78 (0.55)
Indonesia	3,490	0.60	245 (19.1)	2.74 (0.56)
Chile	12,730	3.40	194 (9.0)	2.75 (0.48)
Philippines	3,520	2.90	205 (13.6)	2.73 (0.61)
Morocco	3,310	4.90	205 (22.1)	2.77 (0.57)
South Africa	7,380	7.50	196 (18.3)	2.69 (0.64)

Notes. The countries are listed in descending order of international mathematics scores. SD = Standard deviation; NA = Not applicable because school days a year was reported by the education ministry for all schools in the country.
Educational aspiration: 1 = finish some secondary school; 2 = finish secondary school and 3 = beyond secondary school.
[a]*Source.* World Bank (1999: 134–7). Converted at purchasing power parity (PPP).
[b]*Source.* UNESCO (1999: 490–513).
[c]*Source.* CIA (1999).
[d]*Source.* World Bank (1999: 200–1).

Note

1. Shen (2001) computed an index averaging three measures reflecting students' school attendance based on the TIMSS 1995 school questionnaire: percentage of absenteeism, percentage of students arriving late at school, and percentage of students skipping classes. However, the TIMSS 1999 school questionnaire did not ask about the percentage of students arriving late and skipping classes. Therefore, the present study only includes the average percentage of absenteeism as an indicator of students' school attendance.

References

Bollen, K., Entwisle, B. and Alderson, A.S. (1993) Macrocomparative research methods. *Annual Review of Sociology*, 19, 321–51.

Central Intelligence Agency (CIA) Electronic Document Release Center (FOIA). (1999) *The world factbook 1999*. Retrieved 20 December 2004, from http://www.umsl.edu/services/govdocs/wofact99/51.htm

Elley, W.B. (ed.). (1994) *The IEA study of reading literacy: Achievement and instruction in thirty-two school systems*. London: IEA International Studies in Educational Achievement: Pergamon.

Finn, C.E. (1991) *We must take charge: Our schools and our future*. New York: Free Press.

Howie, S.J. (2003) Conditions in schools in South Africa and the effects on mathematics achievement. In K. Bos, S.J. Howie, and T. Plomp (eds.), Common diversity: TIMSS-1999 findings from a national perspective [Special issue]. *Studies in Educational Evaluation*, 29(3).

Howie, S.J. (in press) Contextual factors on school and classroom level related to pupils' performance in mathematics in South Africa. In S.J. Howie and T. Plomp (eds.), TIMSS findings from an international perspective: Trends from 1995 to 1999 [Special issue]. *Educational Research and Evaluation*.

Johnson, R.A. and Lindblad, A.H. (1991) Effect of mobility on academic performance of sixth grade students. *Perceptual and Motor Skills*, 72, 547–52.

Kawanaka, T., Stigler, J.W. and Hiebert, J. (1999) Studying mathematics classrooms in Germany, Japan and the United States: Lessons from the TIMSS videotape study. In G. Kaiser, E. Luna, and I. Huntley (eds.), *International comparisons in mathematics education* (pp. 86–103). London: Falmer Press.

Kifer, E.W. (2002) Students' attitudes and perceptions. In D.F. Robitaille and A.E. Beaton (eds.), *Secondary analysis of the TIMSS data* (pp. 251–75). Dordrecht: Kluwer Academic Publishers.

Lewis, C.C., Shapes, E. and Watson, M. (1995) Beyond the pendulum: Creating challenging and caring schools. *Phi Delta Kappan*, 76(7), 547–54.

London, B., and Williams, B.A. (1990) National politics, international dependency, and basic needs provision: A cross-national analysis. *Social Forces*, 69, 565–84.

Martin, M.O., Mullis, I.V.S., Beaton, A.E., Gonzalez, E.J., Smith, T.A. and Kelly, D.L. (1999) *School contexts for learning and instruction: IEA's Third International Mathematics and Science Study* (TIMSS). Boston, MA: Boston College, TIMSS International Study Center.

Mullis, I.V.S., Martin, M.O., Gonzalez, E.J., Gregory, K.D., Garden, R.A., O'Connor, K.M., *et al.* (2000) *TIMSS 1999 international mathematics report: Findings from IEA's report of the Third International Mathematics and Science Study at the eighth grade*. Boston, MA: Boston College, TIMSS International Study Center.

Pedhazur, E.J. (1982) *Multiple regression in behavioral research: Explanation and prediction*. New York: Holt, Rinehart, and Winston.

Robitaille, D.F. (ed.). (1997) *National contexts for mathematics and science education: An encyclopedia of the educational systems participating in TIMSS*. Vancouver, BC: Pacific Educational Press.

Robitaille, D.F., and Garden, R.A. (1989) Findings and implications. In D.F. Robitaille and R.A. Garden (eds.), *The IEA study of mathematics II: Contexts and outcomes of school mathematics* (pp. 232–41). Oxford: Pergamon Press.

Schuler, D.B. (1990) Effects of family mobility on student achievement. *ERS Spectrum*, 8, 17–24.

Serpell, R., and Hatano, G. (1997) Education, schooling and literacy in a cross-cultural perspective. In J.W. Berry, P.R. Dasen, and T.S. Saraswathi (eds.), *Handbook of cross-cultural psychology: Vol. 2* (pp. 339–76). Boston, MA: Allyn and Bacon.

Shavelson, R.J., McDonnell, L.M. and Oakes, J. (1987) *Indicators for monitoring mathematics and science education: a sourcebook*. Santa Monica, CA: Rand Publications Corporation.

Shen, C. (2001) Social values associated with cross-national differences in mathematics and science achievement: A cross-national analysis. *Assessment in Education*, 8(2), 193–223.

Shen, C., and Pedulla, J.J. (2000) The relationship between students' achievement and their self-perception of competence and rigour of mathematics and science: A cross-national analysis. *Assessment in Education*, 7(2), 237–53.

Stevenson, H.W., and Stigler, J.W. (1992) *The learning gap*. New York: Simon and Schuster.

Tabachnick, B.G., and Fidell, L.S. (1996) *Using multivariate statistics*. New York: Harper Collins Publishers.

UNESCO (1999). *UNESCO statistical yearbook 1999*. Paris: UNESCO Publishing and Berman Press.

Wood, D., Halfon, N., Scarlata, D., Newacheck, P., and Nessim, S. (1993) Impact of family relocation on children's growth development, school function, and behavior. *JAMA*, 270, 1334–8.

World Bank (1999) *World development report 1998/99*. New York: Oxford University Press.

Wray, H. (1999) *Japanese and American education: Attitudes and practices*. Westport, CT: Bergin and Garvey.

Index

Contexts of Learning

Classrooms, Schools and Society
ISSN 1384-1181

1. Education for All
   Robert E. Slavin
   1996   ISBN 90 265 1472 7 (hardback)
           ISBN 90 265 1473 5 (paperback)

2. The Road to Improvement – Reflections on School Effectiveness
   Peter Mortimore
   1998   ISBN 90 265 1525 1 (hardback)
           ISBN 90 265 1526 X (paperback)

3. Organizational Learning in Schools
   Edited by Kenneth Leithwood and Karen Seashore Louis
   1998   ISBN 90 265 1539 1 (hardback)
           ISBN 90 265 1540 5 (paperback)

4. Teaching and Learning Thinking Skills
   Edited by J.H.M. Hamers, J.E.H. Van Luit and B. Csapó
   1999   ISBN 90 265 1545 6 (hardback)

5. Managing Schools towards High Performance: Linking School Management
   Theory to the School Effectiveness Knowledge Base
   Edited by Adrie J. Visscher
   1999   ISBN 90 265 1546 4 (hardback)

6. School Effectiveness: Coming of Age in the Twenty-First Century
   Edited by Pam Sammons
   1999   ISBN 90 265 1549 9 (hardback)
           ISBN 90 265 1550 2 (paperback)

7. Educational Change and Development in the Asia–Pacific Region: Challenges
   for the Future
   Edited by Tony Townsend and Yin Cheong Cheng
   2000   ISBN 90 265 1558 8 (hardback)
   2000   ISBN 90 265 1627 4 (paperback)

8. Making Sense of Word Problems
   Edited by Lieven Verschaffel, Brian Greer and Eric De Corte
   2000   ISBN 90 265 1628 2 (hardback)

9. Profound Improvement: Building Capacity for a Learning Community
   Edited by Coral Mitchell and Larry Sackney
   2000   ISBN 90 265 1634 7 (hardback)

10. School Improvement Through Performance Feedback
    Edited by A.J. Visscher and R. Coe
    2002   ISBN 90 265 1933 8 (hardback)

11. Improving Schools Through Teacher Development. Case Studies of the Aga
    Khan Foundation Projects in East Africa
    Edited by Stephen Anderson
    2002   ISBN 90 265 1936 2 (hardback)

12. Reshaping the Landscape of School Leadership Development – A Global
    Perspective
    Edited by Philip Hallinger
    2003   ISBN 90 265 1937 0 (hardback)

13. Educational Evaluation, Assessment and Monitoring: A Systemic Approach
    Jaap Scheerens, Cees Glas and Sally M. Thomas
    2003   ISBN 90 265 1959 1 (hardback)

14. Preparing School Leaders for the 21st Century; An International Comparison of
    Development Programs in 15 Countries
    Stephan Gerhard Huber
    2004   ISBN 90 265 1968 0 (hardback)

15. Inquiry, Data, and Understanding: A Search for Meaning in Educational Research
    Lorin W. Anderson
    2004   ISBN 90 265 1953 2 (hardback)